SOME CRITICAL ISSUES FOR INJECTION MOLDING

Edited by **Jian Wang**

Some Critical Issues for Injection Molding
Edited by Jian Wang

Published by InTech

Janeza Trdine 9, 51000 Rijeka, Croatia

Publishing Process Manager Martina Blecic

Technical Editor Teodora Smiljanic

Cover Designer InTech Design Team

Additional hard copies can be obtained from orders@intechopen.com

Some Critical Issues for Injection Molding, Edited by Jian Wang
p. cm.
ISBN 978-953-51-0297-7

For EU product safety concerns:
IN TECH d.o.o., Prolaz Marije Krucifikse Kozulić 3, 51000 Rijeka, Croatia,
info@intechopen.com or visit our website at intechopen.com.

Meet the editor

Dr Jian Wang has been working in the area of polymer properties and injection molding, and has extended into numerical simulation and single-polymer composites. He has published 35 papers, applied 27 national patents, and participated in three books written. Numerous grants and awards support his work including the National Key Technology R&D Program, National Natural Science Foundation, Research Fund for the Doctoral Program of Higher Education, Science Foundation for Young Teachers, and several sponsored research programs. At present, Dr Wang is a lecturer in Beijing Institute of Technology, and has served as the fellow of Society of Plastics Engineers, Society of Rheology, Study & Produce Alliance of CAE Mold, and Plastics Technology Collaboration Committee of China Plastic Processing Industry Association.

Contents

Preface

Plastic products are indispensable in our everyday life. Today, the injection molding is the most used method for producing plastics parts. Using injection molding process, the plastic materials can be melted or remelted and injected at high pressure into the mold cavity to produce parts with desired shape.

I started my study on injection molding seven years ago. When I was an undergraduate student, I got acquainted with mechanics and control. Then, injection molding opened a new and interesting world for me to know mechanical engineering and plastic materials. To research in this direction, I got the opportunity to do a PhD study at Beijing University of Chemical Engineering. In the first of my PhD study, ARBURG, a famous global manufacturers of injection molding machines, supported a machine to my laboratory. It became the first injection molding machine which I used. Then, I used different kind of injection molding machines from LK Machinery, Haitian, GSK, etc. The basic of my research is from the PVT (Pressure-Volume-Temperature) properties of polymers which are important for both engineering and polymer physics. According to the polymer PVT properties, mold design, numerical simulation and process control for injection molding could become easily to realize. The on-line PVT measurement and one application example presented in one chapter of this book had been presented in my two articles.

After graduation, I became a teacher in Beijing Institute of Technology, and I am still doing research on polymer engineering. Fields of my interest include injection molding and innovative plastics manufacturing processes, micro-injection molding, design and process optimization, computer-aided engineering (CAE) and numerical simulation, properties for plastics. I am still working hard on these fields and hoping more research achievements will benefit to our life and society.

This book is composed of different chapters which are mainly related to the subject of injection molding. All the contents not only contain the fundamental knowledge but also have many new research results and techniques by the authors, some leading international academic experts in the field. I define the book title as "Some Critical Issues for Injection Molding" in the context of the research field covered.

Chapter 1 of this book contains introduction on polymer PVT measurements and two main application areas of polymer PVT data in injection molding including numerical simulation and process control. The knowledge on PVT properties of polymers could be the fundamental concepts for the engineers in injection molding. Chapter 2 presents the effective run-in and optimization of an injection molding process guiding readers how to install robust processes or how to review and optimize them. Chapter 3 to 7 are mainly on powder injection molding which was invented and developed in the early 1970s and comprises ceramic injection molding and metal injection molding. Powder injection molding has been accepted as one of the fundamental manufacturing techniques in recent years, and it is leading the way to new prospective market segments. Chapter 8 is about microcellular injection molding which was originally concetualized and invented at the Massachusetts Institute of Technology in 1984. Microcellular technology has already had a significant impact on the worldwide plastics industry. The insert molding process employing vapour chamber is introduced in Chapter 9. Insert molding process is a simplified injection molding method that eliminates secondary processing and assembly. Chapter 10 focused on the natural fiber-thermoplastic composites, reviewed some influence factors on the injection molding process to produce natural fibers thermoplastic compound, and introduced some research on interfacial strength of injection molded composites. Chapter 11 explored and analyzed the properties of injection molded high density polyethylene nanocomposites filled with exfoliated graphene nanoplatelets.

It is now a great pleasure for me to complete one chapter and the review of the other chapters in this book. Finally, I am grateful to all the authors and the support of different foundations or companies for their research. I very much hope that the contents presented here will provide some clear presentation of injection molding process and equipment to direct people in plastics manufacturing to solve problems and avoid costly errors. With useful, fundamental information for knowing and optimizing the injection molding operation, I also hope the readers could gain some working knowledge of the injection molding.

Jian Wang
School of Chemical Engineering and Environment,
Beijing Institute of Technology
China

Part 1

Basics for Injection Molding

1

PVT Properties of Polymers for Injection Molding

Jian Wang
School of Chemical Engineering and Environment, Beijing Institute of Technology, China

1. Introduction

PVT (Pressure-Volume-Temperature) properties of polymers are important for both engineering and polymer physics. Fig.1 shows the typical PVT diagrams of an amorphous (a) and semi-crystalline polymer (b). PVT diagram describes the specific volume as a function of pressure and temperature. Specific volume increases with the temperature increasing. There is a thermal transition in the polymer. The primary amorphous transition of any polymer is known as its glass transition temperature, T_g. While T_g is not a sharp transition, the data from below and above T_g will show an intersection that is generally accepted as being T_g. As shown in Fig. 1, semi-crystalline polymer exhibits a different thermal response than amorphous polymer. For the amorphous polymer, T_g is clearly seen as the temperature where the polymer goes from a solid to a melt. The rate of expansion per temperature increment is much smaller in the solid state than in the melt state. By contrast, the semi-crystalline polymer contains sufficient crystallinity to maintain structural continuity above T_g. While the amorphous content in this polymer exhibits a T_g, the crystal structure allows characterization up to nearly the temperature where the crystals melt.

Polymer PVT data become increasingly important in their value in material science. The excess usage of PVT data can be summarized in at least eight major areas (Berry et al., 1998; Hess, 2004):

- Prediction of polymer-polymer miscibility;
- Prediction of service performance and service life of polymeric materials and components on the basis of free volume concepts;
- Correlation of the reducing parameters of equations of state (EOS) with molecular structures;
- Evaluation of start and progress of chemical reactions in polymer melts in the cases when volume effects accompany the reaction;
- Materials properties of systems in contact with solvents or gases;
- Investigation of the nature of phase transitions;
- Optimizing of processing parameters instead of establishing such parameters by trial and error;
- Calculation of the surface tension of polymer melts.

Injection molding is the most common technique for the mass production of complex shaped products that require accurate dimensions. In injection molding process, some

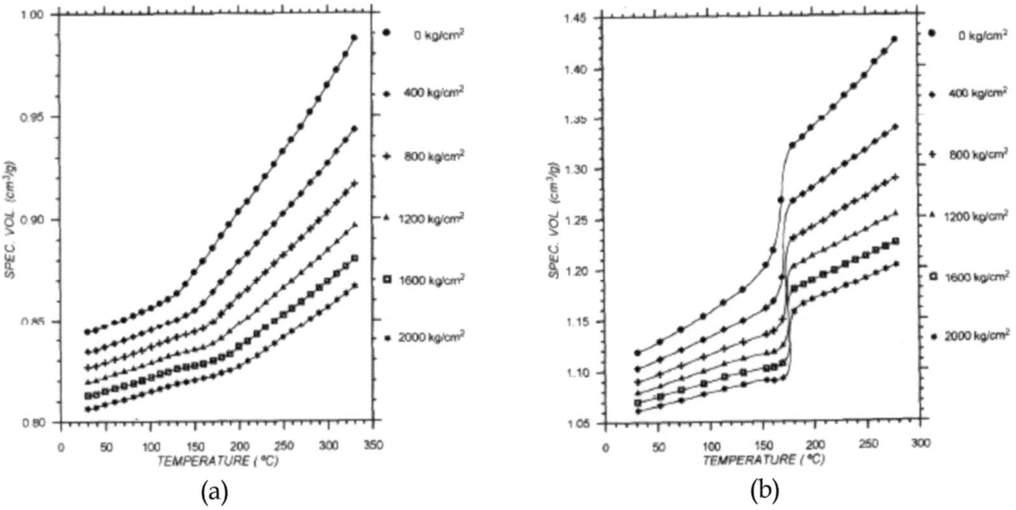

Fig. 1. The typical PVT diagrams of an amorphous (a) and semi-crystalline polymer (b) (Zoller & Walsh, 1995).

defects such as shrinkage, warpage and sink marks will cause thermally induced stresses, and affect both dimensional accuracy and long-term dimensional stability. Software such as Moldflow (Autodesk, Inc., San Rafael, CA) or Moldex 3d (CoreTech System Co., Ltd., Taipei, China) seeks to reduce the risk of producing parts with shrinkage or warpage in plastic components by providing quantitative predictions based on reliable data. In numerical simulation of the injection molding process, one of the important input data set to these software packages is PVT data (Brown & Hobbs, 1998). On the other hand, process analyses have revealed that optimum processing conditions can be achieved by application of PVT diagrams (see Fig.1), particularly if PVT diagrams are used in conjunction with computer control of the process (Menges & Thienel, 1977). Therefore, it is necessary to get more accurate PVT data for more accurate prediction, evaluation, optimization and calculation.

This chapter is an introduction to PVT properties of polymers for injection molding. It is divided into three sections: different measurements of polymer PVT properties are introduced in detail, followed by the testing modes, and then two main application areas of polymer PVT data are illustrated in detail, including numerical simulation and process control.

2. Different measurements of polymer PVT properties

In this section, different measurements of polymer PVT properties are introduced in detail, including conventional measurements (piston-die technique and confining-fluid technique), some improved experimental techniques considering the effect of cooling rate, shear rate and pressure, on-line techniques using injection molding machine or extruder, etc.

2.1 Conventional measurements

Using a dilatometer is the most common technique to measure the bulk specific volume as a function of temperature and pressure of polymers. There are two principally different conventional techniques performing PVT measurements: the piston-die technique and the confining-fluid technique.

2.1.1 Piston-die technique

The piston-die technique (Fig.2a): The material is enclosed and pressurized in a rigid die using a piston which is tightly fitted into the die. During the measuring cycle the volume of the material is recorded by measuring the displacement of the piston. Both temperature and pressure can be varied. The advantage of this technique is the simplicity of the design that can be achieved. The disadvantage is that the pressure applied is not hydrostatic because the material sticks to the wall (He & Zoller,1994). Other problems are the possible leakage between the piston and the die and the formation of voids in the sample when solidifying. The piston-die technique was applied by Chang et al. (1996) who used a PVT-100 apparatus from SWO Germany (see Fig. 3).

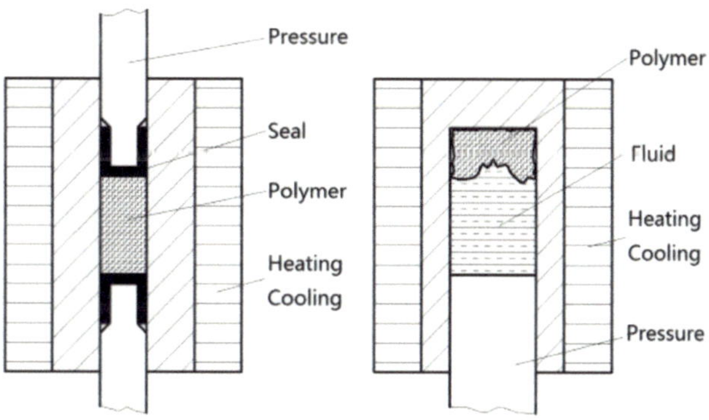

(a) (b)

Fig. 2. Sketch diagrams of piston-cylinder technique (a) and confining-fluid technique (b).

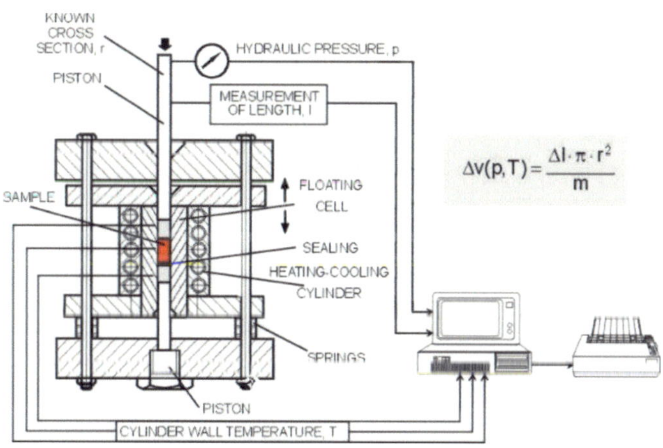

Fig. 3. Principle schematic diagram of PVT100 (SWO Polymertechnik GmbH, 1998).

2.1.2 Confining-fluid technique

The confining-fluid technique (Fig.2b): The sample is enclosed in a rigid sample chamber, and it is submerged into a fluid (mercury or silicon oil). The cell is closed using a flexible wall or bellows. The bellows is used to apply hydrostatic pressure to the fluid and polymer by reducing the sample chamber volume, and sensing the cumulative volume change of fluid and polymer. The absolute specific volume of the polymer can be obtained by correcting the relative volume difference with the specific volume of the confining fluid. Both pressure and temperature can be varied. The advantages of this technique are:

- Pressure is purely hydrostatic as the sample is surrounded by the confining fluid in both a melted and solid state;
- There is an absence of leakage and friction.

The disadvantages are:

- The volumetric changes measured are not that of the polymeric sample only;
- Sealing of the pressurized fluid and reactions may occur between polymers and the confining fluid.

Quach & Simha (1971) constructed an apparatus based on the confining-fluid technique. Mercury was employed as the confining liquid. The operating range is $0 \le T \le 200°C$ and $1 \le P \le 200 MPa$. Calibration with benzene and mercury showed an accuracy of $\pm 2 \times 10^{-4}$ cm^3/g in the measurement of the specific volume change. Zoller et al. (1976) developed an apparatus which was based on the classical confining-fluid technique. The apparatus (Gnomix, Inc., Boulder, CO) shown in Fig. 4 was used by Moldflow to get PVT data. The PVT instrument has some advantages compared to a capillary rheometer if accurate numbers are desired (0.0001 cm^3/g) (Zoller & Fakhreddine, 1994).

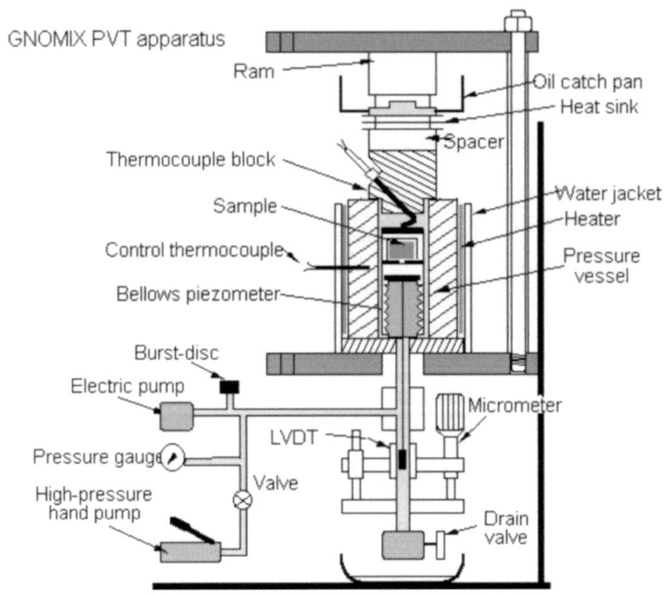

Fig. 4. Gnomix PVT testing device developed by Zoller et al. (1976).

Barlow (1978) developed a bellows dilatometer at pressures up to 280 MPa but temperatures only up to 55 °C. The experimental system used a bellows-type dilatometer for measuring sample volume changes. As shown in Fig.5, the below cell was designed expressly for the purpose of measuring volume changes of solid polymers surrounded by mercury.

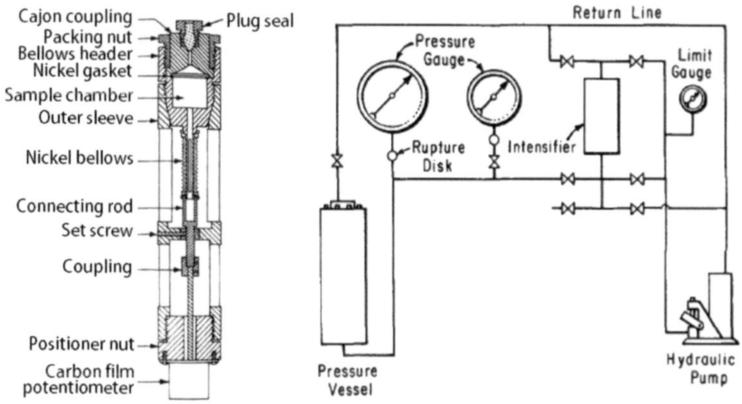

Fig. 5. PVT testing setup developed by Barlow (1978).

Sato et al. (1997) developed an apparatus (Fig. 6) using a metal bellows at temperatures from 40 to 350 °C and pressures up to 200 MPa. The experimental uncertainty of specific volumes was estimated to be within ±0.2%. The effects of a sample cup and sample forms were investigated. The use of the sample cup showed a little effect on the measurments of PVT properties for both samples. The shape (pellet and pillar) of the samples caused a small difference in the specific volumes only under high temperatures and low pressures.

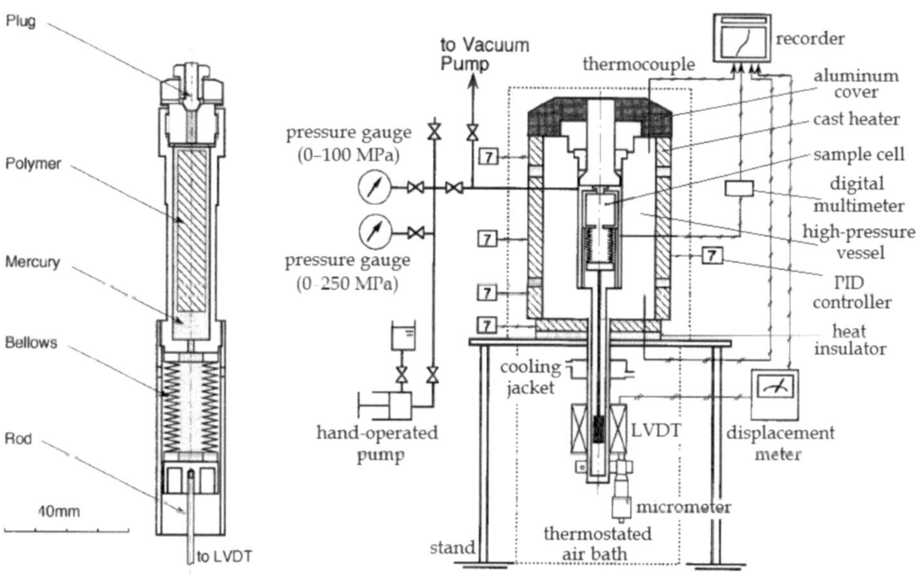

Fig. 6. Schematic diagram of the PVT testing setup developed by Sato et al. (1997).

It has been shown that differences between the piston-die and confining-fluid technique in methods or measurement principles do not lead to differences in measured values bigger than 4% (Sato et al., 1997).

2.2 Improved measurements

The material properties of polymers show a distinct time dependence, i.e., depending on the rate of change of state, they show different thermal material behaviour, particularly in the transition range between melt and solid. Thus, the values of specific volume differ for low and high cooling rates (Fig. 7) (Menges & Thienel, 1977).

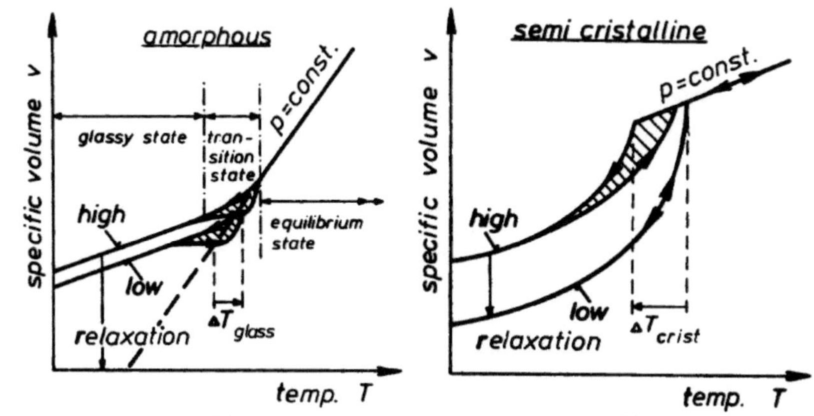

(a) (b)

Fig. 7. PVT diagrams of an amorphous (a) and semi-crystalline polymer (b) under different cooling rates (Menges & Thienel, 1977).

As shown in Fig. 7, amorphous polymer reaches its equilibrium degree of order above the glass temperature T_g within the time of measurement. This is not possible at temperatures below T_g; the melt will then pass over from a state of equilibrium to one of diequilibrium, i.e., to the frozen-in or glassy state. This state is distinguished, in general, by a greater heat content (enthalpy). In this range, there is a smaller coefficient of thermal expansion compared to the molten state. In the glassy state, the molecules retain the state of order they had reached at the glass temperature. At higher cooing rates the glass temperature is shifted to higher values. Since at higher glass temperatures, a lower state of order prevails, a high cooling rate causes a lower density or a greater specific volume than do lower cooling rates.

With semi-crystalline polymer, crystallization takes place below the crystallization temperature (beginning of solidification). If the time available for crystal growing decreases due to an increased cooling rate, crystallization will be slowed down or reduced. This becomes evident in a shift of the crystallization temperature to lower values (melt begins solidifying at lower temperatures) and in a growing reduction in density (increase of specific volume) below this temperature.

Apart from the effects of cooling rates during the transition between melt and solid, specific volume below the glass temperature or crystallization temperature is furthermore influenced by:

- The pressure during cooling.
- The testing modes of operation: isobaric measurement and isothermal measurement.

If specific volume data are to be used to gain better understanding of what is happening during industrial processing of polymers or as input for constitutive modelling used in simulation software, it is necessary to measure specific volume at conditions comparable to industrial processing conditions. Conventional techniques can only be used at relatively low cooling rates. However, plastic processing, such as injection molding, is a rapid, high pressure process where both cooling rate and pressure play critical roles in the final component dimensions. Since high cooling rates are present during injection molding, some improved PVT testing techniques have been designed.

Menges and Thienel (1977) developed a new measuring instrument (see Fig. 8) which made it possible to determine such PVT diagrams using an adjustable barrel chamber under normal processing conditions. Piccarolo (1992) used a different method to measure the specific volume of semicrystalline polymers at high cooling rates above 1200 °C/min. Bhatt and McCarthy (1994) developed a PVT apparatus for computer simulations in injection molding. Imamura et al. (1996) determined PVT relationships at different cooling rates (up to approximately 100 °C/min) and verified their influence on computer simulations. Lobo (1997) attempted to measure the specific volume of semicrystalline polymers at higher cooling rates using a combination of conventional PVT apparatus, differential scanning calorimetry (DSC) measurements and the K-System thermal conductivity apparatus.

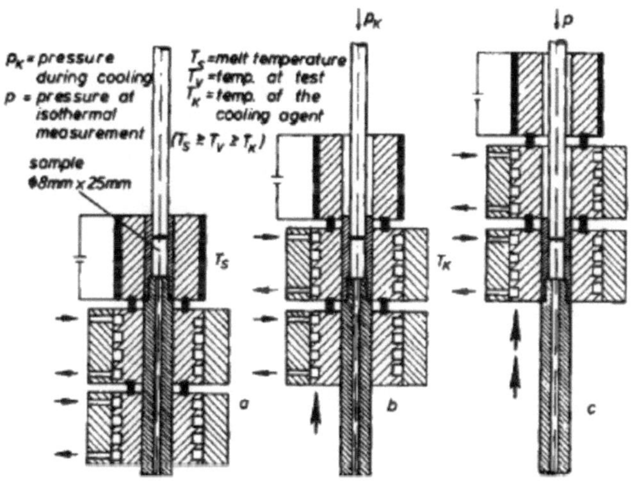

Px: pressure during cooling; P: pressure of isothermal measurement; Ts: melt temperature; Tv: temperature of test; TK: temperature of the cooling agent (Ts≥Tv≥TK).

Fig. 8. PVT testing device at cooling rates developed by Menges and Thienel (1977).

The National Physical Laboratory (NPL) in the UK (Brown & Hobbs, 1998) has demonstrated a method during rapid cooling (up to 250 °C/min) and under high pressure (up to 250 MPa). Chakravorty (2002) in NPL developed the PVT equipment (Fig. 9) for measuring polymer properties at industrial processing conditions based on the piston-die technique. The equipment could reach a maximum cooling rate of 200 °C/min in combination with maximum pressure of 160 MPa. Luyé et al. (2001) discussed the specific volume measurement methods for semicrystalline polymers in order to obtain reliable data and analyzed the effect of the cooling rate taking into account the thermal gradient in a cylindrical sample. They discussed the PVT measurement methods for semicrystalline polymers needed in order to obtain reliable data. Particularly, the effect of the cooling rate was analyzed, taking into account the thermal gradient in a cylindrical sample. Specific-volume evolutions from 220°C to 50°C for different cooling rates and different pressures were analyzed, revealing that when the data are corrected to eliminate the thermal gradient effect, the transition zone is much narrower than the experimental one. Moreover, the effect of the pressure and the cooling rate can be more accurately evaluated. It were obtained with an apparatus (PVT100) manufactured by the German supplier SWO Polymertechnik GmbH (Krefeld, Germany). In addition, using the piston device produced a higher imposed cooling rate at the sample periphery than did the immersion system: typically up to 30°C/min in the former case and a few degrees min^{-1} in the latter case.

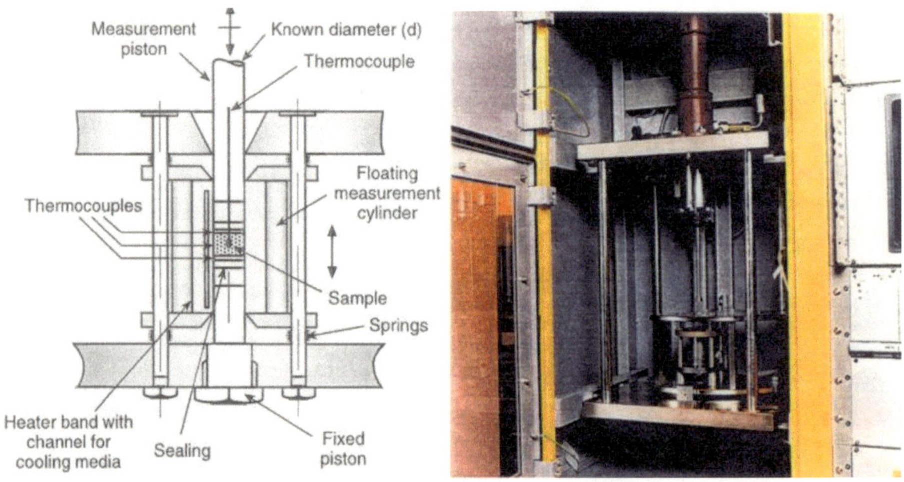

Fig. 9. PVT testing device during rapid cooling and under high pressure developed by NPL (Chakravorty, 2002).

Zuidema et al. (2001) built a setup based on the confining-fluid technique and analyzed the influence of the cooling rate. The setup could reach cooling rates up to 3600 °C/min, but pressure only up to 20 MPa and the accuracy or resolution of the setup was not reported. To minimize the influence of heat on the rest of the measuring equipment, cooling channels were present at the top and the bottom of the heated area to create heat sinks (Fig. 10). When the steady conditions were reached, the vicinity of the sample was quenched with pressurized water via cooling channels positioned close to the sample holding area.

van der Beek et al. (2005a, 2005b) developed a dilatometer to measure specific volume, and the specific volume of polymers as a function of elevated pressure (up to 100 MPa), temperature (up to 260 °C), cooling rate (up to 6000 °C/min) and shear flow in the range of processing conditions as found in injection molding or extrusion. The main goal was not to reach the maximum accuracy possible but rather the ability to analyse specific volume for combination of processing conditions. The dilatometer was based on the piston-die technique in combination with a tensile testing machine with rotation capability. The design

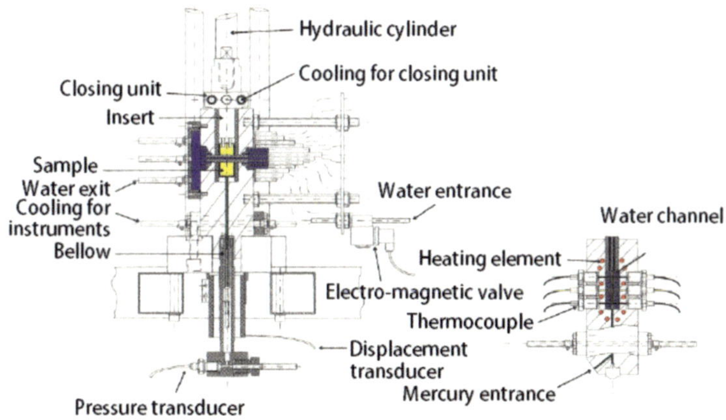

Fig. 10. PVT testing setup developed by Zuidema et al. (2001).

of piston and die was chosen such that an annular shaped sample spacing was created, similar to dilatometers developed by Chakravorty (2002). The particular sampler shape has the advantage that the radial thickness can be chosen small to enable rapid cooling without introducing thermal gradients. Fig. 11 shows a schematic drawing of the pressure cell as mounted to the tensile testing machine.

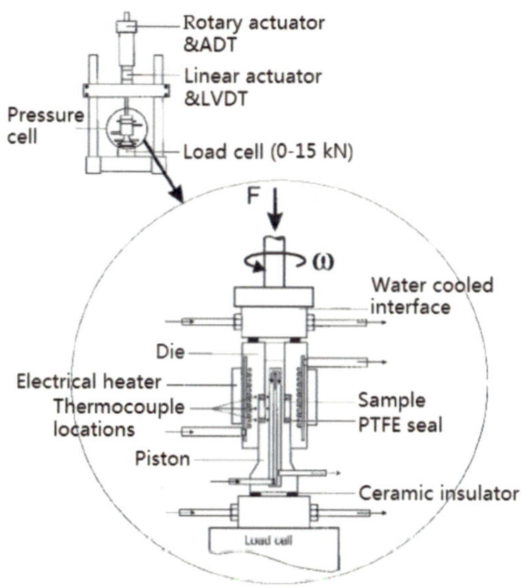

Fig. 11. PVT testing setup developed by van der Beek et al. (2005a).

2.3 On-line techniques

The PVT relationships were almost measured by a special dilatometer. Actually, the technology for measuring PVT relationships using the IMM (injection molding machine) or extruder, which can be called an on-line measurement, is a potentially powerful tool for programming process controllers, because normal process conditions of injection molding or extrusion can be obtained.

Nunn (1989) utilized an IMM to develop PVT properties for a given material by pressurizing the material in the barrel against the blocked outlet nozzle, measuring the volume during pressurization. The injection outlet of the molding machine barrel was blocked and a shot of material was plasiticated in the barrel and then pressurized in the barrel at a certain temperature and pressure by axially advancing the screw toward the blocked outlet. The volume of the pressurized shot was then measured by measuring the axial position of the screw. The pressurized shot was then purged and the purged shot was weighed. The specific volume of the material was calculated as the ratio of the shot volume divided by its weight. These steps were then repeated for a plurality of different temperatures and pressures. However, the testing accuracy was not much good.

Chiu et al. (1995) established a method for generating PVT data for thermoplastics on a microcomputer-controlled IMM, which utilized the injection barrel of the IMM as a pressure chamber for determining the specific volume of thermoplastics at various pressure and temperature conditions. The quantities measured with this apparatus were the hydraulic pressure, the barrel temperature, and the screw position. The hydraulic pressure was controlled by adjusting the openings of the relief valve and the servovalve. The specific volume of the polymer in the barrel was calculated from the response of the screw position sensor. A schematic diagram of the control system is given in Fig. 12.

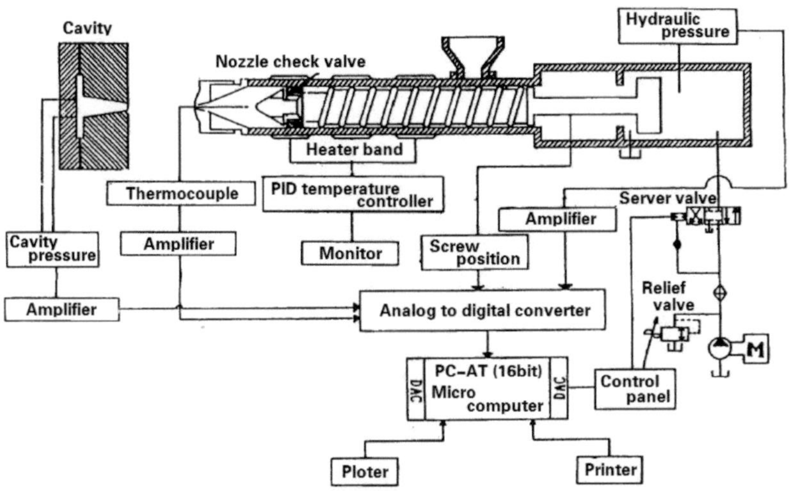

Fig. 12. Microcomputer-controlled injection molding system developed by Chiu et al. (1995).

Park et al. (2004) presented a dilatometer that can measure the PVT properties of polymer melt in a molten state using a foaming extruder and a gear pump. They developed this on-line experimental apparatus, based on an extrusion system. The basic rationale is the determination of the specific volume of the polymer melt by measuring the volume and mass flow rates separately, while controlling the pressure and temperature in the extrusion system independently. Fig. 13 shows a schematic of the designed PVT measurement system. Two extruders were used. The first extruder of the tandem extrusion system plasticated polymer pellets into a melt. The second extruder of a tandem system was used to uniformly lower the melt temperature, while building up the pressure and reducing the fluctuation of pressure. The volumetric flow rate can be determined by measuring the rotational speed of a gear pump due to its positive displacement nature for a polymer melt.

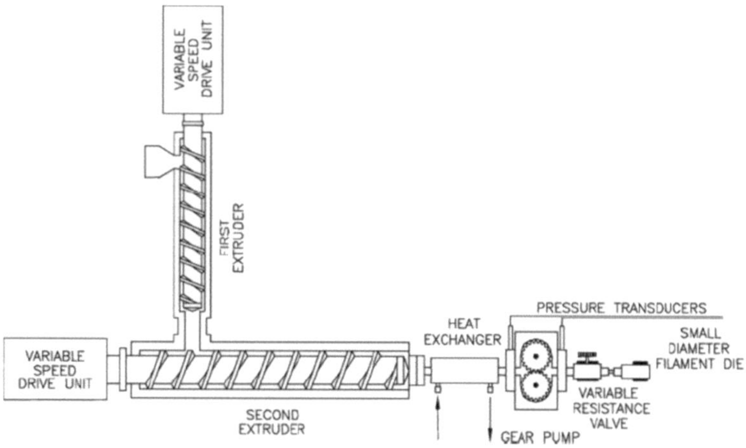

Fig. 13. Testing setup for PVT data of a Polymer/CO_2 solution developed by Park et al. (2004).

All of the experimental results of those above experiments were significantly limited by the machines, especially the maximum pressures were only 9.646 MPa (Nunn, 1989), 96.44 MPa (Chiu et al., 1995) and 28 MPa (Park et al., 2004), respectively. Wang et al. (2009) developed a novel method for testing PVT relationships of polymers based on an IMM. The advantage of this testing approach is that it can be used to obtain PVT data of polymers in mold directly by a special testing mold, however the temperature range was limited below 130 °C, and the pressure range was limited up to 120 MPa.

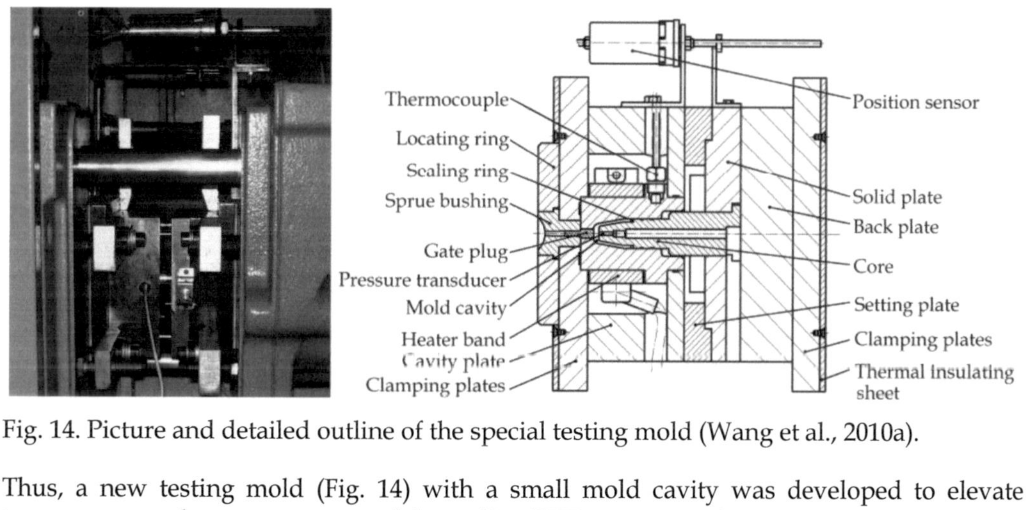

Fig. 14. Picture and detailed outline of the special testing mold (Wang et al., 2010a).

Thus, a new testing mold (Fig. 14) with a small mold cavity was developed to elevate temperature and pressure ranges of the on-line PVT equipment (Wang et al., 2010a), and the heating rate was improved either. This testing mold was assembled onto an IMM to measure the PVT data under normal processing conditions. It used the mold cavity as a pressure chamber for determining the specific volume of polymers under various pressure and temperature conditions. The material was enclosed and pressurized in the rigid mold cavity with a core; the core was close fitting in the mold cavity. During the measuring cycle, the volume of the material was recorded by the measurement of the displacement of the core. The advantages of this method are:

- The use of the clamping system of the IMM to control the pressure of the sample directly;
- The simplicity in design that can be achieved;
- It is feasible to set parameters directly on the IMM and provide industrial processing conditions;
- The sample is molded and tested directly in an IMM.

2.4 Other techniques

There are also some other techniques for measuring the PVT behavior of polymers utilizing ultrasonic technique or X rays. Kim et al. (2004) investigated PVT relationship by ultrasonic technique and its application for the quality prediction of injection molded parts. It proved that the feasibility of ultrasonic response in describing PVT behaviour offered an important basis for monitoring the progress solidification within the mold after injection of polymer melt. Michaeli et al. (2007) developed a new method to determine the specific volume of polymers over a wide range of temperature and pressure based on X-ray attenuation. This method allows the application of different cooling rates enabling the investigation of the density depending on the thermal history.

3. Testing modes of polymer PVT properties

There are several testing modes of operation: isothermal compression taken in order of increasing temperature, isothermal compression taken in order of decreasing temperature, isobaric heating and isobaric cooling. Almost all the PVT measurement apparatuses can be used in these several testing modes.

The different testing modes to obtain a PVT diagram can be listed as: (Luyé et al., 2001)

- Isothermal compressing taken in order of increasing temperature (Mode 1): the specific volume is recorded along isotherms (in order of increasing temperature) and at different pressures. Between each temperature there is a stabilization time of a few minutes before the next isotherm. This procedure is often considered the "standard" one.
- Isothermal compressing taken in order of decreasing temperature (Mode 2): the procedure is the same as mode (1), but the isotherms are in order of decreasing temperature.
- Isobaric heating (Mode 3): The specific volume is recorded with a constant heating rate while a constant pressure is maintained and the temperature is varied. When the temperature scan is completed, another pressure is selected and the temperature is varied again.
- Isobaric cooling (Mode 4): The specific volume is recorded along isobars with a fixed cooling rate.

First, since in injection molding the polymer enters the cavity in the melt state and is cooled in the mold, it seems obvious that the transition that must be considered is crystallization. So modes (1) and (2) appear inconvenient because they show the melting transition even if mode (1) is often used.

Second, modes (1) and (2) exhibit a single transition temperature (respectively, melting and crystallization temperatures) independent of the pressure. Nevertheless, this phenomenon is in total contradiction with thermodynamics because both melting and crystallization temperatures are increased by the pressure. Actually, the single observed transition temperature is explained by the following arguments: When the isotherms are followed in the order of increasing temperature, the polymer melts for a given temperature at the lower pressure. Then, when the pressure is increased, it does not have enough time to recrystallize because the crystallization is very slow in this range of temperature. Therefore, the apparent melting temperature on the PVT diagram corresponds to the lowest pressure used in the procedure. When the isotherms are followed in the order of decreasing temperature, due to the thermodynamics, crystallization occurs at a given temperature for the highest pressure (at the end of the isotherm if the pressures are scanned in increasing order). Then, when measuring the next point—that is, the lowest pressure and the next lowest temperature—the polymer is already crystallized, but generally the temperature is lower than the melting point corresponding to this pressure so it cannot melt. Therefore, the apparent single crystallization temperature coincides with the crystallization temperature for the highest pressure used in the procedure. Moreover, since crystallization can be a very slow phenomenon, it can either occur or not occur depending on the chosen stabilization time between two temperatures.

For all the previous reasons, the best procedure seems to be isobaric measurements in cooling mode, mode (4), because in that case the observed transition is the crystallization, and it depends only on the pressure and the cooling rate. The effect of the cooling rate can be investigated. Nevertheless, as it will be shown further, the analysis of the data still faces a difficulty because of the thermal gradient that occurs in the sample, complicating the effect of the cooling rate on the crystallization kinetics.

4. Application of polymer PVT data

In the introduction section, eight major areas were listed for the application of polymer PVT data. Actually, especially for the injection molding, polymer PVT data could be important in two areas: numerical simulation and process control. So in this part, we firstly introduce the equation of state widely used to describe the polymer PVT data, and then show you an example on numerical simulation using different PVT data, at last the development of the control concepts based on the polymer PVT relationship will be introduced.

4.1 PVT equation of state

The equation-of-state (EOS) is very important in describing the thermodynamic properties of liquids and gas-liquid solutions. Spencer and Gilmore (1949) first developed the PVT relations for polymer using a modified van der Waal's equation. The EOS is correlated using the experimental data in a molten state, far above the glass transition temperature. Among the models used to describe the specific volume of polymers in many literatures, the Tait EOS is the most convenient and the most widely one used for polymers. As it was shown by Zoller and Fakhreddine (1994), Tait EOS represents well amorphous as well as semicrystalline polymer melts. It has only been used to describe the PVT behaviour at temperatures above the melting point for polymers. Nonetheless further developed modified 2-domain Tait Equation has been used.

4.1.1 Tait EOS

The PVT relationships can be represented by the 2-domain Tait EOS:

$$V(T,P)=V_0(T)\{1-C\ln[1+P/B(T)]\}+V_1(T,P) \tag{1}$$

where $V(T,P)$ is the specific volume at temperature T and pressure P, V_0 is the specific volume on the zero gauge pressure, C is 0.0894 (universal constant), and B represents the pressure sensitivity of the material. Two temperature domains are required to model the PVT relationship, because the thermodynamic properties of polymers change at the transition to the solid state. The volumetric transition temperature at zero gauge pressure is denoted by b_5, and the linear increase in the transition with pressure is denoted by b_6. The specific volume obtained by extrapolating the zero-isobar curve to the transition temperature is denoted by b_1. This value is the same for both domains when crossing the glass transition. However, when the material is semi-crystalline, the transition due to crystallization is accompanied by an abrupt change in specific volume, such that b_{1m} (the melt specific volume at b_5 and zero pressure) is greater than b1s. The temperature dependence of the specific volume is measured by b_2, while b_3 and b_4 characterize $B(T)$ in the solid and melt state. The specific volume becomes more pressure sensitive with increasing temperature when b_4 is positive. The constants, b_7, b_8 and b_9 characterize V_1 in the solid state.

For $T<b_5+b_6P$:

$$V_0=b_{1s}+b_{2s}(T-b_5) \tag{2}$$

$$B(T)=b_{3s}\exp[-b_{4s}(T-b_5)] \tag{3}$$

$$V_1=b_7\exp[b_8(T-b_5)-b_9P] \tag{4}$$

For $T>b_5+b_6P$:

$$V_0=b_{1m}+b_{2m}(T-b_5) \tag{5}$$

$$B(T)=b_{3m}\exp[-b_{4m}(T-b_5)] \tag{6}$$

$$V_1=0 \tag{7}$$

b_5, b_6, b_{1m}, b_{2m}, b_{3m}, b_{4m}, b_{1s}, b_{2s}, b_{3s}, b_{4s}, b_7, b_8 and b_9 were determined by fitting the experimental PVT data using a nonlinear regression. Software SPSS (SPSS Inc., Chicago, Illinois) could be used for the nonlinear regression. Before the nonlinear regression, the experimental data should be divided into two phases with the transition temperature. With the transition temperature at different pressure, b_5 and b_6 should be calculated firstly; then b_{1m}, b_{2m}, b_{3m}, b_{4m} in melt state and b_{1s}, b_{2s}, b_{3s}, b_{4s}, b_7, b_8, b_9 in solid state could be calculated separately.

4.1.2 Fitting of PVT data

An example was taken to illustrate the fitting of PVT data. PVT properties of a semi-crystalline polymer,PP (polypropylene), were measured through both the on-line PVT testing mold (Wang et al., 2010a) and a conventional piston-die dilatometer (PVT 100 from SWO Polymertechnik GmbH, 1998). The PVT data of the semicrystalline polymer PP were measured by on-line measurement in the temperature ranges from 17 to 160 °C, pressures from 50 to 200

MPa and heating rate of 15 °C/min; and by piston-die dilatometer in the temperature ranges from 30 to 260 °C, pressures from 20 to 150 MPa and the cooling rate of 2 °C/min. Fig. 15 shows the shape and dimensions of the cup sample used in the on-line measurement. The isothermal compression taken in order of increasing temperatures was applied as the experimental mode of the on-line measurement. The isobaric cooling mode was used by the piston-die dilatometer. The specific volume is recorded along isobars with a fixed cooling rate.

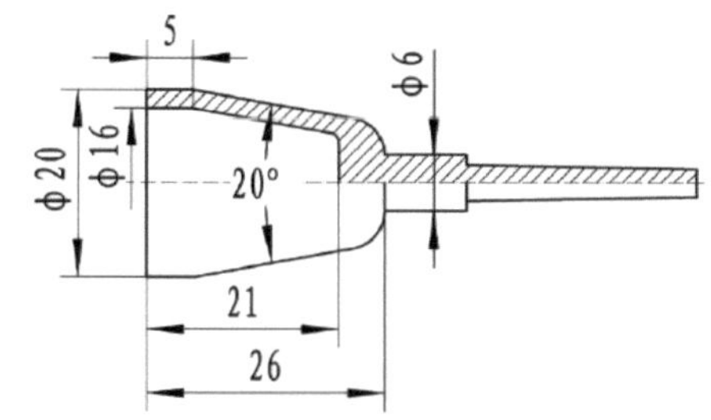

Fig. 15. Shape and dimensions of the cup sample used in the on-line measurement (Wang et al., 2010a).

The experimental results are shown by gray lines and cross dots separately in Fig. 16. The correlation PVT results of both on-line measurement and piston-die dilatometer calculated by Tait EOS are shown by solid lines and dots separately in Fig. 16. The characteristic parameters are listed in Table 1.

Parameter	On-line	Piston-die
b_5 (K)	403	417.64
b_6 (K/Pa)	0.00000015	3.022E-07
b_{1m} (m³/kg)	0.001295	0.001306
b_{2m} (m³/(kg·K))	8.588E-07	9.019E-07
b_{3m} (Pa)	77200000	74300000
b_{4m} (1/K)	0.003487	0.004006
b_{1s} (m³/kg)	0.001242	0.001195
b_{2s} (m³/(kg·K))	9.153E-07	4.929E-07
b_{3s} (Pa)	63200000	131900000
b_{4s} (1/K)	0.006881	0.003355
b_7 (m³/kg)	0.00002616	0.00009623
b_8 (1/K)	0.0714	0.1338
b_9 (1/Pa)	3.355E-09	3.973E-08

Table 1. Correlated parameters for the Tait EOS (Wang et al., 2010a).

From Fig. 16, the specific volume of on-line measurement at 20 °C and 0 MPa is 0.64% higher than that of piston-die dilatometer. The specific volume data of on-line measurement increase more with temperature in comparison with those of piston-die dilatometer in a rubbery state, whereas volume change at the melting point temperature of on-line measurement is less than that of piston-die measurement. Above the melting point temperature, although the specific volume of on-line measurement is slightly higher than that of piston-die dilatometer, a difference in the specific volume between them is not significant; the PVT data of the two measurements give a good agreement within 0.045%.

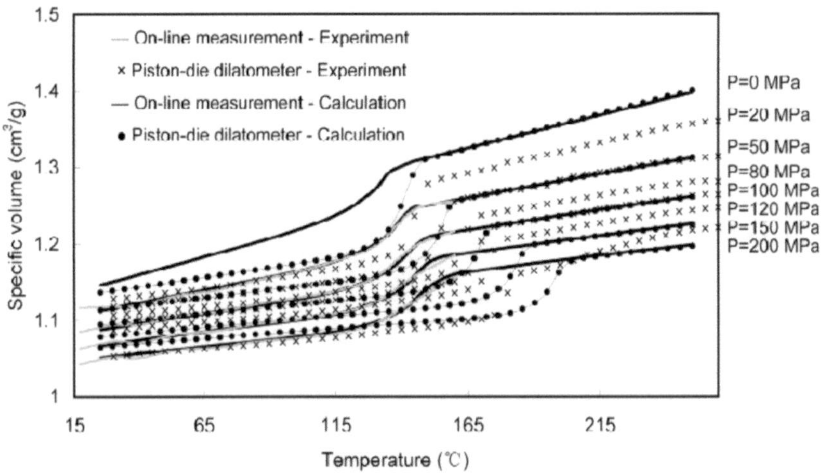

Fig. 16. Experiment and correlation PVT diagrams of PP (Experiment using on-line measurement at T=17-160 °C and P=50, 100, 150, 200 MPa; Experiment using piston-die dilatometer at 30-260 °C and P=20, 50, 80, 100, 120, 150 MPa; Correlation using Tait EOS at T=20-250 °C and P=0, 50, 100, 150, 200 MPa) (Wang et al., 2010a).

The reason for the significant difference in the rubbery state must be related to many factors. Because the principles of both of the two measurements are similar, the main differences between them are the different procedures and sample forms. For the procedures, the influence of cooling rate or heating rate is the most important factor. The heating rate of the on-line measurement was 15 °C/min; while the cooling rate of the piston-die dilatometer was only 2 °C/min. The resulting specific volume in the solid state clearly increases with increasing heating rate.

For the sample forms, the sample of the piston-die dilatometer was a pillar with 7.4 mm diameter and 0.5–1 g weight, and the sample (Fig. 15) of the on-line measurement was a cup with 26mm length, 2 mm thickness and 2.2–2.4 g weight. To measure PVT properties accurately, a large amount of samples should be put into the dilatometer. Actually, a sample with large length diameter ratio is the best one to be measured, for it is more convenient and accurate to get the displacement, even using a position sensor with low precision. The thickness of the sample used by on-line measurement was only 2 mm. The PVT data thus generated would be closer to typical component part thickness. So the cup sample with 26 mm length and 2 mm thickness was better for testing than the pillar sample.

4.2 Numerical simulation with PVT data for injection molding

PVT properties of a semicrystalline polymer, PP, were measured through both the on-line testing mold and a conventional piston-die dilatometer. Both of the PVT data were used in numerical simulations for injection molding process. The simulation was carried out by software Moldflow (Autodesk, Inc., San Rafael, CA), and the results of warpage and shrinkage prediction with different PVT data were compared and investigated.

4.2.1 Numerical simulation process with PVT data

The PVT relationships of polymers are the most important factors influencing the shrinkage and warpage of polymer products. In order to show the application of PVT data in numerical simulation and verify the reliability of the PVT results obtained by using the on-line measurement and piston-die dilatometer, prediction of sample shrinkage and warpage was carried out by using Moldflow Plastics Insight (MPI) 6.1. The approach implemented took advantage of the Finite Element (FE) analysis to simulate component fabrication and investigate the main causes of defects. The basic idea is to create a model of the geometry or mold to be analyzed as Fig. 17 shows.

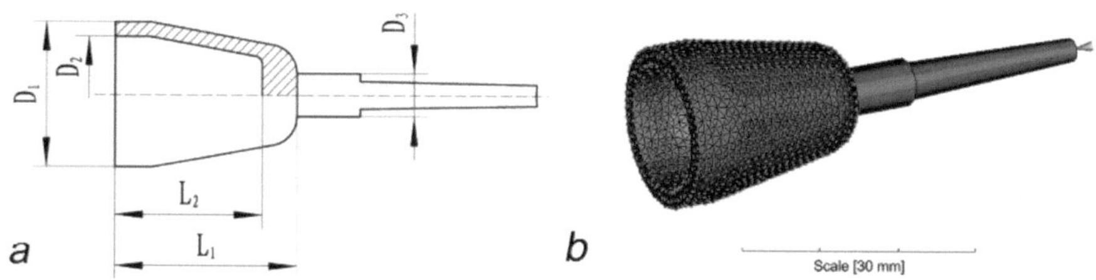

Fig. 17. Principle dimensions (a) and FE mesh (b) of the sample (Wang et al., 2010a).

In order to get an accurate prediction, 3D numerical simulation was used to simulate injection molding process. Fig. 17(a) shows the location of the five characteristic dimensions (D_1, D_2, D_3, L_1 and L_2) where results have been analyzed, and Fig. 17(b) shows the FE mesh of the sample. The calculation mesh number is 54338 and the mesh type is four-node tetrahedral element.

Efforts were made to reduce differences in prediction, except the different PVT data, the other input material data such as rheological, thermal and mechanical properties were all measured through the appropriate instruments and kept identical in the simulation. For viscosity, the Cross-WLF model was used, and the flow curves are given in Fig. 18. For PVT data, the modified Tait EOS was used. These other input material data can be referred to Table 2, which were provided by Beijing Yanshan Petrochemical Co., Ltd., SINOPEC.

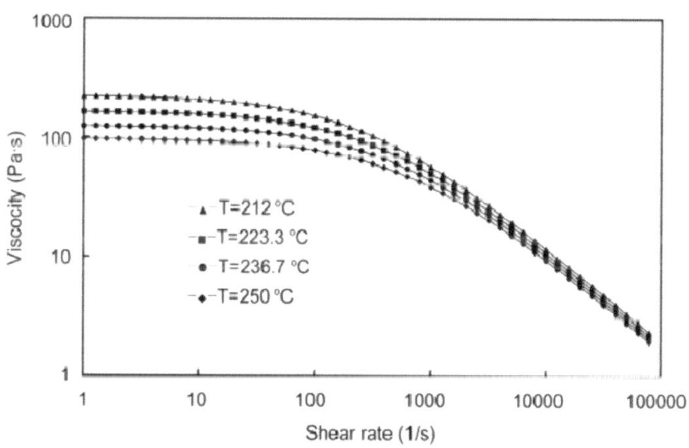

Fig. 18. Flow curves of PP using Cross-WLF model (Wang et al., 2010a).

Mechanical Properties	
Elastic modulus (MPa)	1480
Poissons ratio	0.38
Shear modulus (MPa)	570
Transversely isotropic coefficient of thermal expansion (CTE) data	
Alpha1 (1/°C)	1.07×10^{-4}
Alpha2 (1/°C)	1.07×10^{-4}
Stress at yielde (MPa)	24.8
Strain at break (%)	45
Thermal Properties	
Specific heat data at heating/cooling rate -0.3333 °C/s	
Temperature (°C)	Specific heat (J/kg·°C)
60	1757
100	2029
150	3158
160	5368
170	4886
200	2259
230	2364
Thermal conductivity data	
Temperature (°C)	250
Thermal conductivity (J/kg·°C)	0.26
Rheological Properties	
Cross-WLF viscosity model coefficients	
n	0.2065
Tau* (Pa)	57861
D1 (Pa·s)	2.25304E+14
D2 (K)	263.15
D3 (K/Pa)	0
A1	34.12
A2 (K)	51.6
Transition temperature (°C)	135
Melt mass-flow rate (MFR)	
Temperature (°C)	230
Load (Kg)	2.16
Measured MFR (g/10min)	45.0

Table 2. Mechanical, thermal and rheological properties of the material (Wang et al., 2010a).

Similarly, the processing conditions were all identical. As a result, it is possible to focus solely on the PVT data differences. According to the recommended and experimental processing, the process parameters used are reported in Table 3.

Parameter	Value
Volumetric injection rate (cm^3/s)	30
Mold temperature (°C)	70
Melt temperature (°C)	230
Hold pressure (MPa)	1.8
Hold time (sec)	10
Cooling time (sec)	20

Table 3. Process parameters (Wang et al., 2010a).

4.2.2 Comparison of results with different PVT data

The influence of the PVT data on shringkage and warpage prediction of the sample was examined. Results for shrinkage and warpage, using Moldflow, are given in Fig. 19 where displacements have been amplified by a factor of 5. The behavior of shrinkage and warpage is observed from the change in dimensions of the sample.

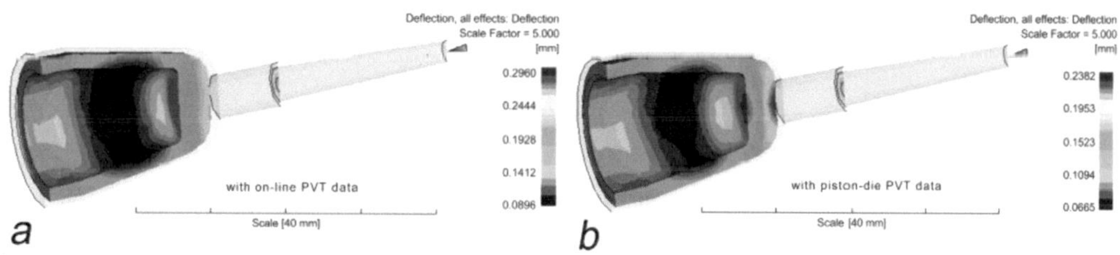

Fig. 19. Numerical simulation of shrinkage and warpage of the sample with on-line PVT data (a) and piston-die PVT data (b) (Wang et al., 2010a).

Table 4 indicates the principle dimensions of the sample measured by micrometer. The experimental shrinkage is 5.3038%, the shrinkage (s) calculation equation is:

$$s = 1 - V_{\text{molding}} / V_{\text{cavity}} \tag{8}$$

In Eq. 8, V_{molding} is calculated with the five average dimensions of Table 4, V_{cavity} is calculated with the dimensions of the mold cavity.

The experimental results for shrinkage and warpage of final parts are compared with the analysis results of software Moldflow. Shrinkage of the sample is shown in Fig. 20. The quantitative results of shrinkage, 4.557%, calculated by Moldflow with on-line PVT data are observed as being somewhat underestimated in this study, and it is larger than 2.183% analyzed with piston-die PVT data, but much closer to the experimental shrinkage 5.3038%.

Warpage of the sample measured by micrometer and analyzed by Moldflow with both on-line and piston-die PVT data is shown in Fig. 21, and Table 5 indicates the experimental and analyzed deviation value of the principle dimensions. The qualitative description of warpage in this study is sufficiently comparable with the experimental results. The quantitative results are found to be slightly underestimated since the factors which result in warpage have not all been considered. Several factors affect the shrinkage and warpage of molded parts, including molecular orientation, the nonequilibrium state of the polymer PVT behavior, the effect of the flow field on polymer crystallization, anisotropy of material properties etc., which were considered to accurately describe the real conditions of injection molding as well as to predict the real shrinkage and warpage of molded parts.

	D_1 (mm)	D_2 (mm)	D_3 (mm)	L_1 (mm)	L_2 (mm)
Sample 01	19.60	15.52	5.66	25.42	20.38
Sample 02	19.44	15.50	5.72	25.60	20.50
Sample 03	19.42	15.44	5.68	25.60	20.56
Sample 04	19.48	15.46	5.72	25.50	20.60
Sample 05	19.42	15.44	5.64	25.40	20.50
Sample 06	19.60	15.56	5.66	25.44	20.50
Sample 07	19.42	15.52	5.62	25.62	20.62
Sample 08	19.62	15.58	5.56	25.44	20.34
Sample 09	19.52	15.54	5.70	25.50	20.58
Sample 10	19.40	15.40	5.68	25.60	20.70
Sample 11	19.46	15.50	5.62	25.50	21.00
Sample 12	19.50	15.40	5.62	25.12	20.48
Sample 13	19.48	15.50	5.70	25.50	20.50
Sample 14	19.58	15.54	5.50	25.16	20.74
Sample 15	19.46	15.50	5.78	25.56	20.46
Sample 16	19.46	15.50	5.70	25.52	20.54
Sample 17	19.50	15.50	5.72	25.58	20.48
Sample 18	19.46	15.50	5.64	25.50	20.60
Maximum	19.62	15.58	5.78	25.62	21.00
Minimum	19.40	15.40	5.50	25.12	20.34
Average	19.49	15.49	5.66	25.48	20.56
Cavity	20.00	16.00	6.00	26.00	21.00

Table 4. Principle dimensions of the sample measured by micrometer (Wang et al., 2010).

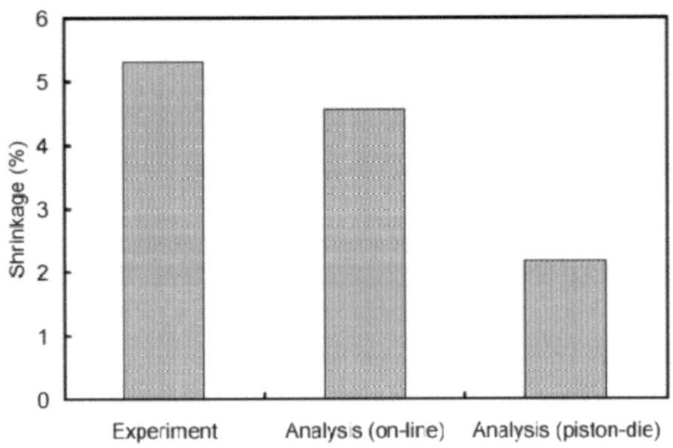

Fig. 20. Shrinkage of the sample (Wang et al., 2010a).

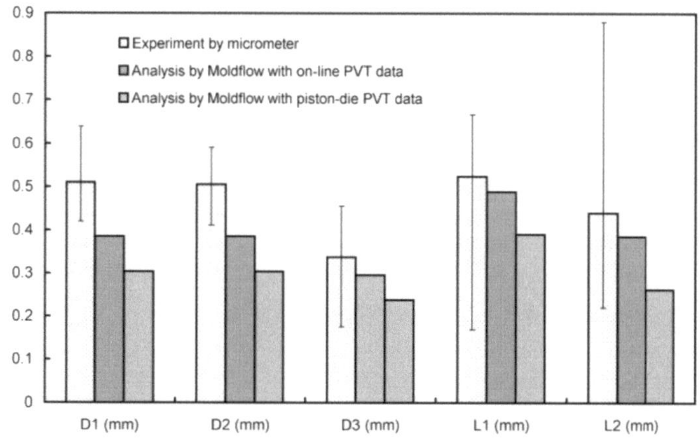

Fig. 21. Comparison of the deviation value of the principle dimensions (Wang et al., 2010a).

	D_1 (mm)	D_2 (mm)	D_3 (mm)	L_1 (mm)	L_2 (mm)
Experiment by micrometer	0.5100	0.505556	0.337778	0.524444	0.4400
Analysis with on-line PVT data	0.3856	0.385600	0.296000	0.488800	0.3856
Analysis with piston-die PVT data	0.3046	0.304600	0.238200	0.390500	0.2617

Table 5. Deviation value of the principle dimensions (Wang et al., 2010a).

For both PVT data, the warpage predictions are very different. Overall, as Fig. 21 shows, the warpage prediction with on-line PVT data is closer to the experimental results than the piston-die testing data.

4.3 Process control with PVT properties for injection molding

The PVT diagram describes the dependence of the specific volume on melt temperature and pressure. It is the basis for a constant quality with the same degree of orientation, residual stresses, and shrinkage which is the goal of isochoric process control (Johannaber, 1994). The application of PVT properties of polymer in process control for injection molding has been studied for years. In this section, injection molding process is described with PVT diagram, and then the development of some process control methods based on the polymer PVT properties are presented.

4.3.1 Description of injection molding process with PVT diagram

Injection molding is a cyclic process consisting of four phases: filling, melt compressing (or packing), holding and cooling, as shown by the typical PVT diagram in Fig.22(a), cavity pressure profile in Fig. 22(b), and cavity temperature profile in Fig.22(c). The filling process starts at Point A. The cavity pressure signal begins at Point B – where the melt plastics touch the pressure sensor for the first time – and then the pressure increases steadily as the filling proceeds. The filling phase is complete at Point C, where the cavity is only volumetrically filled by the melt without being compressed. The packing process then embarks and the pressure rises rapidly to the peak value at Point D. At point D, the injection pressure switches over to the holding pressure and the holding pressure control sets in. Thereafter, the melt within the cavity is maintained at an assigned pressure during the holding phase, when additional plastic melt can be packed into the cavity to compensate for the plastic shrinkage caused by cooling, so as to have the mold completely filled. This process continues until the gate is frozen, as marked at Point E. Point E is the end-point of the holding phase. Beginning at Point E, a phase of constant volume is maintained (isochoric phase). This isochoric phase is especially important because one strives for a minimum of orientation, residual and distortion. This phase is decisive for the dimensional accuracy of the molding. Reaching Point F in a uniform way is decisive for the constancy of weight and dimensions of the molding. After Point F, the molding cannot be influenced anymore. It shrinks unaffected, usually down to ambient temperature.

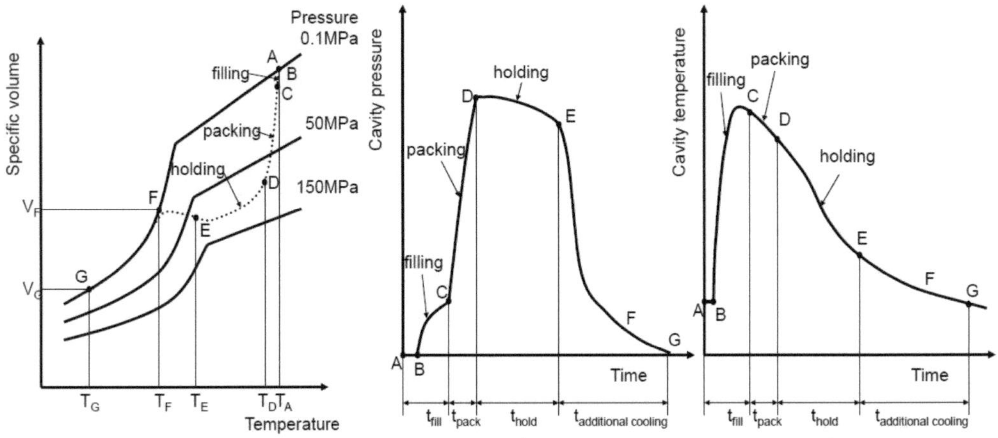

(a) (b) (c)

Fig. 22. PVT diagram (a), cavity pressure profile (b) and cavity temperature profile (c).

Therefore, Point D and E are the important transfer points to be controlled in order to obtain optimized holding phase control. There are many methods and theories on the best way to control the transfer of plastic from the injection unit to the mold cavity. Point D is referred to as the filling-to-packing switchover point. The filling-to-packing switchover control during injection molding plays an important role in ensuring the quality of the molded parts (Huang, 2007). There are many filling-to-packing switchover modes that machines are capable of using today, and these different filling-to-packing switchover modes were investigated and compared by some researchers (Edwards, 2003; Chang, 2002; Kazmer et al., 2010). However, it only affects one critical control point in the whole injection molding process. Wang et al. (2011) investigated the filling-to-packing switchover mode based on cavity temperature, and it was proved that the switchover mode based on cavity temperature can be used to accommodate the product weight change due to the variation of mold temperature. Point E is referred to as the holding phase end-point. While the transfer from fill to pack is particularly crucial, transfers from pack to hold and hold to screw recovery also significantly affect part quality. In previous work, authors also studied the end-point control of the holding phase and found that end-point control with cavity temperature can be used to adjust the holding time or cooling time to produce parts with optimum qualities (Wang et al., 2010b). Besides the control of these key transfer points, according to the typical PVT diagram (Fig 22(a)), one way to maintain a high yield rate from molding is to reproduce the trace from Point A to Point G (especially from Point D to Point E) in every injection molding cycle. Therefore, based on the PVT behaviour of polymer, the qualities of the parts could be optimized.

4.3.2 Development of process control methods with polymer PVT properties

Process variations in the injection molding process can be attributed to a wide variety of possible causes, including process pressure and temperature variations. The cavity pressure profile and its repeatability remarkably influence the quality of the molded part, especially on its weight, dimensional stability, mechanical behavior, and the surface quality (Huang, 2007). Many studies have proposed that the cavity pressure profile can be used to maintain high quality product and help to control the machine in the injection molding process (Huang, 2007). Gao et al. (1996) studied the dynamic behavior and control of cavity pressure for the filling and packing phases. It indicated that reproducing the cavity pressure curve in every shot is one way to maintain the best shot to shot consistency. However, the cavity temperature could not be negligible.

The influence of the cavity temperature on the production rate and the stability of the injection molding process, as well as on the quality of moldings, has been being investigated for a number of years. The cavity temperature influences the following: surface quality, after-shrinkage, orientation, residual stresses and the morphology of polymers (Čatić, 1979). The lower the cavity temperature, the higher the orientation, residual stresses, density of the plastic products, and the lower the surface quality (Menges et al., 1974). Irregularities in cavity temperature profiles from shot to shot can result in defects in the product such as poor mechanical behavior due to residual stresses and other defects causing warpage and differential shrinkage. The weight of the final product and the time required to cool the product are also affected by these cavity temperature profiles (Manero, 1996). However, the cavity temperature changes very fast from melt temperature to the ejected temperature during injection molding, it is difficult to control the cavity temperature exactly in a short time because of the slow response of most thermocouples or temperature sensors.

Therefore, a robust process control technique for achieving high process capabilities which is related to both the cavity pressure and temperature should be developed. It needs to take into account the actual behaviors of the melt plastics being injected into the cavity. The PVT diagram describes the dependence of the specific volume on melt temperature and pressure. It is the basis for a constant quality with the same degree of orientation, residual stresses, and shrinkage which is the goal of isochoric process control (Johannaber, 1994). The application of PVT properties of polymer in process control for injection molding has been studied for years. As the above section, based on the PVT behaviour of polymer, the qualities of the parts could be optimized. The quality of molded parts is primarily determined by the process inside the cavity. The variables of state, such as the pressure, temperature, and specific volume, play an important role (Menges, 1974). Yakemoto et al. (1993) had described an adaptive holding pressure control based on the prediction of polymer temperature in the mold cavity. In their study, they regarded temperature variations as a primary cause of fluctuations in product quality. Their results indicated a strong correlation between the temperature variations and product quality. Sheth and Nunn (1998, 2001) investigated an adaptive process control system to compensate for melt temperature variations that can occur from shot to shot. The pressure compensation phase of the molding process is modified to compensate for the melt temperature (barrel temperature) change in an effort to achieve dimensional consistency. Kamal et al. (1999) proposed two methods for the control of part weight in injection molding by controlling the peak melt pressure in the cavity and the estimated bulk temperature of the polymer in the cavity at the time when the gate freezes. Both control methods reduced variance in part weight. Michaeli et al. (2004, 2009) developed an online cavity pressure controller, which realized a constant gradient of the cavity pressure during the filling phase and calculated the cavity pressure with a PVT optimization in the holding pressure phase. They indicated that the specific volume displays a significant dependence on the viewed disturbance variables, in particular the melt/mold temperature.

5. Summaries

This chapter provided an introduction to different measurements of polymer PVT properties and the application of polymer PVT data for injection molding. Different measurements of polymer PVT properties includes conventional measurements (piston-die technique and confining-fluid technique), some improved experimental techniques considering the effect of cooling rate, shear rate and pressure, on-line techniques using injection molding machine or extruders, etc. Several testing modes of operation were discussed, including: isothermal compression taken in order of increasing temperature, isothermal compression taken in order of decreasing temperature, isobaric heating and isobaric cooling. Almost all the PVT measurement apparatuses can be used in these several testing modes. For injection molding, polymer PVT data could be important in two areas: numerical simulation and process control. So the 2-domain Tait EOS which is used widely in injection molding was introduced in this chapter, then an example on numerical simulation using different PVT data was shown, at last the development of the control concepts based on the polymer PVT relationship was introduced.

From these different research fields, we can see that the PVT properties of polymers play the most important role in both numerical simulation and process control for injection molding. The knowledge on PVT properties of polymers could be the fundamental concepts for the engineers in injection molding.

6. Acknowledgments

Some contents of this chapter originally appeared in the References. The author gratefully acknowledge the support of the National Natural Science Foundation of China (Grant No. 50973009), the Basic Research Fund of Beijing Institute of Technology (Project No. 3100012211108) and the InTech.

7. References

Barlow, J.W. (1978). Measurement of the P-V-T behavior of cis-1,4-polybutadiene, *Polymer Engineering and Science,* Vol. 18, No. 3, (February 1978), pp. 238-245. ISSN 0032-3888

Berry, J.M.; Brostow, W.M.; Hess, M. & Jacobs, E.G. (1998). P-V-T relations in a series of longitudinal polymer liquid crystals with varying mesogen concentration. *Polymer,* Vol. 39, No. 17, (August 1998), pp. 4081-4088. ISSN 0032-3861

Bhatt, S.M. & McCarthy, S.P. (1994). Pressure volume and temperature (P-V-T) apparatus for computer simulations in injection molding, *Society of Plastics Engineers Annual Technical Conference*, ISBN 9995283220, 9789995283223, San Francisco, May 1994, pp. 1831-1832.

Brown, C. & Hobbs, C. (1998). Pressure-Volume-Temperature behaviour of polymers during rapid cooling. *NPL Measurement Notes*: CMMT (MN) 033, November 1998.

Čatić, I. (1979). Cavity temperature—an important parameter in the injection molding process. *Polymer Engineering and Science*, Vol.19, No. 13, (October 1979), pp. 893-899. ISSN 0032-3888

Chakravorty, S. (2002). P-V-T Testing of Polymers under industrial processing conditions. *Polymer Testing*, Vol. 21, No. 3, (2002), pp. 313-317. ISSN 0142-9418

Chang, R.Y.; Chen, C.H. & Su, K.S. (1996). Modifying the Tait equation with cooling-rate effects to predict the pressure-volume-temperature behaviors of amorphous polymers: modeling and experiments, *Polymer Engineering and Science*, Vol. 36, No. 13, (July 1996), pp. 1789-1795. ISSN 0032-3888

Chang, T.C. (2002). Robust process control in injection molding - process capability comparison for five switchover modes. *Journal of Injection Molding Technology*, Vol. 6, No. 4, (December 2002), pp. 239-246. ISSN 1533-905X

Chiu, C.P.; Liu, K.A. & Wei, J.H. (1995). A method for measuring P-V-T relationships of thermoplastics using an injection molding machine, *Polymer Engineering and Science*, Vol. 35, No.19, (October 1995), pp. 1505-1510. ISSN 0032-3888

Edwards, R.; Diao, L. & Thomas C.L. (2003). A comparison of position, cavity pressure, and ultrasound sensors for switch/over control in injection molding. *Society of Plastics Engineers Annual Technical Conference*, ISBN , Nashville, USA, May 2003, pp. 586-590.

Gao, F.; Patterson, W.I. & Kamal, M.R. (1996). Cavity pressure dynamics and self-tuning control for filling and packing phases of thermoplastics injection molding. *Polymer Engineering and Science*, Vol. 36, No. 9, (May 1996), pp. 1272-1285. ISSN 0032-3888

He, J. & Zoller, P. (1994). Crystallization of polypropylene, nylon-66 and poly(ethylene terephthalate) at pressures to 200 MPa: kinetics and characterization of products, *Journal of Polymer Science Part B: Polymer Physics*, Vol. 32, No. 6, (April 1994), pp. 1049-1067. ISSN 0887-6266

Hess, M. (2004). The use of pressure-volume-temperature measurements in polymer science. *Macromolecular Symposia*, Vol. 214, No. 1, (August 2004), pp. 361-379. ISSN 1022-1360

Huang, M.S. (2007). Cavity pressure based grey prediction of the filling-to-packing switchover point for injection molding. *Journal of Materials Processing Technology*, Vol. 183, No. 2-3, (March 2007), pp. 419-424. ISSN 0924-0136

Imamura, S.; Mori, Y.; Kaneta, T.; Kushima, K. & Kobunshi, R. (1996). Influence of accuracy of P-V-T measurement on the simulation of the injection molding process, Japanese Journal of Polymer Science and Technology, Vol. 53, (1996), pp. 693-699. ISSN 0023-2556

Johannaber, F. (1994). *Injection molding machines: a user's guide* (4th edition), Hanser Gardner Publications, ISBN 1569901694,Western Europe

Kamal, M.R.; Varela, A.E. & Patterson, W.I. (1999). Control of part weight in injection molding of amorphous thermoplastics. *Polymer Engineering and Science*, Vol. 39, No. 5, (May 1999), pp. 940-952. ISSN 0032-3888

Kazmer, D.; Velusamy, S.; Westerdale, S.; Johnston, S. & Gao, R. (2010). A comparison of seven filling to packing switchover methods for injection molding. *Polymer Engineering and Science*, Vol. 50, No. 10, (October 2010), pp. 2031-2043. ISSN 0032-3888

Kim, J.G.; Kim, H.; Kim, H.S. & Lee, J.W. (2004). Investigation of pressure-volume-temperature relationship by ultrasonic technique and its application for the quality prediction of injection molded parts. *Korea-Australia Rheology Journal*, Vol. 16, No. 4, (December 2004), pp. 163-168. ISSN 1226-119X

Lobo, H. (1997). New approaches for PVT measurements, Toronto meeting of the CAMPUS/ISO, 1997, Toronto, Canada.

Luyé, J.F.; Régnier, G.; Le, B.P.; Delaunay, D. & Fulchiron, R. (2001). P-V-T measurement methodology for semicrystalline polymers to simulate injection molding process, *Journal of Applied Polymer Science*, Vol. 79, No. 2, (January 2001), pp. 302-311. ISSN 0021-8995

Manero, F. (1996). Cavity temperature and control in thermoplastics injection moulding. 1996, Master's Thesis, McGill University

Menges, G.; Wübken, G. & Čatić, I. (1974). Thermal balance, temperature distribution, and thermal design of injection molds. Plastverarbeiter, Vol. 25, No. 3, (1974), pp. 155-158. ISSN 0032-1338

Menges, G. (1974). Findings for normal operations from studies into process control. Plastverarbeiter, Vol. 25, No. 5, (1974), pp. 265-274. ISSN 0032-1338

Menges, G. & Thienel, P. (1977). Pressure-specific volume-temperature behavior of thermoplastics under normal processing conditions, *Polymer Engineering and Science*, Vol. 17, No. 10, (October 1977), pp. 758-763. ISSN 0032-3888

Michaeli, W. & Gruber, J. (2004). Increasing quality by online control of the cavity pressure. *Society of Plastics Engineers Annual Technical Conference*, ISBN 0975370707, 9780975370704, Chicago, USA, May 2004, pp. 688-692.

Michaeli, W.; Hentschel, M. & Lingk, O. (2007). A novel approach for measuring the specific volume of (semi-crystalline) polymers at elevated cooling rates using X-rays. *Society of Plastics Engineers Annual Technical Conference*, ISBN 1604232145, 9781604232141, Ohio, USA, May 2007, pp. 1608-1612.

Michaeli, W. & Schreiber, A. (2009). Online control of the injection molding process based on process variables. *Advances in Polymer Technology*, Vol. 28, No. 2, (Summer 2009), pp. 65-76. ISSN 0730-6679

Nunn, R.E. (1989). Adaptive process control for injection molding, U.S. Patent 4850217, July 1989.

Park, C.B.; Park, S.S.; Ladin, D. & Naguib, H.E. (2004). On-line measurement of the P-V-T properties of polymer melts using a gear pump, *Advances in Polymer Technology*, Vol. 23, No. 4, (Winter 2004), pp. 316-327. ISSN 0730-6679

Piccarolo, S. (1992). Morphological changes in isotactic polypropylene as a function of cooling rate. *Journal of Macromolecular Science - Physics*, Vol. 31, No. 4, (1992), pp. 501-511. ISSN 0022-2348

Quach, A. & Simha, R. (1971). Pressure-volume-temperature properties and transitions of amorphous polymers; polystyrene and poly (orthomethylstyrenes). *Journal of Applied Physics*, Vol. 42, No. 12, (November 1971), pp. 4592-4606. ISSN 0021-8979

Sato, Y.; Yamasaki, Y.; Takishima, S. & Masuoka, H. (1997). Precise measurement of the P-V-T of polypropylene and polycarbonate up to 330 °C and 200 MPa, *Journal of Applied Polymer Science*, Vol. 66, No. 1, (October 1997), pp. 141-150. ISSN 0021-8995

Sheth, H.R. & Nunn, R.E. (1998). An adaptive control methodology for the injection molding process. Part 1: Material data generation. *Journal of Injection Molding Technology*, Vol. 2, No. 2, (June 1998), pp. 86-94. ISSN 1533-905X

Sheth, H.R. & Nunn, R.E. (2001). An adaptive control methodology for the injection molding process. Part 2: Experimental application. *Journal of Injection Molding Technology*, Vol. 5, No. 3, (September 2001), pp. 141-151. ISSN 1533-905X

Spencer, R.S. & Gilmore, G.D. (1949). Equation of state for polystyrene. *Journal of Applied Physics*, Vol. 20, No. 6, (June 1949), pp. 502-506. ISSN 0021-8979

van der Beek, M.H.E.; Peters, G.W.M. & Meijer, H.E.H. (2005a). A dilatometer to measure the influence of cooling rate and melt shearing on specific volume. *International Polymer Processing*, Vol. XX, No. 2 (2005) 111-120. ISSN 0930-777X

van der Beek, M.H.E.; Peters, G.W.M. & Meijer, H.E.H. (2005b). The influence of cooling rate on the specific volume of isotactic poly (propylene) at elevated pressures, Macromolecular Materials and Engineering, Vol. 290, No. 5, (May 2005), pp. 443-455. ISSN 1438-7492

Wang, J.; Peng, J. & Yang, W. (2011). Filling-to-packing switchover mode base on cavity temperature for injection molding. Polymer-Plastics Technology, Vol. 50, No. 12, (August 2011), pp. 1273-1280. ISSN 0360-2559

Wang, J.; Xie, P.; Ding, Y. & Yang, W. (2009). On-line testing equipment of P–V–T properties of polymers based on an injection molding machine. *Polymer Testing*, Vol. 28, No. 3, (May 2009), pp. 228-234. ISSN 0142-9418

Wang, J.; Xie, P.; Ding, Y. & Yang, W. (2010a). Online pressure–volume–temperature measurements of polypropylene using a testing mold to simulate the injection-molding process. *Journal of Applied Polymer Science*, Vol. 118, No. 1, (October 2010), pp. 200-208. ISSN 0021-8995

Wang, J., Xie, P., Ding, Y. & Yang, W. (2010b). Study on the end-point control of holding phase during injection molding. *Advanced Material Research*, Vol. 87-88, (2010), pp. 222-227. ISSN 1662-8985

Yakemoto, K.; Sakai, T.; Maekawa, Z. & Hamada, H. (1993). Adaptive holding pressure control based on the prediction of polymer temperature in a mold cavity. *Society of Plastics Engineers Annual Technical Conference*, ISBN 9993108979, 9789993108979, New Orleans, USA, May 1993, pp. 2192-2202.

Zoller, P.; Bolli, P.; Pahud, V. & Ackermann, H. (1976). Apparatus for measuring pressure–volume–temperature relationships of polymers to 350 °C and 2200 kg/cm². *Review of Scientific Instruments*, Vol. 47, No. 8, (August 1976), pp. 948-952. ISSN 0034-6748

Zoller, P. & Walsh, D.J. (1995). *Standard pressure volume temperature data for polymers*. Technomic Publishing Company. Inc., ISBN 97815667663288, U.S.A., Lancaster

Zoller, P. & Fakhreddine, Y.A. (1994). Pressure-volume-temperature studies of semicrystalline polymers. Thermochimica Acta, Vol. 238, (June 1994), pp. 397-415. ISSN 0040-6031

Zuidema, H.; Peters, G.W.M. & Meijer, H.E.H. (2001). Influence of cooling rate on pVT-data of semicrystalline polymers. *Journal of Applied Polymer Science,* Vol. 82, No. 5, (October 2001), pp. 1170-1186. ISSN 0021-8995

Part 2

Optimization of Injection Molding Process

2

Effective Run-In and Optimization of an Injection Molding Process

Stefan Moser
Moser Process Consulting,
Germany

1. Introduction

This chapter is written to give developers and machine operators a better idea how to install robust processes or how to review and optimize these. The clear structure moves from basic introduction to in-depth application of the methods and tools, thus guiding readers through these processes. While this paper cannot replace a further deepening in this matter, it can assess its usefulness.

Since the publication of my article "Effective Run-In of an Injection Molding Process", (Moser & Madl, 08/2009) I have noticed that both the start phase of an optimization process and the end phase ("verification / validation") are the most critical parts. Due to this problem, I have decided to extend the upcoming article with the following chapters. Increasingly, "Processes Capability" is a necessary basis for accomplishing design transfer with the customer on a valid foundation. Also "Quality by Design" and "Design Space Estimations" are no longer foreign words within the injection molding business. Especially, the medical and automotive businesses call for process validation. This new chapter will, therefore, be divided into the following sections:

- Familiarization
- Screening
- Optimization
- Robustness
- Validation
- Summary

This procedure will help process' manager move through the setup or optimization process. Most students who joined, for instance, a "Process Capability Statistics"- or a "Design of Experiments" course, have difficulties finding the fulcrum or lever to complete the first steps. Consequently, they often just invest in "trial and error methods" to get their process to work. Also, common paradigms like "change one parameter at a time" will not help accelerate optimization or enable the improvement team to map the whole process, including interactions or nonlinear behaviours. Therefore, this chapter will outline tools to collect the main process factors, identify the disturbance factors and also some more special tools to interpret the impact of these on the process. The best way to get a run in or on optimization process started is to get a "complementary" team of experts at the table. Within in this team, it is important to lift the members into a mode where they are willing and motivated to work on the problem in an open, cooperative, and productive atmosphere. Beside this, because of its rules and structure, "team leading", "mediation skills", and "creativity tools" comprise an indispensable base to build up a mutual attitude towards the improvement process. Furthermore, these tools will lead the team from reflecting to describing a problem and to joint agreement on supported decisions, work-methods and actions. An additional advantage will be a clear structure, such as, for instance, the "DMAIC Cycle".

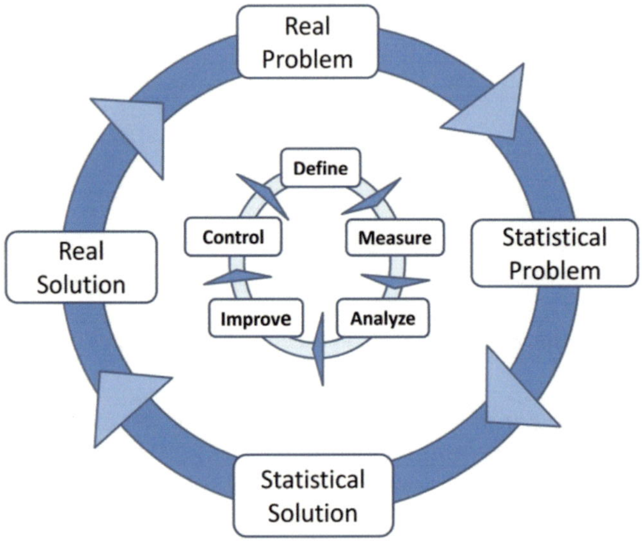

Fig. 1. DMAIC Cycle (Lunau, 2006, 2007).

The DMAIC Cycle „which as a logical further development of the Deming cycle, provides a good structure to get into the "problem solving process". Within this approach, the question "what is the 'real problem'?" is asked. Two different symptoms form cause and effect, so it is helpful to discuss this in the team of experts, for instance, with the following tools and methods. After the "Real Problem" has been defined, it is necessary to find a way to measure cause and effect of the problem. This might sound straight forward and logical, but in most cases, it is not done. This means, for instance, that a check of the capability of the measuring equipment is often not requested for measuring the whole process variation range of the "Process working space". Measurement methods and also the equipment calibration and capability should be validated *(each time)* prior to execution of the experiments. Otherwise, it may happen that a lot expensive, time- consuming experiments are preformed and also a lot of measurements are taken, but these are not adequate to describe cause and effect. (Process space)

The next step to get factor settings and systems- or product-attributes measurably defined is to analyze their correlation with a structured approach. Design of Experiments is a very powerful tool to do this. During the experiments, it is recommended to request every step in planning, such as:

- It has to be verified whether the latest setup (factor variation) of the worksheet is adequate for focusing on the desired responses of the targets (Fig. 18)

- if some responses are not measurable or quantifiable, continuing without adaption leads to fuzziness, respectively to bad "goodness of prediction models"
- It is not beneficial for the experimental room to extend far beyond the realm of objective target functions. Because this will automatically lead to more experiments and fuzziness due to more complex mathematics which is needed to describe cause and effect. (Fig. 18)

After the "test and analyzing phase" the gap between cause and effect could be closed with a mathematical prediction model. The purpose of the model is to reflect how factors and responses are related. On the basis of this, "model contour plots" (Fig. 29) can be generated and potential optima could be calculated and visualized.

Within in the oncoming Improve Phase, the optimum should be verified. After this verification, the robustness of this optimum setting could be rechecked with a reduced factor variation around this optimum to ensure the model-based calculations. In a last or parallel step to the robustness testing, the capability of the optimal setting, including the naturally given process variation, can be determined by using "Monte Carlo Simulations". The output will be, for instance: "Cpk"-value or "defects per million„ within an estimation of the work-point design space. These, "key process indicators" (which will be explained later) will then be a base for validating the process and making it comparable to other sub-processes. *(Cf.3)*

2. Familiarization

But again where to start? The following small collection of tools is a good start to get the first steps done, to reflect and research the process setup or process problem.

2.1 Ask why 5 times! (Michael L. George, 2005)

One of the easiest and most straightforward tools for getting familiar with a process setup or process problems is just to ask why, why and why again. Inquire if tree- or bubble diagrams can be used to document the root cause analysis. This and the following tools should be performed in a team only after it is certain that it follows the basic rules of good brainstorming / communication practice. This means: no direct "pointers" or school assignments should take place. Also there should be no criticism during the creative phase but rather only at the right time and then only constructively expressed.

Small Example of constructive, drill-down questioning:

Why is the injection part not of sufficient quality?
→ Because it contained some color strikes and dells.
Why does the part contain color strikes and dells?
→ Because the filling and cooling process are not as robust as they should be.
Why are the cooling and filling process not robust?
→ Because the density of the melted polymer is not homogenous.
Why is the melted polymer-density-distribution not as it should be?
→ Because the polymer granulate was not dry enough.
Why did the drier not work as it was supposed to?
→ Because the service hadn't been properly done.

2.2 Cause effect diagram

In addition to the "5xWhy", the cause effect diagram (Fig. 2), is very effective if there are people who already know a lot about the process. If they are used to it, they like thinking within cause and effect but still will be affected by suggestions from other team members to review their thought against the inputs from others and be motivated to think out of the box to find even more detailed reasons for product-rejects or process-failures. In combination with Brainstorming or the "5xWhy", it is even more focused and powerful.

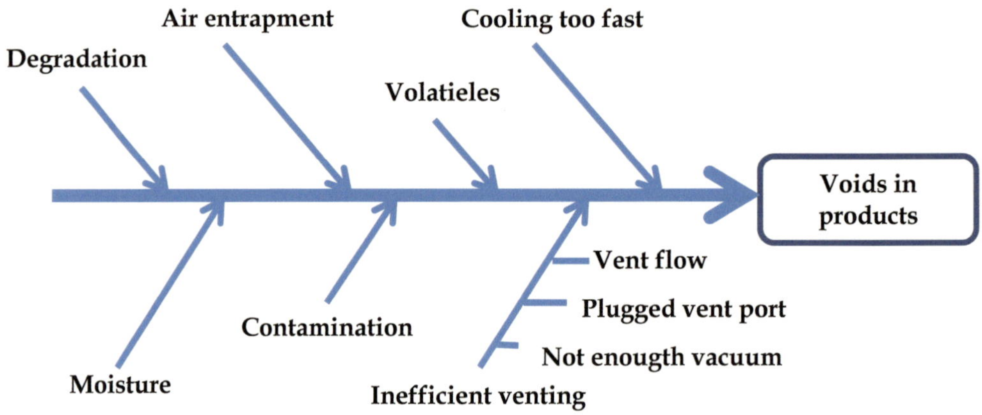

Fig. 2. Cause and effect diagram(Rauwendaal).

2.3 Ishikawa diagram

The Ishikawa diagram (Fig. 3) is the standard diagram to summarize cause and effect when concentrating on root cause and process-influence analysis. It is a good tool for discussing issues beyond the first impressions of why a process did and does not work or a product will not fulfill quality requirements. This is because this tool will guide the focus from different sources to a correlation of sources each time with a focus on a different reject-reason or process-failure. When meetings get stuck (because thoughts are spinning around) focus can be easily reset to another M-block[1]. In many cases, the output can be transferred to a FMEA "*Failure Mode and Effects Analysis*" or vice versa. The source of costs and process rejects are always defined by the product specification. Lowering these by adapting tolerances will instantly guide to lower costs, but may be critical for the next customer in a process line or the end customer who buys the product. The factor and its correlated target tolerances should be as wide as possible and as narrow as necessary.

2.4 AHP: Analytical Hieratical Process

Things are not always easy to interpret. Therefore, the AHP (combined with a grid analysis) is a very powerful tool to extract insights from a complex or fuzzy process. With process diagrams, Ishikawa diagrams or mind maps, the most influential factors are collected and can now be ranked according to Pareto's 20/80 law. This can be done by giving every team

[1] M-blocks are: Man, machine, management, measurement, method, material, milieu

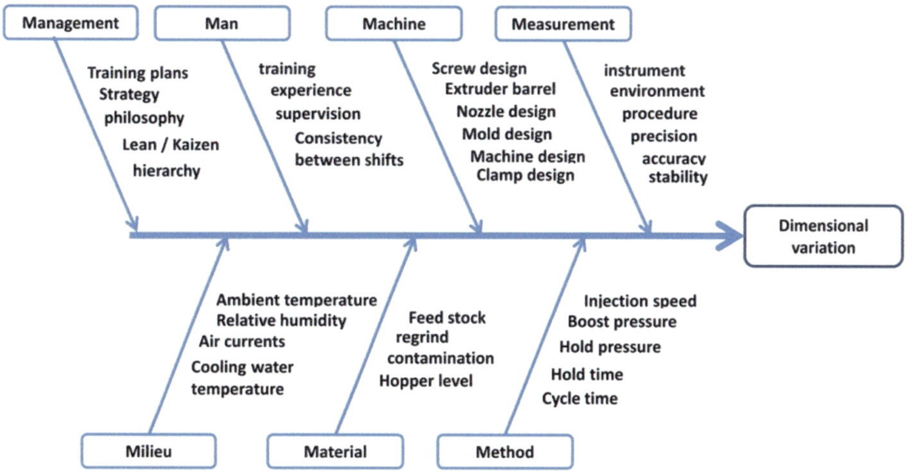

Fig. 3. Ishikawa diagram, (Rauwendaal).

member the chance to write the numbers from one (very important) to ten (less important) behind the collected factors on a white board, or perform a simple hand-up voting (*each number only one time)*.

After this is done, a maximum of ten factors of importance should be placed in the following matrix and weighted pairwise according to the importance of their influence. The direction of the questions is row versus column (Fig 4). Within the AHP, importance is leveled after the following scheme. (Vester, 2002) (Klein, 2007)

Weighting	Weights	Weighting counterpart	Weights
Extremely more important	9	Extremely less important	1/9
Significantly more important	7	Significantly less important	1/7
More important	5	Less important	1/5
Somewhat important	3	Somewhat less important	1/3
Equal important	1	Equally important	1

Table 1. Analytical hieratical process Weight basis.

In general, I recommend doing the weighting vice versa instead of filling the counterpart question automatically *(space below grey diagonal)*. Asking the questions "how much more is "Factor A" important than "Factor B"" and asking the opposite question staggered "how much more is 'Factor B' important than 'Factor A'" - again, will show the uncertainty of the knowledge and make it possible to reflect this fact within the grid diagram. In most cases, it is helpful to visualize the customers' demands prior to the factor ranking. A valuable input for defining customer values against product/process costs is for instance the "Kano model. If the prioritized quality criterion is not available, it needs to be developed because, in most cases, some target functions are more important to achieve then others, so compensating factor settings need to be developed after ranked target functions. On the grid-diagram (Fig. 5) the factors "holding pressure" and "nozzle temperature" are displaced a little bit laterally; this is because of the contradictorily ranked answers summarized in the matrix. Because of this inconsistency, influence of these factors should be discussed again. At some places, due

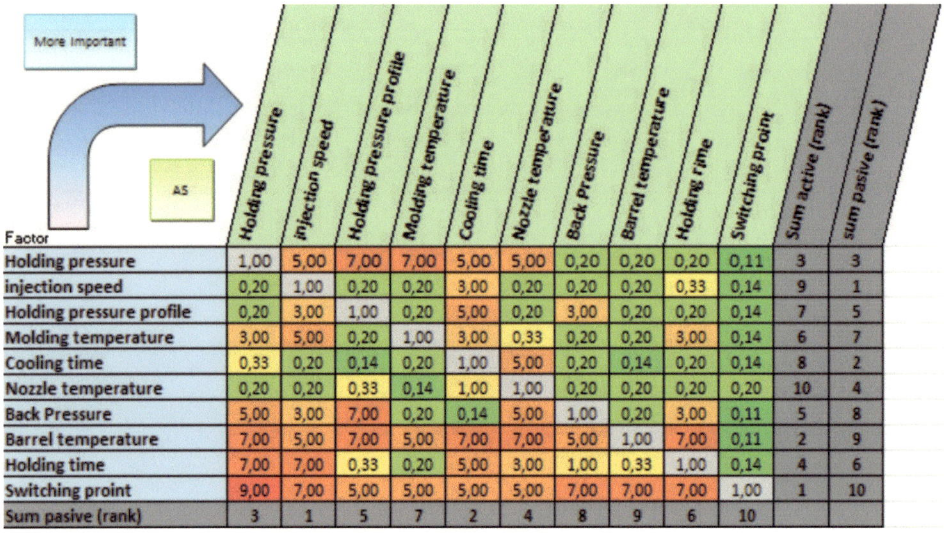

Factor	Holding pressure	injection speed	Holding pressure profile	Molding temperature	Cooling time	Nozzle temperature	Back Pressure	Barrel temperature	Holding rime	Switching proint	Sum active (rank)	sum pasive (rank)
Holding pressure	1,00	5,00	7,00	7,00	5,00	5,00	0,20	0,20	0,20	0,11	3	3
injection speed	0,20	1,00	0,20	0,20	3,00	0,20	0,20	0,20	0,33	0,14	9	1
Holding pressure profile	0,20	3,00	1,00	0,20	5,00	0,20	3,00	0,20	0,20	0,14	7	5
Molding temperature	3,00	5,00	0,20	1,00	3,00	0,33	0,20	0,20	3,00	0,14	6	7
Cooling time	0,33	0,20	0,14	0,20	1,00	5,00	0,20	0,14	0,20	0,14	8	2
Nozzle temperature	0,20	0,20	0,33	0,14	1,00	1,00	0,20	0,20	0,20	0,20	10	4
Back Pressure	5,00	3,00	7,00	0,20	0,14	5,00	1,00	0,20	3,00	0,11	5	8
Barrel temperature	7,00	5,00	7,00	5,00	7,00	7,00	5,00	1,00	7,00	0,11	2	9
Holding time	7,00	7,00	0,33	0,20	5,00	3,00	1,00	0,33	1,00	0,14	4	6
Switching proint	9,00	7,00	5,00	5,00	5,00	5,00	7,00	7,00	7,00	1,00	1	10
Sum pasive (rank)	3	1	5	7	2	4	8	9	6	10		

Fig. 4. AHP Matrix[2].

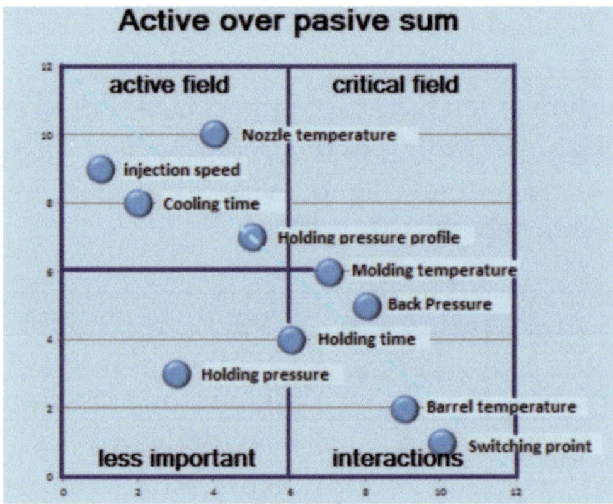

Fig. 5. AHP Grid.

to the complexity and a time reduction approach, it is sometimes recommended to do just a bilateral comparison-matrix with the part either below or on top of the grey diagonal. The other counterpart could also be filled out by asking or by being automatically computed. Often, therefore, the simplified schematic "2" = "more important"; "1" = "equal" and "0" = "less important" is used. This is also a good approach but will not be as differentiated as the previous AHP method.

In both cases, the factors can be weighted after importance by calculating the ranked row numbers at column "sum active ranked". This number and the calculated counterpart "sum passive rank" have to be plotted into the grid to visualize the factors' influence.

Now a new level of information has been extracted from the discussion. And the factors' importance can now be documented with the support of the whole team.

[2] Software Excel 2010, software operator S. Moser

Field	Meaning
Field "active":	factors with strong influence
Field "critical" :	factors with ambivalent influence
Field "less important" :	factors with less impact
Field " interactions" :	factors with possible interactions

Table 2. Interpreting of the grid (Fig. 5).

2.5 Contradictions and correlation

Besides the factor ranking, it is also important to get an understanding how these factors influence each other. For this reason, a similar matrix can be used in addition to a new question structure. Now the question should be: "Does factor 'A' reinforce the influence of factor 'B'?". In posing this question, one can get a better understanding of how the factors are correlated to each other or, in other words, how strong the interactions between these factors are. Thus the impact of potential contradictions can be exposed and documented. This is one of the most important project-management steps because of the necessary risk assumption. If the contradictions in the requirements are too stark to be compensated, the team needs to discuss whether the project should be stopped because of these "show stoppers" or "scope creepers". In any case, the risks should be displayed in a diagram which illustrates the probability of occurrence over the importance/impact of the risks. If there are any show-stoppers (i.e. risks with a large negative influence on the project and a high potential to occur) and they cannot be prevented, tools like TRIZ[3] may be helpful in resolving the contradictions. "TRIZ is problem-solving, analysis and forecasting tool derived from the study of patterns of invention in the global patent literature". In English the name is typically rendered as "the Theory of Inventive Problem Solving", and occasionally goes by the English acronym TIPS.

2.6 System modelling

A more recent approach to understanding process and complexity is to model the system interaction or dynamic. One interactive, easy-to-use software is the Consideo Modeler, which was used for Fig. 6, 7, 8, 9. At the beginning, this approach works similarly to mind mapping but can calculate feedback loops later on in order to visualize the system's dynamic. After the most influential factors have been collected and ranked due to importance, those factors can be connected with arrows to describe their impact. These arrows can be defined with the intensity of the factor-effect, the cause-direction *(enhancing, reducing)*, and the time-dependence of their effect (Fig. 6). One other advantage of this software-approach is that also "attributive" and "qualitative" factors can be embedded into the net-diagram. These factors are treated mathematically equally to quantitative factors in a first step. This is possible because the impact of the feedback loops of each factor *(factor→arrows→factor loops)* will be calculated iteratively. From this, the influence of the factors to a response can be interpreted. This method is also useful for visualizing what has been worked out in a "team problem discussion" by displaying the result of the extracted process on a net diagram (Fig. 6). After this, plots such as a "weighting matrix "(Fig. 9), in addition to the AHP (see 2.4) "root cause" and "cause and

[3] TRIZ / TIPS for more information see http://en.wikipedia.org/wiki/TRIZ

effect" diagrams (Fig. 8) can be easily generated, as well as the "insight matrix" (Fig. 7), to show how the factors affect responses.

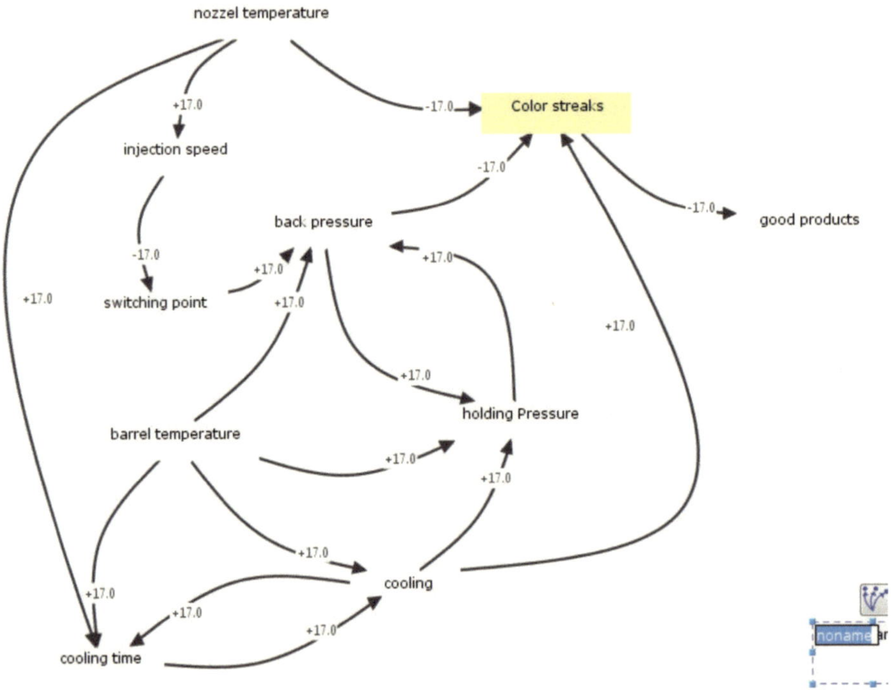

Fig. 6. Example "net diagram"[4].

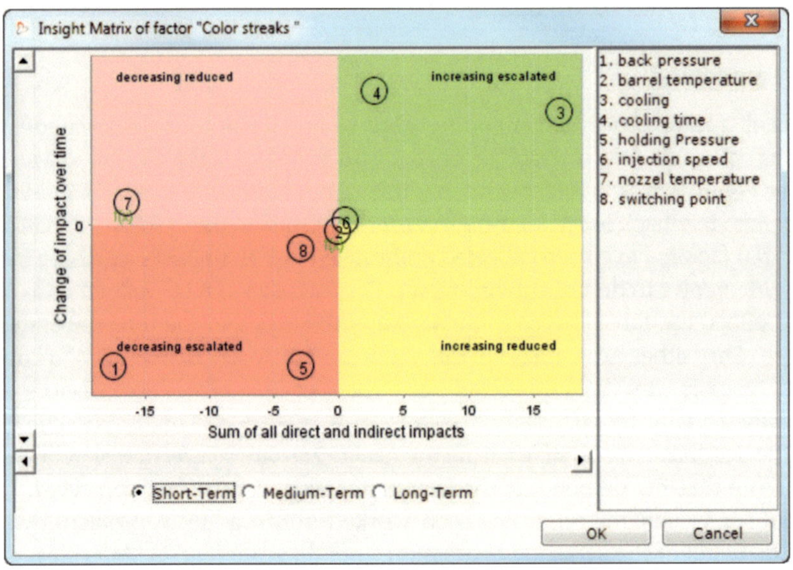

Fig. 7. Example Insight Matrix[4] of factor "color steaks".

[4] Software Consideo Modeler, www.consideo.com , software operator S. Moser

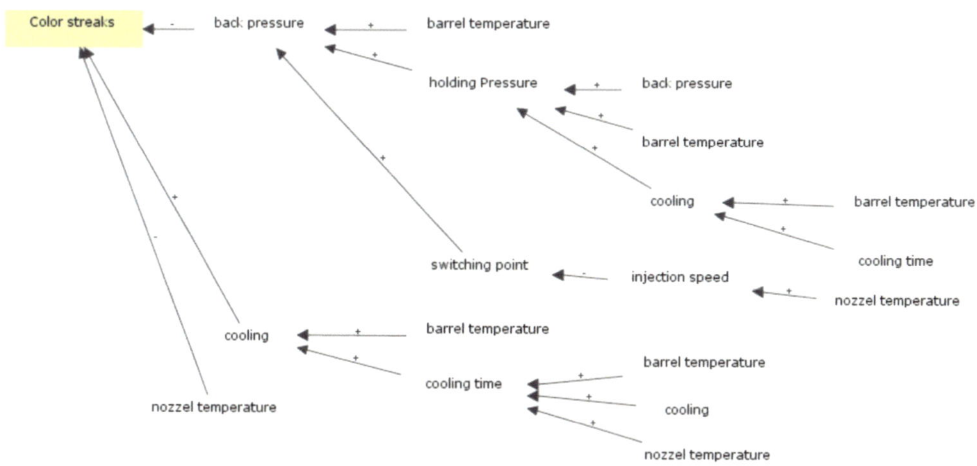

Fig. 8. Cause tree [4].

		1	2	3	4	5	6	7	8	9	10
1	injection speed						17				
2	good products										
3	cooling				17			17			17
4	holding Pressure									17	
5	barrel temperature			17	17			17		17	
6	switching point									17	
7	cooling time			17							
8	nozzel temperature	17						17			17
9	back pressure				17						17
10	Color streaks		17								
	Sum	17	17	34	51	0	17	51	0	51	51

Fig. 9. Weighting matrix [4].

2.7 Restrictions within familiarization steps

This approach *(and also the shortly described preceding methods)* will not deliver a whole picture of the process or an ideal setup, but it will help to concentrate on the really important factors. Furthermore, these tools will help to document the problem-solving process and to support the team in working results oriented and step by step -- in order to do the right things right.

So within difficult problems, tools are capable of raising the creative nouveau of generating innovative ideas or solutions. Instead of only endless problem-focused discussions, which only lead to "questions of power", "influence", "the problem history" and "particular blame" of team members, using supporting tools means the team can concentrate on solving problems by using the tools right! This helps to minimize distracting, time-consuming and conflicted meetings.

2.8 Reflecting the familiarization steps

After the "root cause analysis", factor prioritization and response target definition are done, it is necessary to review these values with a focus on "good project management practice".

Therefore, it has to be considered that doing a research on all desired functions will take time and will consume money and resources. In most projects, there is the sword of Damocles over the project team, which means that there is always not enough time, money and resources. In this context, one often hears a contradiction in terms such as, "We do not have the time for experiments ". This interaction is visualized in Fig. 10.

To briefly illustrate, it can be assumed that the functions to be examined need too much time. These circumstances can only be compensated by moving the timeline or tapping additional resources. Both options will impact the budget. Thus, it is always important to check at regular intervals if any (planned) actions are still result-related and necessary.

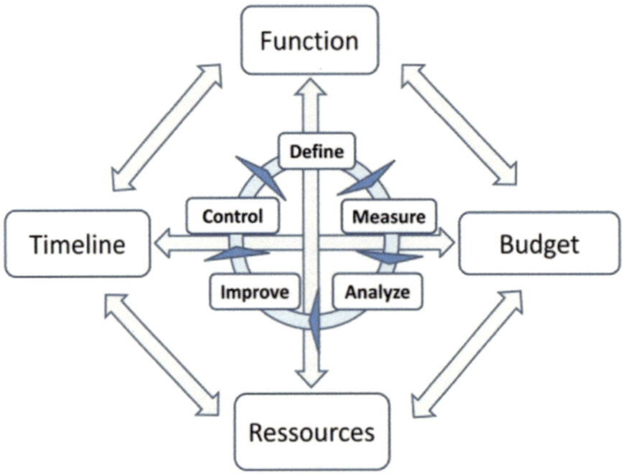

Fig. 10. Interaction between the main components of project management.

3. Screening

The challenge for machine operators is to get the run-in process done as quickly as possible. This means a minimum number of experiments with a maximum ability to describe cause and effect. The operators should be sensitized to the fact that small adjustments in the setup of the machine can have a great impact on the quality of the injection molding parts. Therefore, a well-structured approach and high quality data are needed. To reach these goals, the method of "Design of Experiments" will be introduced on the basis of the software "Modde"[5]. This is done because further steps "optimization" and the "Design Space Estimation" are incorporated tools within the software "Modde".

In general, the software is used when there is a lack of knowledge how cause and effect are related.

The use of "Design of Experiments" is an admission that the correlation of factors and effect could not be fully captured. This condition is depicted as a black-box (Fig. 11). By varying the factors within and according to a structured design, a regression model can be derived. From this model, the effect of the factors can be calculated. Since the experiments and thus the factors are varied to an estimated optimal range, some of the results are, of course, likely to deviate from the optimal targets. Nevertheless, all experiments and their results are very

[5] Modde is a software product of Umetrics, a company of MKS Instruments Inc.

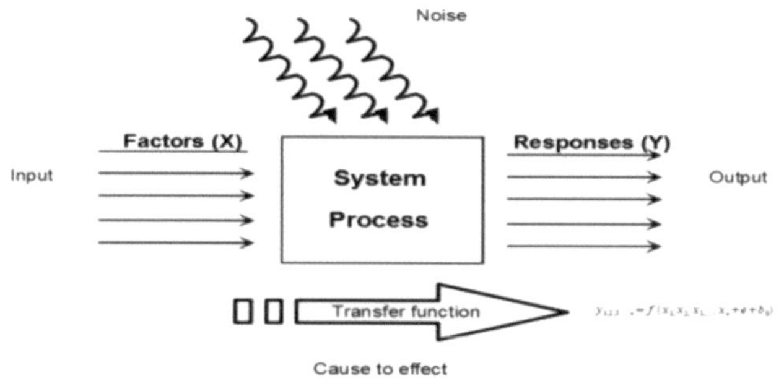

Fig. 11. Process Black-Box.

important because the basis of an entire image of it can be mapped on the work-space. The screening-process starts to extract the most influential factors from the familiarization process. These factors will be used to start with the Design of Experiments method. But not necessarily all factors must be examined in light of variability. So, some of these less important factors should be frozen at a certain level, which still ensures a good product quality. This is because the number of factors substantially affects the sum of the experiments; the number of evaluation criteria (responses) is of secondary importance. In the best case, the factors are quantitative, and so simple geometric designs can be generated. It is more difficult when they are qualitative, e. g. "machine 1" or "machine 2". Such qualitative or attributive parameters increase the number of experiments because they hamper the generation of the design. Once the factors have been identified, it is necessary to assess their effect. The effect is that which is exerted on the target variable when the factor is varied from its minimum to its maximum setting. Since all the factors are changed simultaneously in a factorial design, this effect is difficult to estimate.

It is therefore useful to debate and determine the factor variations within a group of experienced staff. Some factors are even trickier to formulate than the qualitative factors, such as temperature or pressure profiles. Just as in the machine, in the factorial experiment, the profiles can be programmed with some nodes such as (initial value + 9 nodes). The start and end values of the profile are known. Moreover, the process specifies a sloping curve (Fig 12, 13). If the profiles were programmed with real numbers, the sloping profile would necessitate the use of a great many programmed extra constraints. These factor restrictions limit the choice of experimental models and greatly increase the number of necessary experiments. For this reason, a mathematical formulation of the profiles is recommended which allows restrictions to be dispensed with entirely. Thus, the pressure profile is calculated, for instance, from the given initial value and the maximum decrease in pressure (in bar) per node:

$$\Delta p = \frac{\text{initial value max.} - \text{end value min.}}{\text{number of nodes} - 1} \qquad (1)$$

The following variation thus arises for each node in (2):

$$\text{Node } (i+1) = \text{node } (i) - (\text{min. } 0, \text{ max. } \Delta p) \qquad (2)$$

Another way to represent the profile is the use of a simple two point (FU 3).

$$\text{value of factor setting}_{(x)} = \text{mx} + b_0 + \varepsilon \qquad (3)$$

m = *Cf.* (4) ; x = node of factor profile; b_0 = bias ; $\varepsilon = noise$

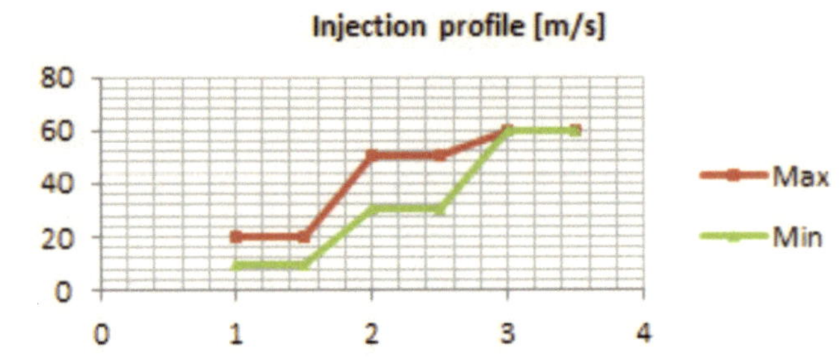

Fig. 12. Injection profile [2].

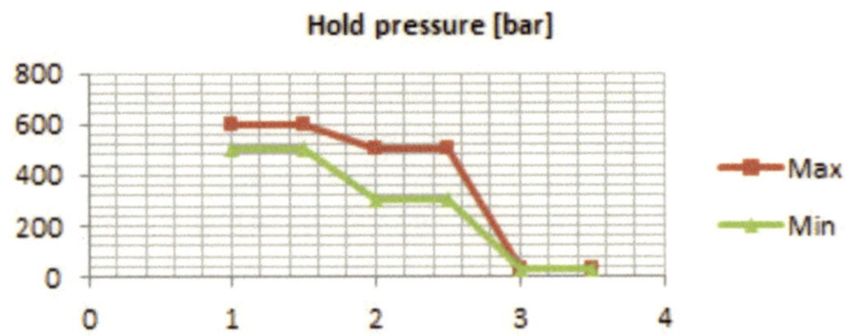

Fig. 13. Hold pressure [2].

For this, the initial value and the maximum slope (FU 4) of "initial $value_{max}$" and "end $value_{min}$" is required. From this data the increasing/decreasing constant slope/node can be described with two factors instead of several nodes (Fig.14, 15). These factors are the "start-value" and the constant amount to increase/ decrease per node, both must have a min./max. variation. Therefore, a constraint (FU 4) needs to be defined that, if decreasing, or increasing with a bigger constant amount beginning from a varied start level cannot lead to exceeding the final max. or min. final-profile-levels.

$$m = \left(\frac{\text{initial value}_{\max} - \text{end value}_{\min}}{2}\right) \qquad \Delta p = \text{const. for each node, note (5)} \tag{4}$$

$$\text{Node (i +1)} = \text{node (i)} - \Delta p \qquad \text{Value}_{\min} < \bigcup_{i=1}^{i=9} \text{node}_i < \text{Value}_{\max} \tag{5}$$

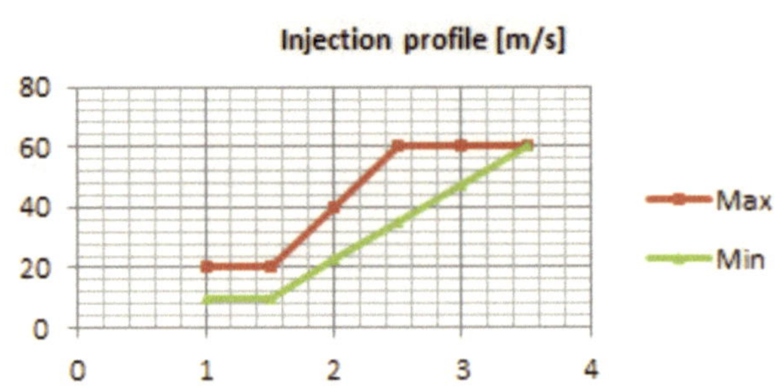

Fig. 14. Injection profile [2].

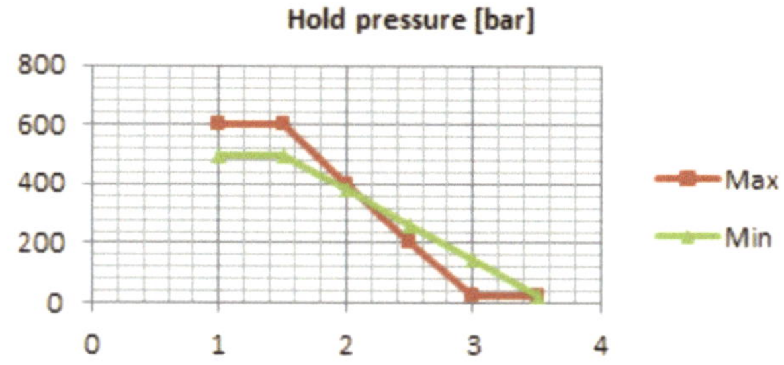

Fig. 15. Hold pressure 2.

Considering that the individual nodes from the first approach can be quantitatively described independently of each other, the experimental scope and so the following experiments are extremely reduced due to freezing of these node factors at a promising level or due to explaining them with the second formula approach (constant amount/node) and smaller variance space. The formulation with the second approach (slope) is also a good approach to describe the constraints with a very limited number of experiments, but not as independent and individual an approach as the first described approach. In the most circumstances, the more effective second approach is recommended. The profile of the injection values could be formulated in the same manner.

3.1 Responses and targets

The target variables could be geometrical variables, various criteria pertaining to surface quality, as well as some measured process parameters. Because the model which will be calculated later on could only be as good as the quantified quality of the test runs, it is very beneficial if the response or target values could be measured as quantitative numbers. If this is not the case, a qualitative ranking method should be discussed and implemented which contains at least 3 graduations or even better several more to support a better predictive model. If it is necessary to set up new qualifications-judgments, one should ensure that the ranking is symmetrical for instance "1" = too hard, "5" = optimum, "10" = too soft. Otherwise, if for instance "1" is optimal-filling and "5" could be more easily less-filled or over-filled, two sources of failure are mixed up, which is suboptimal to the predictability model. The mathematical reason is that the distribution model will be skewed (Fig. 16). Also

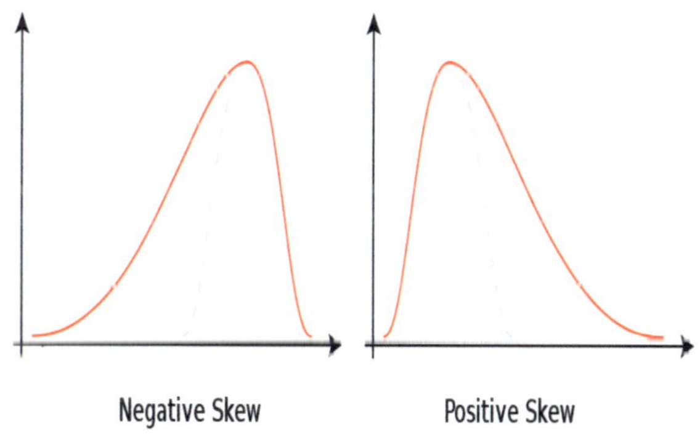

Fig. 16. Skew distribution (From Wikipedia).

the special responses (as results of product-life-tests) tend to deliver a Weibull distribution (Fig. 17) or skew distributed data; so, to handle this, additional knowledge of data, very sensitive factor-settings and possibly transformation of the response data are required in order to achieve good prediction models. For more detailed mathematical information see (From Wikipedia).

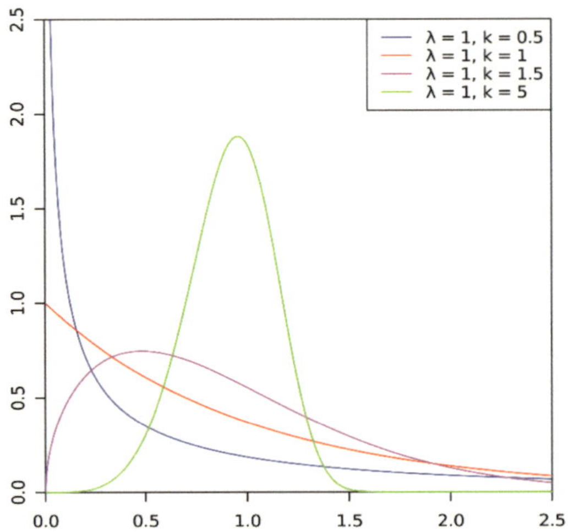

Fig. 17. Weibull probability distribution (From Wikipedia).

Also it should be verified that the quality judgments are reproducible, as well as have been measured with same accuracy over the whole experimental space (Fig. 18). Before any experiments are performed, the capabilities of test methods and test equipment need to be verified. The minimum of such tests should be a test of linearity and reproducibility, representative at the extreme factor settings within the experiment space. If the experimental procedure and the measurement tasks of more than one person are carried out, it should also be verified that the tasks are equally well conducted and reproducible. Optionally, the participants of the experimental design are documented as block variables *(uncontrollable factors)*. In an ideal case, the person-dependent influence can later be repudiated in a hypothesis test. The influence of the employee should then in such a case appear in a very small bar annonation in the diagram "coefficient plot" and therefore assessed as "not significant".

3.2 Safeguarding experiment design space/worksheet

The factorial design is the foundation upon which all further analyses are developed. Because the experimental scale is kept to a minimum, it is essential to measure the target variables of almost all experiments, as otherwise it is difficult to model reality from the results. The circumstances in which the experiments are preformed should as equal as possible *(same raw material, same machine, same operator and room condition);* otherwise, the occurring side effects will be represented by the model noise or modeled fuzzy into the terms of the model. Because some of all possible factors need to be assumed as constant in order to keep the number of varied factors small, the number of varied lasting factors and their variation will always describe a reduced reality. Factors which cannot be controlled,

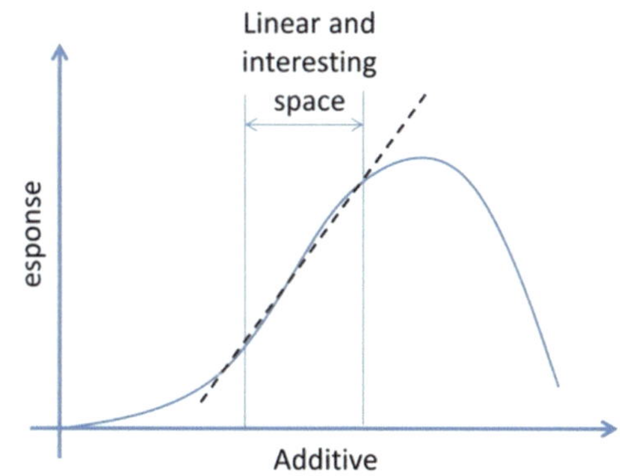

Fig. 18. Response-factor interaction.

like relative humidity or changes in raw-material, should be, if potentially influential, documented as uncontrolled factors. Later on, if the disturbance variation affects the responses, the correlation can be calculated and interpreted to a certain degree. The calculations want be as cause and effect but could be analyzed as a trend to work out a plan of verification or compensation if necessary. The reason why the factor influence could not be ideally assessed is because the uncontrolled factors vary randomly and not as geometrically organized within the design (Fig. 19).

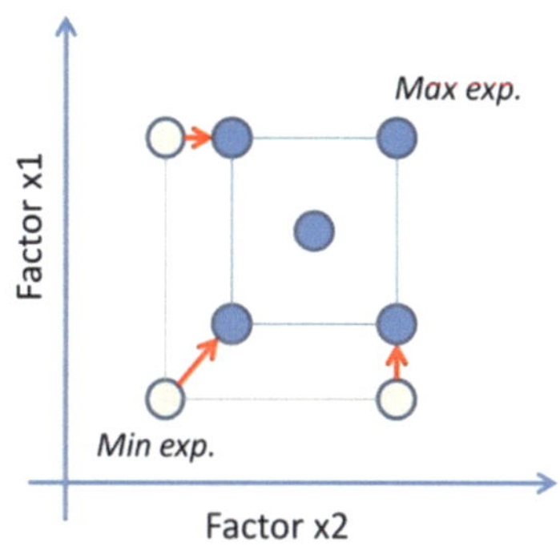

Fig. 19. Example of two factor design adjustment.

To specify the minimum and maximum of the target factor values, only a few preliminary tests integrated into the factorial design are needed -- usually, two successful experiments are commonly sufficient in which all factors are set to the lowest or the highest levels. This ensures that variation in the factors is still measurable in the outcome (responses). Otherwise, the variation in individual factors would have to be reduced as displayed in Fig. 19 or extended in a modified factorial design. Once the min., max. and one-center-point experiments have been performed, it should also be considered if the range of the factor variation is necessary because if these variations are too big, non- linear behavior of some factors will probably occur. In most cases only a small linear area is of interest (Fig. 18). At the start, because of the lack of knowledge about the factors' effects and especially because of the reinforcing interactions, the factor ranges are often sub-optimally set. If this is the case, the factors' impact could be non- linear. Usually, factorial experiments commence with a large number of factors which are initially studied only for their linear influence. This approach has the advantage of requiring just a few experiments to find out which factors influence the targets, and to what extent. It is also possible to obtain an initial indication of the suitability of the evaluation criteria and the measuring technology. Since the focus of the investigation is on linear correlations, experiments are not conducted on the interaction between the factors. Were all factors and their interactions to be investigated, the experimental effort would be much greater. A complete linear description of the factors is given by the following rule of thumb. (FU 6)

$$\text{Number of experiments} = 2^{\text{number of factors}} \tag{6}$$

within full factorial designs. It is recommended that designs with more than four factors should be examined with advanced designs in order to keep the number of experiments small. A comparison of the linear design is in (Tab. 3).

No of Factors	Full Factorial	Frac. Factoriall
2	4	4
3	8	4
4	16	8
5	32	16
6	64	16
7	128	16
8	256	16
9-16	>512	32

Table 3. Summary of fractional factorials designs, excluding replicates (AB, 2009).

Now one could say "With this many experiments, I can do it without a plan/ design." This is probably right and therefore, there are more efficient designs in Tab. 4. But the difference will be that, in case there is no statistic software available, results cannot be visualized adequately; hence, most of standard spreadsheet programs are limited to 2 parameter diagrams. Besides, the whole statistic has to be calculated by hand, which makes this approach highly susceptible to calculation errors[6]. In the case of evolutionary-by-hand-experiments, some time is always necessary between the experiments to discuss and decide what has to be done next. In contrast to this approach, the structured Design of Experiments method enables more experimentation within shorter intervals. From the perspective of identical-process-conditions, the time-consuming, by hand process is also more susceptible to errors and must be more critically conducted. Fig. 20 represents the "by-hand" versus the "statistically-structured" process. The Fig. 20 also highlights another benefit of the structured approach: after each set of experiments, the number of factors to be discussed could be reduced on the basis of the regression models. This will further decrease the number of factors and so the experimental design experiments by freezing the unimportant factors to promising levels.

[6] One option could also be to not calculate anything but just perform experiments. Not performing any statistics at all, will never enable experimenters to judge the quality of their process-setup. And even more important is the fact that if the process is running less sufficiently, the room for improvement cannot be described and project sense and definition cannot be estimated.

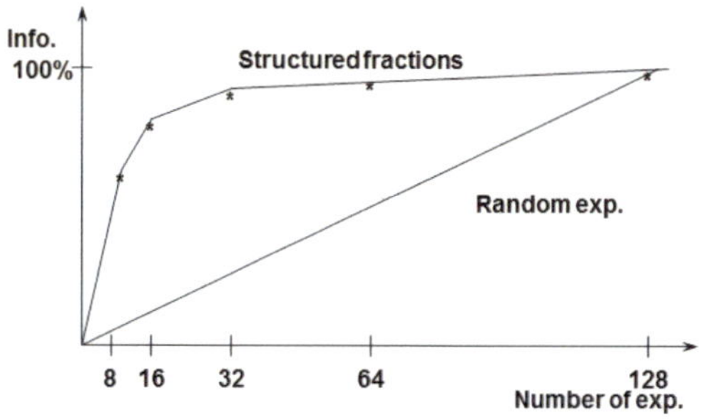

Fig. 20. Structured versus random experiments (AB, 2009).

The solid line in Fig. 21 visualizes, that within the structured Design of Experiments approach, the largest proportion of experiments is done at the project beginning. This ensures a better and faster process-knowledge growth and a more robust production launch. While in the other case (dotted line) a lot of resources are needed for improvements after the production launch, which block resources needed for instance for developing the next product- or process generation.

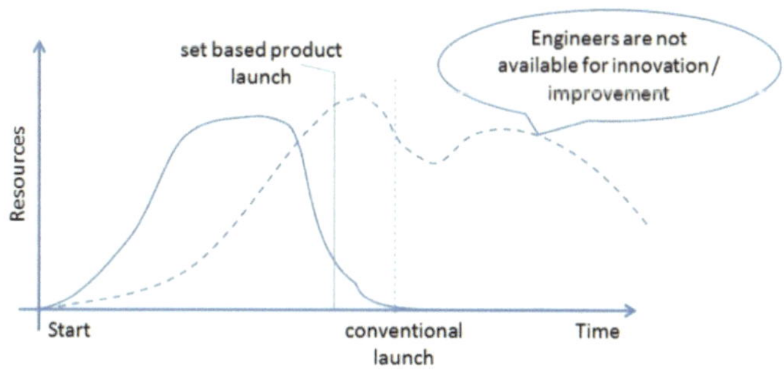

Fig. 21. DoE vers. COST (Kennedy, 2003).

Table 4 shows number of runs (excluding replicates) and alternative supported models. With six or more factors for instance, "Rechtschaffner" designs (L. Eriksson) may constitute a viable alternative to the fractional factorial designs (for the interaction models).

Number of Factors	Factorial / fractional factorial Design	Plackett Burmann design	Rechtschaffner design	L-design
2	4	8_l	n/a	9_l(q)
3	8	8_l	7_i	9_l(q)
4	16	8_l	11_i	9_l(q)
5	16	8_l	16_i	18_l(q)
6	16_l or 32_i	8_l	22_i	18_l(q)
7	16_l or 32_i	8_l	29_i	18_l(q)
8	16_l or 32_i	12_l	37_i	18_l(q)
9	32_l or 64_i	12_l	46_i	27_l(q)
10	32_l or 64_i	12_l	56_i	27_l(q)

Table 4. Summary of screening design families l= linear; i= interaction; q= quadratic (AB, 2009).

3.3 Reproducibility

So- called center-points experiments serve to evaluate the reproducibility of the measuring method and the process relative to the variation within the factorial experiment. In order for the influence of the respective factors to be determined, these are varied about a mean value upward and downward in equal measure. The software "Modde" visualizes these relationships with the replicate plot (Fig. 22) and the 4th bar within the summary plot (Fig. 27, 28). If the center points are not close together within the replicate plot, the reason must be analyzed.

Deviating center points could, for instance, be potentially caused by missing factors, incapable measurement equipment or measurement methods, different machine operators, different machines, different batches of raw material and much more. Also it must be assumed that other experiments' results fluctuate the same as the center- point experiments, otherwise the quality of the prediction model would be weak.

3.4 Design

A common geometric factorial design is chosen to create the following exemplary but actually performed experiment (Moser & Madl, 08/2009). This fractional factorial design entails far fewer experiments than full factorial design. For comparison: Were a full dual level factorial design to be applied to this factorial experiment, 227 experiments would be needed. Although other designs, such as "Plackett Burmann" designs (L. Eriksson), would entail fewer experiments (28+), they would rule out the possibility of studying interactions or quadratic terms at a later date.

This screening design consists of two profiles, each of which has ten nodes and seven quantitative factors. This yields 27 factors from which, with the aid of fractional factorial design, an experimental scale of 64 + 3 center-point experiments are generated. This number is derived from the experimental design in which all the factors are studied independently and without interaction at two levels (min./max.). A feasibility study was conducted in which the part is injected in a 1-cavity mold. To ensure constant melt and mold temperatures in the process, 30 moldings were initially produced. Two molded parts were then produced and characterized. After the experiments, the moldings were measured in a coordinate measuring table.

The values were transferred to the software. Before a model can be generated out of the data, it is very important to look at the raw data. Therefore, common statistical software supplies the replicate, histogram and correlation plots. The first plot to look at is the replicate plot (Fig. 22), which plots all the experiments in a row in order to see if experiments with equal factor settings (center- points and replicates) cause equal results. The second plot is the histogram (Fig. 23), which is important for reviewing the data and ensuring a symmetrical Gaussian distribution (Fig. 34, bell distribution). Because of the limited number of experiments it is very important to have close to normally distributed data in order to get a good regression model, fitted with the method of least squares.

After this, the "black box" (Fig. 11) between factor setting (cause) and targets (effect) can be graphically modeled. The model itself is a tailor polynomial, the mathematical formulation of the relationships being derived from the results of the factorial experiment. Since

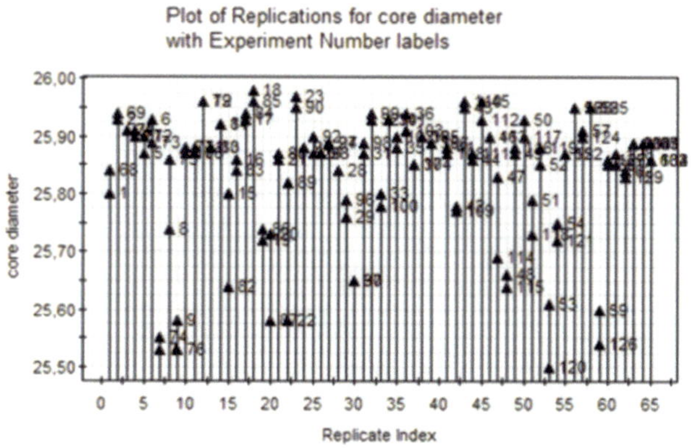

Fig. 22. Replicate plot [5].

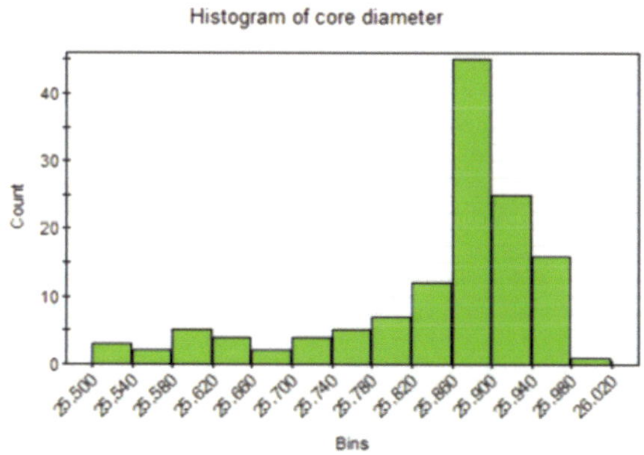

Fig. 23. Histogram [5].

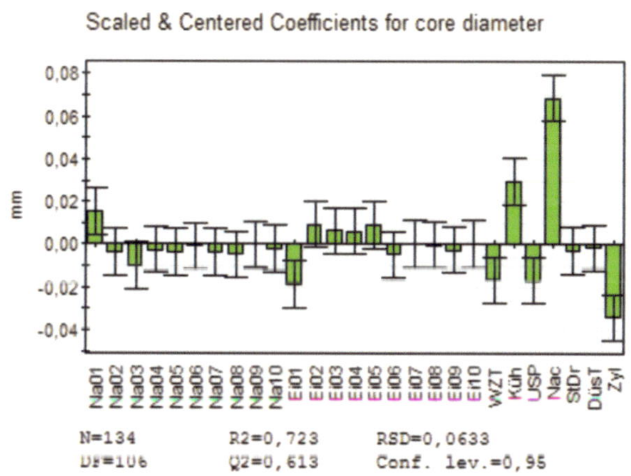

Fig. 24. Coefficient plot [5].

screening examines only the linear correlations, only the linear effects of all factors are plotted in a coefficient bar graph (Fig. 24).The next set up to study is the "N- residual probability plot" (Fig. 25).

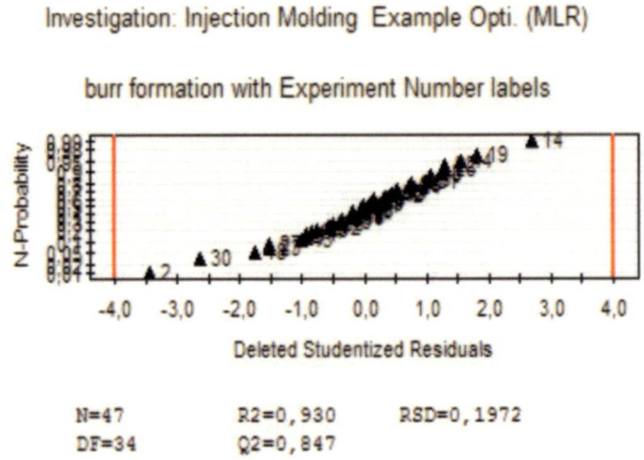

Fig. 25. N-Residual Plot [5].

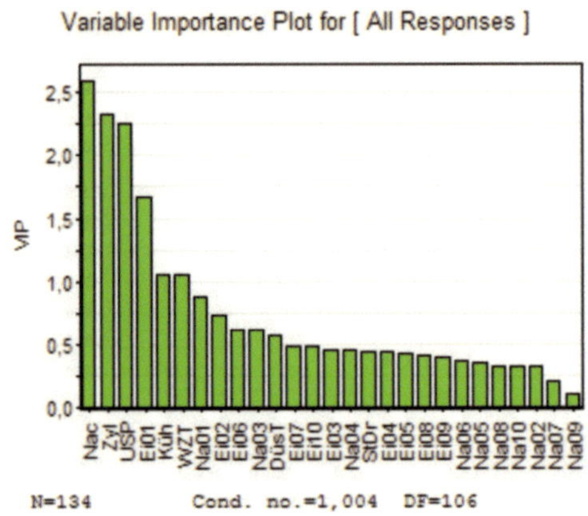

Fig. 26. Variable importance plot [5].

The residuals are plotted on a cumulative normal probability scale. This plot makes it easy to detect normality of the residuals. If the residuals are normally distributed, the points on the "N- residual probability plot" follow close to a straight line and will also support detection of outliers.

Points deviating from the normal probability line with large absolute values of studentized[7] residuals, i.e. a larger than 4 standard deviations indicated by red lines on the plot.

According to the Pareto Principle, about 20 % of the factors account for 80 % of the effect. In line with this hypothesis, any factor without influence (significance) can be graphically removed from the coefficient plot, while the software works in the background to calculate the updated model from the remaining terms[8]. In the next step, the experimenter can focus on the significant factors. To see what factors are important for all considered responses, the "variable importance plot" (Fig. 26) could be analyzed. This is crucial because the number of factors decisively determines the extent of further experiments. The quality of the test series and its calculated model are also illustrated as several four bar plots (Fig. 27, 28). These

[7] For more details see Student's t-distribution at (From Wikipedia)
[8] For this reason, the order in which non-significant terms are removed is important.

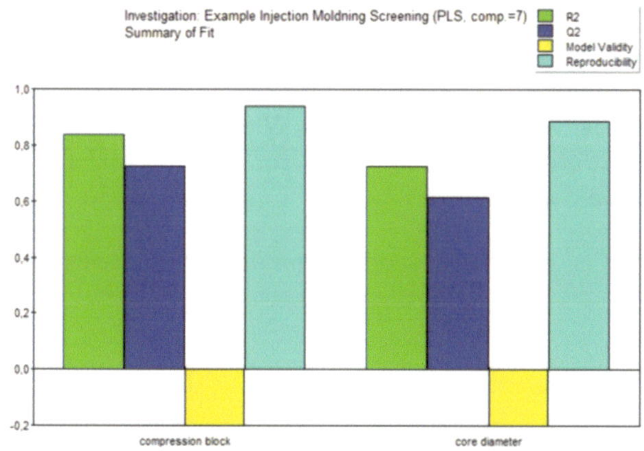

Fig. 27. Summary plot after screening [5].

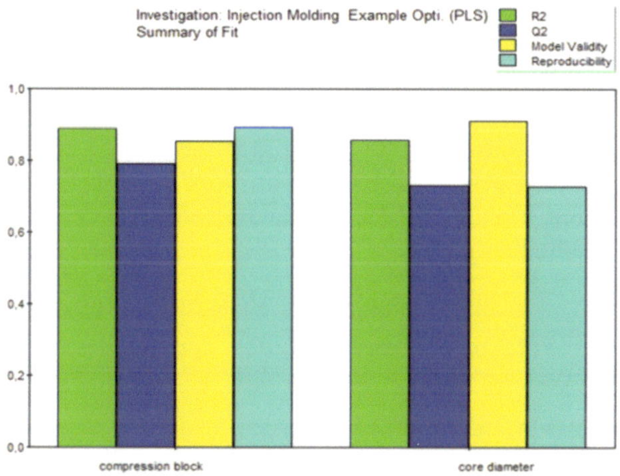

Fig. 28. Summary plot after optimization [5].

stand for the quality of the results (1th bar), the prognosis for targets in new experiments (2th bar), validity (3th bar) and the reproducibility of the model (4th bar). The bars are shown normalized (0 to 1), i. e. the closer the values are to 1, the better. The model calculated after the screening phase (Fig. 27) was of surprisingly good quality. Fig. 28 is the summary of the enhanced "optimization model". In detail the "summary plot" is calculated as follows: (L. Eriksson) (AB, 2009)

R^2 (Fig.27, 28; 1St bar): Quality of results or the Goodness of fit is calculated from the fraction of the variation of the response explained by the model (FU 7): The R2 value is always between 0 and 1. Values close to 1 for both R2 and Q2 indicate a very good model with excellent predictive power.

$$R^2 = \frac{SS_{REG}}{SS} \tag{7}$$

SSREG = the sum of squares of the Response (Y) corrected for the mean, explained by the model.
SS = the total sum of squares of Y corrected for the mean.

The second column in the summary plot is **Q2** (Fig.27, 28) and is the fraction of the variation of the response predicted by the model according to cross validation and expressed in the same units as R2. Q2 underestimates the Goodness of fit (FU 8). The Q2 is usually between 0 and 1. Q2 can be negative for very poor models. With PLS negative Q2 are truncated to zero for computational purposes. Values close to 1 for both R2 and Q2 indicate a very good model with excellent predictive power.

$$Q^2 = 1 - \frac{\text{PRESS}}{\text{SS}} \tag{8}$$

PRESS= the prediction residual sum of squares
SS = the total sum of squares of Y corrected for the mean.

The third column in the summary plot is the "model validity" (Fig.27, 28) and a measure it. (FU 9). When the model validity column is larger than 0.25, there is no lack of fit of the model. This means that the model error is in the same range as the pure error. When the model validity is less than 0.25 there is a significant lack of fit and the model error is significantly larger than the pure error (reproducibility). A model validity value of 1 represents a perfect model.

$$\text{Validity} = 1 + 0.57647 * \log(\text{plof}) \qquad \text{where plof} = \text{p for lack of fit.} \tag{9}$$

The forth column in the summary plot is the Reproducibility (Fig.27, 28) which is the variation of the response under the same conditions (pure error) (FU 10), often at the center points, compared to the total variation of the response. A reproducibility value of 1 represents perfect reproducibility.

$$\text{Reproducibility} = 1 - \frac{\text{MS(Pure error)}}{\text{MS(total SS corrected)}} \tag{10}$$

MS = Mean squares, or Variance

4. Optimization (of process parameters)

The screening experiments for this project revealed the region in which the profiles for holding pressure and injection speed must lie. This enables the nodes to be dispensed with in favor of a description of the profiles with a varying initial value that decreases constantly Formula (FU 3). Every experimental approach also got less significant factors, such as in this case: The back pressure and the temperature of the hot runner nozzles. These factors were frozen at a calculated optimal value. This reduced the number of factors to be varied from twenty-seven to six. These six factors holding pressure, injection speed, mold temperature, cooling time, switching point and barrel temperature have been studied further in a multilevel geometrical experimental design ("Central Composite Face") with 44 experiments and three center-points. After executing the additional experiments the raw data analysis has to be performed again, such as checking the data for normal distribution and possibly transformation of responses, checking the reproducibility and pruning the model terms at the "coefficient plot" in order to get the best possible prediction model. The summary plot is visualized within (Fig. 28):

Now with regard to the predictive quality of the model, contour plots (Fig. 29) can be generated. These plots are like a map of a process; the ordinate and abscise represent factors, and the contour lines visualize response values.

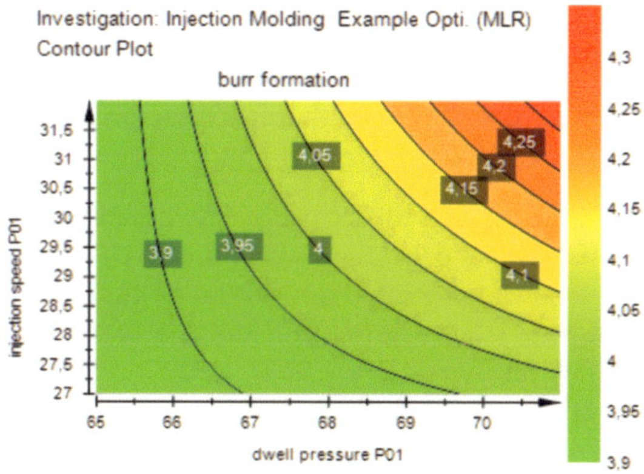

Fig. 29. Contour plot [5].

4.1 Finding an optimal process-setup

While the color shading is value-neutral, it is predefined by default with blue as low and red as high response values.

Since humans are incapable of thinking in more than three dimensions, it is difficult to display more than two factors, plus a target. However, it is possible for software to calculate the degree of fulfillment of any number of evaluation criteria as a function of several factors. This can be shown in a sweet-spot plot (Fig. 30).

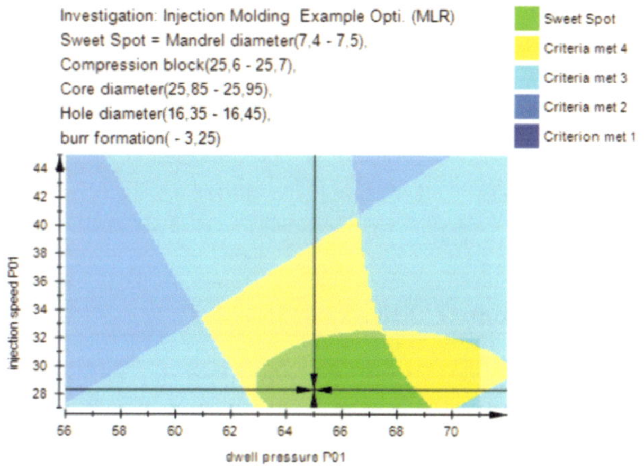

Fig. 30. Sweet spot plot [5].

As in set theory, the target functions obtained are displayed on top of each other in different colors. The region in which all targets are met is called the sweet spot (green).

The injection prototype tool was created to ascertain the process capability of the tool. The determination of the process optimum from the model proved to be so good and reproducible that the project team eschewed a study of the robustness of the optimum. On production tools, and especially when it comes to large piece numbers of very high requirements to produce, further verification steps are inevitable.

5. Robustness testing

In most cases, some of the target definitions could be fully met, others only to a certain degree. Sometimes a compromise has to be defined to outline actionable new demands of targets and specifications to be derived from this updated knowledge.

After this important step of reflecting the possible and defining new process specifications in the light of costs over benefits, a potential optimum can be calculated with the optimizer and visualized with the sweet spot plot. The Modde-Optimizer is a software tool which cannot optimize anything but is a very helpful tool for searching in a multidimensional space for a setting in which all targets are met max. To do this, the target limits and the factor settings have to be taken or updated.

At this point, it is also possible to narrow the factor limits or to expand them. This inter- or extrapolation is combined with the possibility of estimating the accuracy of the factors setting; this means how accurately this factor can be adjusted to a certain limit. From this data, a reduced linear design will be used to place small, mathematical isosceles triangles into the multidimensional working space. From there, these triangles will be mirrored on their flanks in order to check if the new additional peaks of the mirrored triangles (Fig. 31) are better positioned to fulfill the predefined targets' requirements.

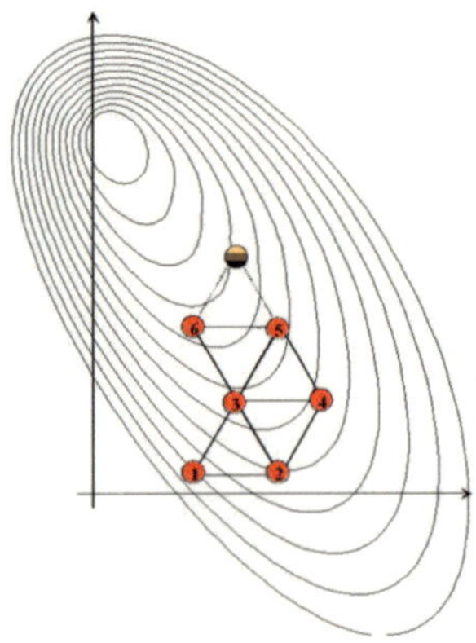

Fig. 31. Sketch: simplex algorithm, (L. Eriksson).

This process will be repeated iteratively till no better solution can be found. Because one of the small triangles could be easily trapped at local maxima or minima, a small proportional number of triangles are used to form a different position of the factor space. This can be seen with the ending lines in Fig. 32. This so-called "simplex algorithm" is a very simple but is powerful tool to check to which degree targets' values can be simultaneously fulfilled. The previously discussed target priorities can be weighted within the response target settings. The process model serves as a basis for finding a setting in which the molded part can be produced within the required quality. For the purpose of evaluating robustness in the region of the calculated optimum, it often takes only a few experiments, so that the effect of the factor variations may be adequately described. To do this in fine-tuning or a robustness check, a new small set of linear experiments with a factor variation similar to realistic process conditions should be performed. After executing the experiments and doing the repetitive data analysis, the results can be summed up in one of these four cases:

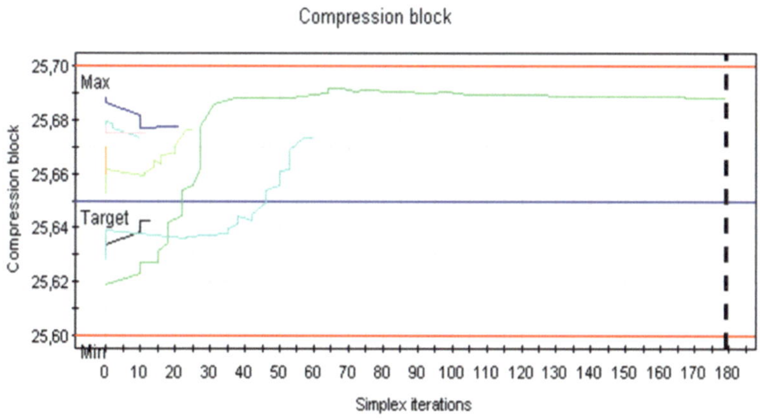

Fig. 32. Response simplex evaluation [5].

1. **Case one:** All target values are fulfilled; a good predictive model can be developed. This means the process is robust, but the variation of the factors is still influential on the target functions. It should be discussed if some of the process conditions should be improved.
2. **Case two:** All target values are fulfilled; no model can be achieved. This is the best outcome because factor variations seem to be too small to affect the results. This case should also be discussed if the process setup can be simplified in order to make the process more cost efficient
3. **Case three:** No target values are fulfilled; a model can be developed. This means that the model prediction was not as accurate as expected. The quality of the "predictive basis model" should be rechecked; maybe the Q2 and the validity are not as good as supposed, or there are outliers in the experiments. By implementing these test experiments in the Modde-File, the model-quality can be checked as to whether it can be enhanced or whether these results deviate and if so, for which the reasons.
4. **Case four:** No target values are fulfilled; no model can be developed. Typical reasons for this outcome are that process conditions have been changed among the experimental blocks, likewise different raw materials, machines, operators, tools etc. Another common reason is that due to the models' quality, the requirement has been underestimated.

6. Validation/process capability

Finally after a process setup has been found and verified to produce parts with a sufficient quality, the process capability should be checked. This is a very important step for summarizing the quality and robustness of the process. Within this step, a variation caused by realistic production conditions can be attached to the calculated optimum of the regression model. This is done by performing a Monte Carlo Simulation. This means that 100,000 simulations are computed while the uncertainty of factor settings is randomly applied. One way is to define a narrow factor range, such as in a robustness model or a certain factor setting spot within an optimization design. To explain the concept, a more practical Fig. 33 is shown, where the process target borders are displayed by the road shoulders and the process itself as a car on this road.

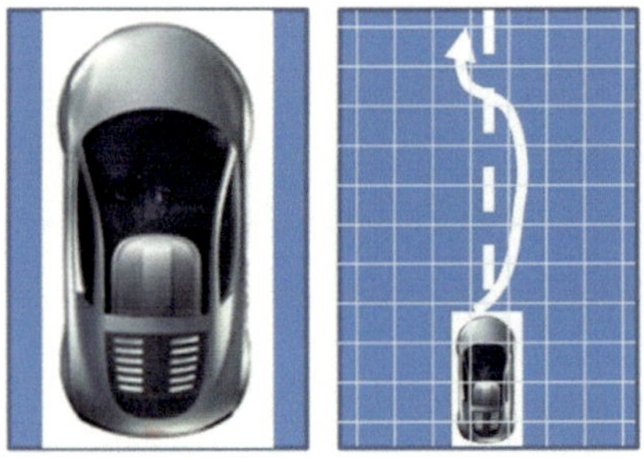

Fig. 33. Car example.

The smaller the car, in comparison to the road width, the more easily the process can be controlled. If a grid is applied on the street and the horizontal car positions during driving is added up, the positions within the grid intervals can be plotted as bars in a histogram (Fig. 23) or distribution plot (Fig. 34). Again, it is easy to interpret how close the car is driving right-handed relative to the middle lane marking and thus how safe the driving is. If the standard deviation to the "ideal driving line" is checked as a critical indicator between the expected value *(ideal driving line)* and each process border *(road shoulder)*, then the process can be described in terms of quality and robustness. Consequently, it can be assumed that process quality can be described as a function of tolerance and process variation. This number is called Cpk (FU 9), the process capability or capability index and has its origin in "SixSigma" statistics. Within in Fig. 35 normal distributions with differed sigma levels are plotted. At higher CpK levels the underlying data is centered closer to the expected value (μ).

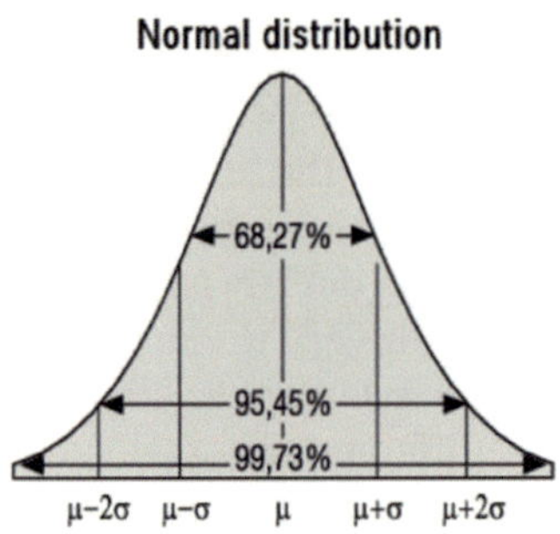

Fig. 34. Normal distribution.

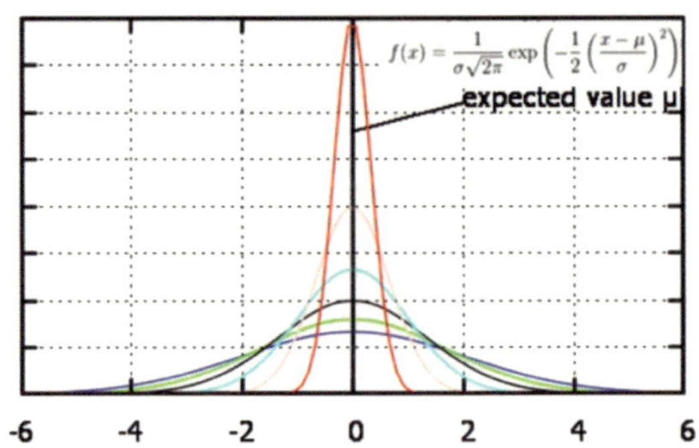

Fig. 35. Sigma distributions 1-6 sigma.

$$Cpk = \min\left[\frac{USL - \mu}{3 * \sigma}, \frac{\mu - LSL}{3 * \sigma}\right] \quad (9)$$

USL = Upper specification limit, LSL = Lower specification limit and μ = estimated standard deviation for predictions.

$$DPMO = Ho * \frac{1\,000\,000}{NS} \quad (10)$$

Ho = Hits outside specifications, Ns = Number of simulations based on an infinite number of predictions.

Another way to describe the process quality is to calculate the DPMO which is short for "Defects per Million opportunities outside specifications" and is used as stop criteria in the design space estimation (FU 10).

Within the Table 5 higher the Cpk is, the better the process capability and robustness are. This can be seen also within the numbers of "DPMO" or the "%Outside" specification. To calculate these numbers from the Design of Experiments experiment data, the "Monte Carlo simulation" (MCS) can be computed round the previously calculated optimal. The factor variance is adjusted to the approximated variance of the process settings within normal working conditions (Assumption 5%). The response variance calculated by the MCS is caused then by the factor setting -- and the model uncertainty. At Fig. 36 The black T-bars represent the space in which one factor can be varied while freezing the other factors and still keeping the calculated response fulfillment (Fig. 37). This is a very important information to set the process tolerances as closely as necessary and as wide as possible.

Cpk	DPMO	% Outside
0,4	115070	11,51
0,6	35930	3,59
0,8	8198	0,82
1,0	1350	0,13
1,2	159	0,02
1,5	3,4	7,93328E-05
1,8	0,03	3.33204E-06
2,0	0,0010	9.86588E-08

Table 5. Six Sigma, source: (From Wikipedia).

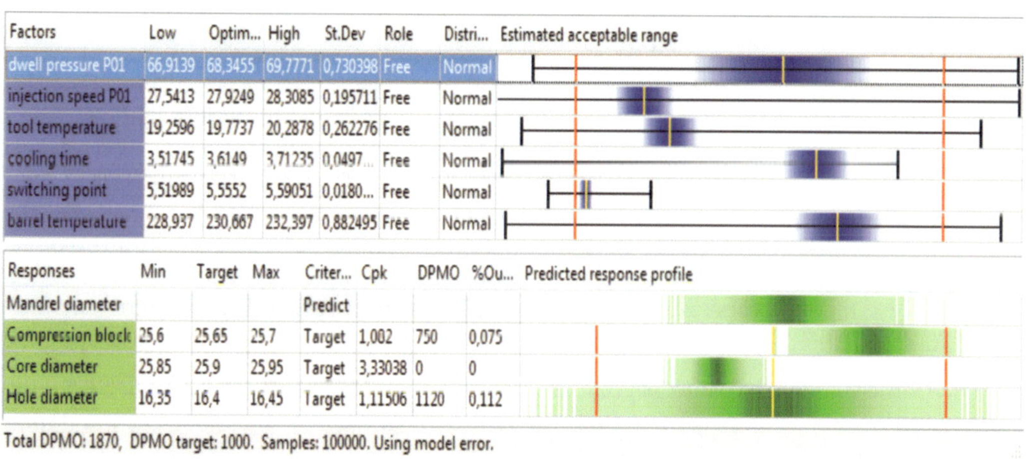

Fig. 36. Design space estimation [5].

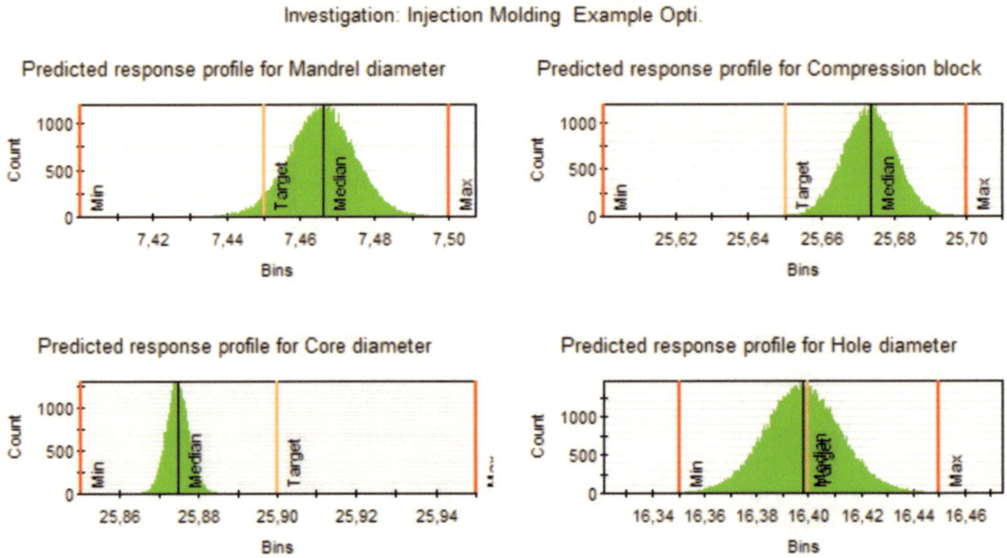

Fig. 37. Predictive design space histogram [5].

Chart legend of Modde-predicted response profile

- The yellow line represents target value for the responses as specified in the Optimizer.
- The red lines are the specification limits for each response as specified in Optimizer.
- The faded green region represents the probability of a prediction for a random distribution of factor settings in the given ranges (low-optimum-high), the design space.
- The black T-bars represent the space in which one factor can be varied while freezing the other factors and still keeping the calculated response fulfillment.

The histograms of Fig. 37 represent the response targets based on the regression model at an optimal factor setting including the factor uncertainty variation calculated by the MCS.

7. Success with restrictions

Assuming the predictive model is of good quality, the responses and its degree of achievement can be evaluated against priorities of the projects' definitions. Within the most experimental setups, responses targets are optimistically defined and with a high degree of safety. Sometimes not all of these targets can be fully achieved. Therefore, and as described in the optimization, responses can be weighted again according to their priorities and their targets, in order to find a sufficient compromise. This compromise could mend not only the factors and responses that need to be adjusted but could also mean that conditions that act as disturbances need to be compensated. Potential disturbance factors are, for instance: Relative air humidity, temperature, water content in raw materials, using different machines or bad machine calibration.

Also predictive models are only as good as their data. Even if factors are sufficiently enough arranged to describe the target functions, there are still a lot of things which could impair the prediction model such as bad pruning of the model terms, bad distributions or deviating experiments, missing factors, bad measurement *(calibrated)* equipment.

In addition to this, the approach is always to minimize the work, time and budget with efficient designs *(fewer experiments)* and fewer factors, so that not all important factors are implanted or the effect of factors are underestimated, and are, for instance, of a higher order than assumed. So, if a process is not linear and linear designs are used, the predictive capability of this model is very limited. If necessary due to interaction-, squared- or cubic factor-terms, the design could be complemented step-wise. The design of higher-order processes, complex designs are not recommended at the beginning, since these drastically increase the number of experiments. Complexity can always be reduced by focusing only on a small process space (Fig. 18).

8. Conclusion/summary

After reading the chapter, the readers should now have a good understanding how far the combined methods could help them to achieve the predefined requirements. In addition, they should also be sensitized to the fact that non-structured approaches are weak and time-consuming. It is also important to understand that while the design of experiments "DoE" does not necessarily lead to good results or capable processes, they can help to describe and document the potential of a process. Even if the targets could not be achieved, it is still possible to derive useful, cost-effective and robust knowledge with this structured approach in order to identify and assess possible disturbance factors or possible process constraints. This can provide beneficial clues for fine-tuning the factors and conditions in order to ensure and optimize process capability and success.

By following this good "DoE" practice recommendation, the iterative difficulties in finding the fulcrum or lever at the beginning of an optimization process first hand can now be reduced--- if not eliminated. And by following this consistently structured approach, the right things can be done in the right way with the right tools. Thus, Pareto's law can be intelligently leveraged, and finally, the optimization team can operate in the most efficient and effective fashion.

9. References

AB, U. (2009). Software & Help File Modde 9.0. ISBN-10 91-973730-4-4, Sweden.

From Wikipedia, t. f. (n.d.). *Wikipedia.* Retrieved 09 15, 2011, from: http://en.wikipedia.org

Kennedy, M. N. (2003). *Product Development for the Lean Enterprise.* USA, Richmond Verginia, ISBN 1-892538-09-1: The Oaklea Oress.

Klein, B. (2007). *Versuchsplanung-DoE.* Oldenbourg Wissenschaftsverlag GmbH, ISBN 978-3-486-58352-6: München, Germany.

L. Eriksson, E. J.-W. (n.d.). *Design of Experiments:Principles and Applications,.* Sweden, Umea, ISBN 10 91-973730-4-4.

Lunau, S. (2006, 2007). *Six Sigma + Lean Toolset.* Berlin, Heidelberg, ISBN 978-3-540-69714-5: Springer Verlag.

Michael L. George, D. R. (2005). The Lean Six Sigma Pocket Toolbook. ISBN 0-07-144119-0, USA: McGraw-Hill.

Moser, S., & Madl, D. (08/2009). *Effective Run-In of an Injection Molding Process.* Kunststoffe interantional: Carl Hanser Verlag Munich.

Rauwendaal, C. (n.d.). *SPC Statistical Process Control in Injection Molding end Extrusion.* München 2008, ISBN 978-3-446-40785-5: HanserVerlag.

Vester, F. (2002). *Die Kunst vernetzt zu denken.* dtv, ISBN 3-423-33077-5: München, Germany.

Part 3

Powder Injection Molding

3

Powder Injection Molding of Metal and Ceramic Parts

Joamín González-Gutiérrez, Gustavo Beulke Stringari and Igor Emri
Center for Experimental Mechanics, University of Ljubljana, Ljubljana, Slovenia

1. Introduction

Powder injection molding (PIM) is a technology for manufacturing complex, precision, net-shape components from either metal or ceramic powder. The potential of PIM lies in its ability to combine the design flexibility of plastic injection molding and the nearly unlimited choice of material offered by powder metallurgy, making it possible to combine multiple parts into a single one (Hausnerová, 2011). Furthermore, PIM overcomes the dimensional and productivity limits of isostatic pressing and slip casting, the defects and tolerance limitations of investment casting, the mechanical strength of die-cast parts, and the shape limitation of traditional powder compacts (Tandon, 2008).

Due to the demand of high performance materials and the miniaturization of complex components in various fields, PIM market has exceeded the $ 1 billion mark in 2007, becoming approximately six times larger than 15 years before (German, 2008). This impressive growth rate is not expected to slow down in the next few years, as a recent report from Global Industry Analysts announced that together, world metal and ceramic PIM market is forecast to reach $ 3.7 billion by the year 2017 (Global Industry Analysts [GIA], 2011). Metal powder injection molding (MIM) is still considered the largest segment of this market, accounting for more than 70% of global output. Although PIM is globally widespread, Europe and Asia-Pacific account for a major share of MIM segment, while USA is still the largest market for Ceramic Injection Molding (CIM) (GIA, 2011).

In Europe, the MIM production is dominated by automotive applications and the so called consumer market (which includes watches and eyeglasses), while the North American production is mainly applied to the medical/healthcare field. On the other hand, the Asian production, considered the largest one, is dominated by consumer electronics and information technology applications. The consumer electronics market is, indeed, one of the drivers behind MIM, whose growth is largely taking place in Asia, specifically in Taiwan, Malaysia, Thailand, China, Singapore and South Korea. Another growth factor is the expansion of medical component production also in Asia, as a larger population gains access to improved health care (German, 2008).

A recent increase in MIM sales has generated a need for new equipment, with a simultaneous investment in research and development. Typically, leader companies invest an average of 10.5% of sales in the combination of capital expansion and research. Besides this, several contract electronic assembly firms that used to purchase MIM components from custom molders have now elected to make from MIM a captive operation (German, 2008). This could lead to rapid growth for those operations since the relationship between design and production can be improved.

Stainless steel continues to dominate MIM applications, accounting for around half of the global production, reflecting the capture of components that would have otherwise gone to investment casting. Nevertheless, other materials as copper, nickel alloys, bronze and more recently tungsten alloys and titanium also represent important markets (German, 2008). On the other hand, the CIM market is mainly dominated by products made out of alumina, zirconia, and silicon or aluminium nitride powders (Ruprecht et al, 2002).

Typical components produced by PIM, either MIM or CIM, range from 0.002 to over 100 g (German, 2008). Furthermore, in both segments it is clear the progressive move towards smaller products, demanding improved technologies regarding machinery, process, and materials, hence leaving a still open field for applied research.

2. Overview of process

The PIM process presents countless variations which are used in the industry today. Invariably, it consists of four steps (Gonçalves, 2001; Tay *et al*, 2009):

- Feedstock preparation;
- Injection molding;
- Debinding;
- Sintering.

In Fig. 1 a flow chart illustrating the main stages of PIM is presented.

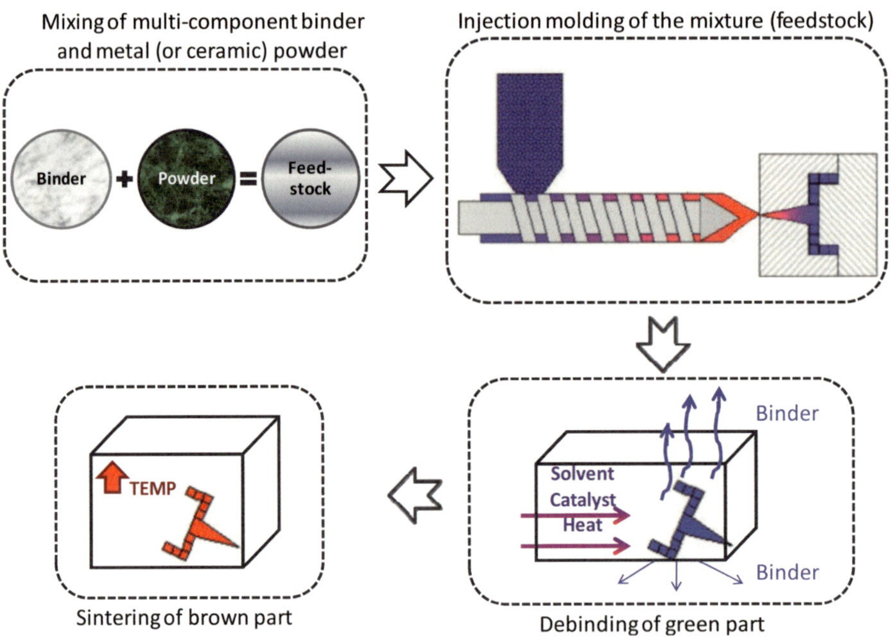

Fig. 1. Flow chart illustrating the main stages of PIM.

The starting material for PIM, usually termed feedstock, is a homogeneous pelletized mixture of metal or ceramic powder and an organic multi-component binder. The binder and the powder are combined in a variety of compounding equipment, such as extruders and mixers. The mixture is then pelletized to an appropriate shape for feeding into the molding machine. The binder is simply a carrier medium for the powder and once a part is molded, the binder is removed in a subsequent step.

The injection molding process is mainly identical to conventional plastic injection molding. Nevertheless, some machine hardware changes are usually required to process a specific feedstock based on their compressibility and viscosity. Control of the molding process is vital for maintaining tight tolerances in subsequent steps. Most design advantages of PIM technology are captured during molding by relying on the flexibility of incorporating complexities in the tool. A molded part is called a "green part" and is oversized to allow shrinkage during sintering (Tandon, 2008).

Before sintering, it is necessary to remove the binder from the molded part. The debinding is the most expensive and time-consuming stage in the PIM technology. This removal process should be based on a progressive opening of the surface channels to facilitate the removal of vapors inside the nucleus (Gonçalves, 2001). Three main methods can be applied depending on the composition of the binder: thermal, solvent, and catalytic. In thermal debinding, the binder is removed by degradation, evaporation, or liquid extraction, at temperatures ranging from 60 to 600 °C. The relatively long time associated with thermal debinding is greatly reduced using an organic solvent or in some cases even water to dissolve the soluble components of the binder, in the so called solvent debinding (Tandon, 2008). The catalytic debinding, in turn, focus in a solid-to-vapor catalytic degradation, as it is the case of exposing acetal-polyolefin-based feedstocks to acid vapors, resulting in much faster binder removal and superior handling strength when compared to thermal or solvent debinding (Krueger, 1996; Mathew & Mastromatteo, 2003). Nevertheless, it is worth to point out that in all debinding methods, a skeleton of backbone binder often remains to impart adequate strength and shape retention up to the onset of sintering. This remaining backbone is thermally removed between 200 °C and 600 °C in a pre-sintering step (Tandon, 2008).

Sintering is the last stage of the process, providing the inter-particle bonding that generates the attractive properties from otherwise loose powder mass. Depending on the material, debound parts or "brown parts" are sintered at temperatures ranging from 1200 to 1600 °C. It is essentially a removal of pores, accompanied by growth and strong adhesion among the adjacent particles, causing the retraction of the product whose dimensions usually reduce between 14 and 20% (Gonçalves, 2001; Krug, 2002; Tandon, 2008). Therefore, green parts are oversized to compensate for the sintering shrinkage. The fine particle size used in the PIM process results in high sintered density ranging from 95% to 99.5% of theoretical, thus providing superior mechanical and corrosion properties as compared to press and sinter technology.

The following sections of this chapter have the goal to describe in more detail the different steps of the PIM and the desired characteristics of raw materials to be used during this process.

3. Feedstock preparation

The first step in the powder injection process is the preparation of feedstock materials. The powder and binder are hot mixed above the softening point of the binder constituents to provide a uniform coating on the powder surface (Fig. 2A). The feedstock is prepared by compounding polymeric binders with fine metallic or ceramic powders. Commercially available feedstock material is generally supplied in the shape of pellets (Fig. 2B), so that is easy to handle before and during the injection molding step.

(A) (B)

Fig. 2. (A) Micrograph of feedstock material showing metallic particles surrounded by polymeric binder, (B) Pellets of commercially available feedstock material Catamold® by BASF.

The powder content usually ranges from 50 to 65% in volume, although there are claims of optimized commercial formulations in which even more than 80% is used. If the powder content is found to be lower than 50 vol.%, the sintering ability of the feedstock and the final density of the part are significantly lowered. From another standpoint, it is also important to keep the viscosity of the feedstock as low as possible in order to facilitate the injection molding process, reason for why a powder content higher than 65 vol.% should be handled with care (Merz *et al*, 2002).

One of the most important properties of the feedstock is certainly its homogeneity. A homogeneous distribution of powder particles and binder in feedstock is important as it helps to minimize segregation during the injection molding stage and later on to obtain isotropic shrinkage after debinding and sintering (Quinard *et al*, 2009). Avoiding segregation of feedstock components is necessary to prevent visual defects, excessive porosity, warpage and cracks in the sintered part (Thornagel, 2010).

The technique used for mixing binder and powder can influence the homogeneity of feedstock materials. Feedstock materials can be either produced in a batch process or continuously. Four different types of machines are generally used: high-shear mixers, roll mills, screw extruders and shear rolls. The first two are examples of batch operations while the last two are continuous. Which approach to take depends on the details of the application and the materials to be used to prepare the feedstock (Clemens, 2009). When using fine particles, which have a tendency to agglomerate batch mixing in planetary or z-blade mixers (Fig. 3) is preferred, even though the process can take a couple of hours. In high volume productions, twin-screw extruders or shear rolls (Fig. 4) are employed for feedstock preparation (Hausnerová, 2011).

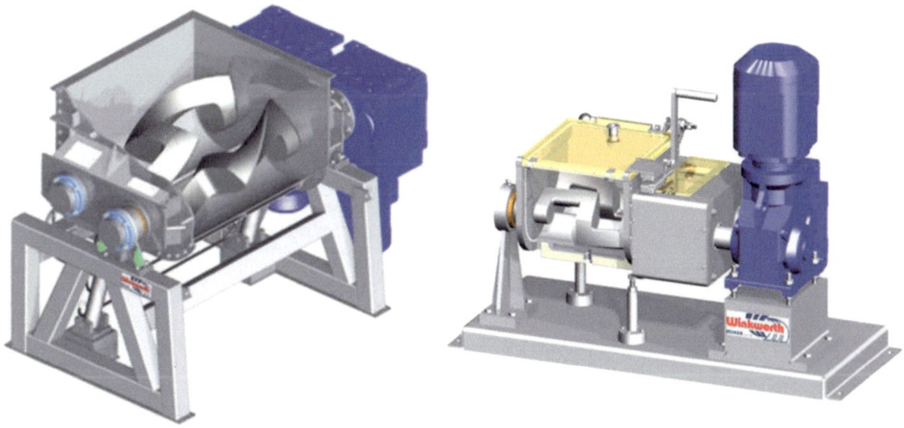

Fig. 3. Z-blade mixers for batch production of feedstock materials (courtesy of Winkworth Mixer Co., UK, www.mixer.co.uk).

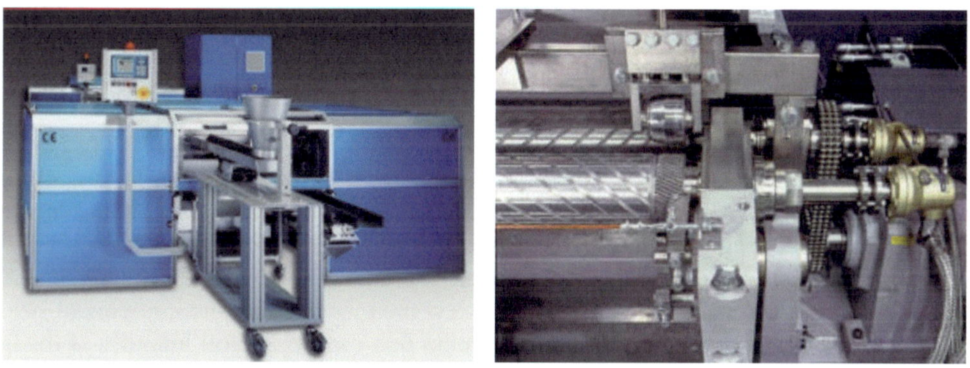

Fig. 4. Shear rolls for continuous production of feedstock materials (courtesy of Bellaform GmbH, Germany, www.bellaform.com).

The following sections have the purpose to further describe the two main components of the feedstock material: binder and powder. The binder formulation, powder synthesis processes, and their desirable properties are indicated below.

3.1 Binder formulation

Binder vehicles used for PIM are usually designed as multi-component systems. One of the main components is termed backbone, which is a thermoplastic polymer that supports and maintains the shape of the molded part until the last stages of debinding (Thomas-Vielma *et al*, 2008). As examples of currently used backbones, it is possible to mention ethylene vinyl acetate (EVA), polyethylene (PE), polypropylene (PP), polystyrene (PS), polyethylene glycol (PEG), polymethyl methacrylate (PMMA) among others (Ahn *et al*, 2009; Chuankrerkkul *et al*, 2007; Krug *et al*, 2002; Thomas-Vielma *et al* 2008; Yang *et al*, 2002) (Fig. 5).

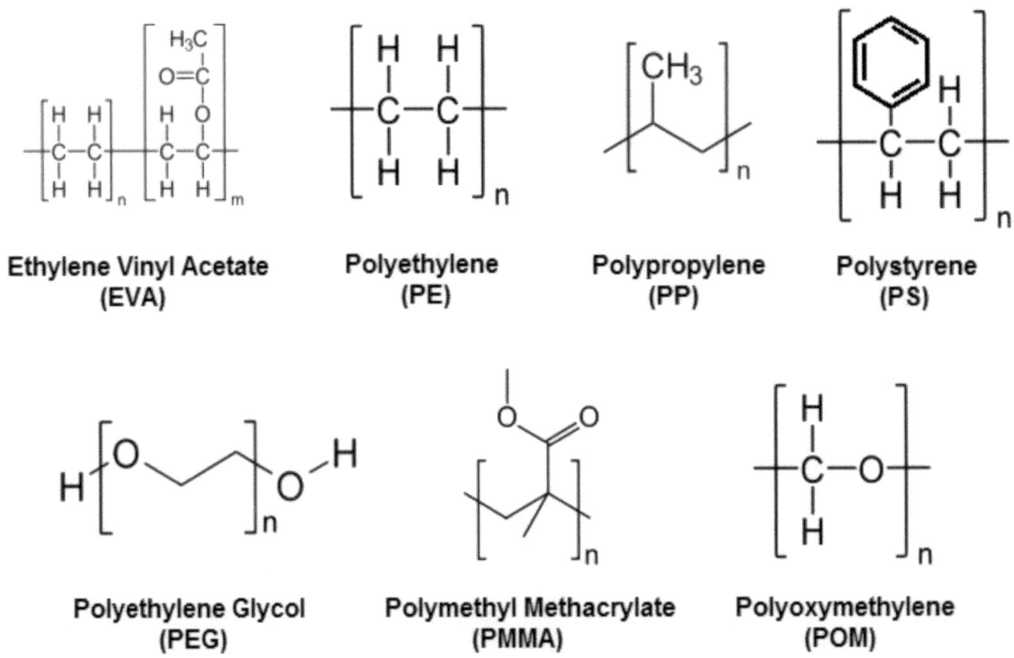

Fig. 5. Examples of polymers used in different binder formulations.

The second component, usually in a proportion similar to the backbone is commonly a wax, as paraffin or carnauba wax, or in some cases even agarose, that improves the material flowability (Ahn *et al*, 2009). Besides improving flowability, such component should be easily removed in early stages of debinding, in general via solvent methods, leaving open pores that will allow the gaseous products of the remaining polymer to diffuse out of the structure (Thomas-Vielma *et al*, 2008). Even though this low-melting temperature component has an important role in the process, it is worth to mention that the mechanical integrity of the final product is reduced as its proportion increase after certain limits (Tseng & Hsu, 1999).

The importance of each of these two main binder components can be better understood with a further description of the debinding mechanism. It is worth to remember that at the beginning of debinding no pores or free space are shown in the molded part, hence the backbone component has a crucial role retaining the shape of the part, and avoiding cracks while the low-melting component leaves this molded structure (Thomas-Vielma *et al*, 2008). In the last stages of debinding, it is due to the open porous created by this second component that the backbone can diffuse out without damaging the structure of the product. If not by these pores, an excessive pressure would easily build up within moldings from the degradation species during burnout, causing distortions and cracks (Tseng & Hsu, 1999).

However, the emergence of a POM-based binder system for PIM has made it possible to remove the polymer vehicle from up to 35 mm thick sections without the use of any wax or low molecular weight component (Krug *et al*, 2000). As previously described, POM (Fig.5) decomposes predominantly to formaldehyde in the presence of an acid vapor (as oxalic or nitric acid) well below its softening point, that is, in the solid state, avoiding the cracks and bloating that can be caused by the boiling of the binder (Krug *et al*, 2001). It is also important to mention that the polymer is not penetrated by the gaseous acid and the decomposition proceeds exclusively at the gas-binder interface with a nearly planar debinding front moving through the compact. In this sense, gas exchange is limited to the already porous shell and the buildup of an internal pressure is avoided. Nevertheless, POM-based binder systems often contains up to 30% of polyethylene which does not react with acid vapors, acting as a backbone until being burned out during the sintering cycle.

Finally, additives as surfactants can compose the binder, being stearic acid the most common example of them. These surface-active dispersants normally present a low melting temperature and affinity to preferentially adsorb onto powder surfaces, forming a densely thin outer layer on a particle surface which leads to a more homogeneous packing structure (Chan & Lin, 1995). However, bubbles and cracks were reported to occur as the amount of the surfactants increases, presumably owning to the reduced vaporization temperature since the surfactants are composed of mostly short molecules (Tseng & Hsu, 1999).

3.2 Powder manufacturing

It is from the powder material that the final product will be constituted, and its selection often involves the combination of a tailored particle size distribution to maximize packing densities. Powders for ceramics and metals can be obtained from a variety of methods; the following section will describe some of the methods used for obtaining ceramic and metallic powders of various shapes and sizes.

3.2.1 Ceramics

The methods used for synthesis of ceramic powders range from mechanical methods that involve grinding or milling (commination) for size reduction of a coarse, granular material to chemical processes involving chemical reactions under carefully controlled conditions. Generally speaking, mechanical methods are considerably cheaper than chemical methods. However, chemical methods offer better control of the powder characteristics, such as shape and size (Rahaman, 2003).

Mechanical methods are generally used to prepare powders from naturally occurring raw materials. Operations such as crushing, grinding and milling are classified as mechanical methods. Machines like jaw, gyratory and cone crushers are used to produce powders in the size range from 0.1 to 1mm. In order to achieve particles with less than 1 mm milling is generally used. Mills used today include high-compression roller mills, jet mills and ball mills (Rahaman, 2003).

Chemical methods are generally used in the synthesis of powders of advanced ceramics from synthetic materials or from naturally occurring materials with a significant chemical refinement. The most common chemical methods are summarized in Table 1. It is important to mention that many chemical methods require a milling step to break down agglomerates and for determining the average particle size and particle size distribution (Rahaman, 2003).

Method	Advantages	Disadvantages
Solid-state reactions: - Decomposition - Reactions between solids	Inexpensive, simple equipment used	Agglomerated powders, limited homogeneity for multicomponent powders
Liquid solutions: -Precipitation -Solvent vaporization -Gel routes	Small particles, chemical homogeneity, high purity, composition control	Agglomerated powders, poor for non-oxides, expensive
Non-aqueous liquid reaction	Small particles, high purity	Limited to non-oxides
Gas-solid reactions	Inexpensive for large particle sizes	Low purity, expensive for fine powders
Gas-liquid reactions	Small particles, high purity	Expensive, limited applicability
Reaction between gasses	Small particles, high purity, inexpensive for oxides	Agglomerated powders, expensive for non-oxides

Table 1. Commonly used chemical methods for manufacturing of ceramic powders (Rahaman, 2003).

3.2.2 Metals

Most metal powders can be produced by comminution of the refined ore. Milling and grinding are methods to produce powders of any degree of fineness from friable or malleable metals (e.g. titanium or steel). However not all particle sizes are sinterable, in general particles have to be below 45 μm. Currently, there is a tendency to use submicron and nanoparticles, since a better packing density can be achieved, their activation energy is higher and their sintering temperature is lower (Shearwood *et al*, 2005). Nevertheless, care must be taken when handling nano powders since they can be explosive or can easily oxidize; therefore most of the time powder metallurgy utilizes particles in the 1 to 45 μm range (Hartwig *et al*, 1998). In particular, for metal injection molding the mean particle size ranges from 5 to 15 μm (Krug *et al*, 2002). This type of size particles can generally be achieved in ball mills, rotary mills, planetary mills, jet mills, vibrating grinders, stampers and crushers (Schrader *et al*, 2000).

Nickel or iron can react with carbon monoxide to form metal carbonyls; carbonyl vapors undergo decomposition by instantaneous mixing with a large volume of hot inert gas followed by quenching the aerosol formed by diluting and cooling. Metal powders from carbonyls have high purity, small and uniform grain size, and particles that are dense and round (Neikov *et al*, 2009; Schrader *et al*, 2000). The most commonly used powders in today's MIM production are carbonyl powders. Simple mixtures of iron and nickel carbonyl powders are the basis of powder mixtures for metal injection molding; other metals included in these mixtures include heat treatable steels and stainless steels (Hartwig *et al*, 1998).

Metals or alloys that may be homogeneously melted can also be atomized in a stream or air or an inert gas (often argon). Some metals are melted first and later injected through an orifice into the stream to later be dropped in water, such process is called shotting. Other metals like iron and stainless steel may be fused in an electric arc and refractory metals in a plasma arc. For example, titanium droplets freeze to powder after being thrown from the end of a rapidly rotating bar heated by a plasma arc in a helium atmosphere (Schrader *et al*,

2000). Water atomization leads to irregular particles with a good yield of particles below 45 μm. Gas atomized powders leads to mostly spherical particles (Fig. 6) but the yield of particles below 45 mm is limited (Hartwig *et al*, 1998).

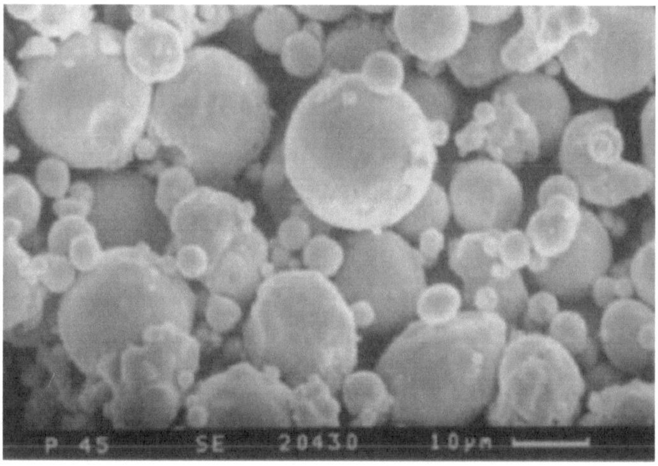

Fig. 6. SEM micrograph of gas atomized stainless steel powder (Hartwig *et al*, 1998).

Metal powder can also be electrolytically deposited under the right conditions. The deposited material may have to be broken up and ground to achieve the desired fineness, heated to be annealed, drive off hydrogen, sorted and blended. Electrodeposited powders are among the purest powders available, but they are pricier (Schrader *et al*, 2000).

Other less common methods of manufacturing metallic powders include vapor condensation, chemical decomposition, ordinary machining, impacting of chips and scrap, and granulation by stirring vigorously during solidification (Schrader *et al*, 2000). The most common methods applied for obtaining powders of different metals can be found in Table 2.

Production method	**Used for**
Chemical	Tantalum, tungsten
Electro-chemical	Copper, iron, nickel
Thermo-chemical	Carbonyl-iron, nickel
Mechanical (comminution & atomization)	Steels, titanium, intermetallics

Table 2. Common production methods for metal injection molding powders (Hartwig *et al*, 1998).

3.2.3 Submicron powders

With the growing demand for microparts (Fig.7) to be used in consumer electronics, automotive parts and in the biomedical field, there is also a requirement to use powders with smaller particles and perform what is called micro-PIM. It is generally accepted that the smallest feature of a part can only be ten times larger than the mean particle diameter, thus in order to obtain microparts with good edge definition and shape retention, submicron or nanoparticles need to be used. Also, submicron powders have an enhanced sintering activity, which is beneficial for attainment of high-density bodies at lower sintering temperatures. Such powders are still under development since there are many issues that need to be resolved.

Fig. 7. Ceramic microparts produced by PIM (Pioter *et al*, 2001).

Submicron powders tend to be pyrophoric and therefore it is essential to avoid any oxidation during handling. One way to avoid oxidation is to handle submicron powders in a glove box under argon; inside this box powders are coated with a binder constituent, in order to avoid oxidation in the subsequent processing steps of feedstock compounding and injection molding (Zauner, 2006). Another issue that arises as the particle size decreases below 1 µm is the greater tendency for particle-particle interaction, which results in agglomerates. If many agglomerates are present, the final sintered part could have a non-uniform distribution of particles and little benefit is achieved compared to coarse particles. Therefore, the use of submicron powders requires special handling and mixing procedures in order to minimize detrimental effects due to the presence of agglomerates (Rahaman, 2003).

3.3 Desired properties of feedstock

A homogeneous distribution of powder particles and binder in feedstock is important as it helps to minimize segregation during the injection molding stage and later on to obtain isotropic shrinkage after debinding and sintering (Quinard *et al*, 2009). Avoiding segregation of feedstock components is necessary to prevent visual defects, excessive porosity, warpage and cracks in the sintered part (Thornagel, 2010). The technique used for mixing binder and powder can influence the homogeneity of feedstock materials. As shown by Quinard *et al* (2009), using a twin-screw extruder yielded a better binder distribution. However, when using a z-blade mixer a better binder volume distribution around the powder particles was achieved, that would lead to an isotropic shrinkage after sintering.

Understanding which the desired properties are for a binder is a crucial step in the optimization of PIM regarding both productivity and quality of the final product. Only a few authors have been working in this direction and not all of them agree in their conclusions. For example, regarding the flow characteristics, German (1990) highlighted the importance of low viscosity at the molding temperature, a characteristic stressed by Liu *et al* (2001) as even more important in the case of micro moldings. Ahn *et al* (2008) mentioned that a high drop in viscosity at the high shear rates (shear-thinning behavior) is also a desired property for cavity filling with less energy, especially for complicated geometries. However, stability of the mixture should be taken into account in order to avoid powder-binder separation.

The binder should also be strong and rigid after cooling and present small molecules to fit between particles (German, 1990). No agreement is found regarding the viscosity-temperature dependence, which is suggested to be the least as possible by German (1990) but high by Thomas-Vielma *et al* (2007). Nevertheless, Ahn *et al* (2009) shown that contrary to the viscosity-shear rate dependence, the viscosity-temperature dependence is more influenced by the powder selection than the selection of the binder itself.

High thermal conductivity, low thermal expansion coefficient, short chain length, no orientation, low contact angle and good adhesion with powder, capillary attraction of particles and be chemically passive with respect to the powder are also desired properties that should be taken into account when selecting binder components (German, 1990).

Thomas-Vielma *et al* (2007) reported that defects as cracking (Fig. 8), slumping and sagging are partly related to the swelling of the binder component during solvent debinding, a problem that can be reduced by increasing the crystallinity of the polymer acting as a backbone. Besides this, the decomposition of such polymer should be non-corrosive, non-toxic, and having low ash content (German, 1990).

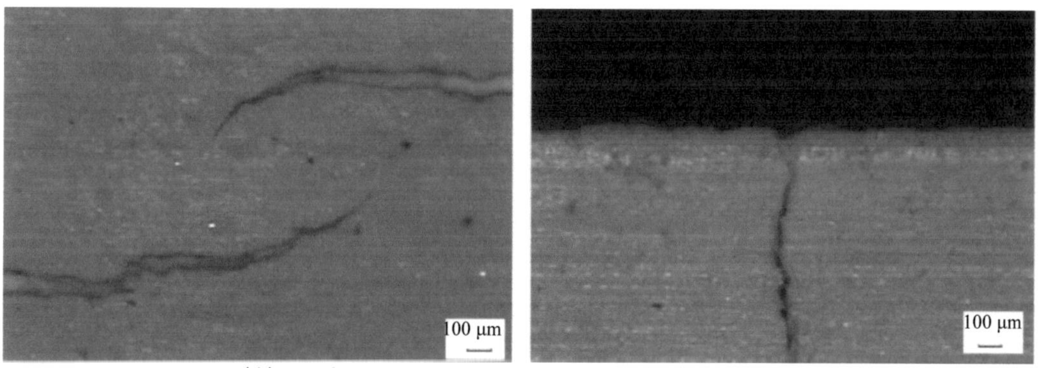

(A) (B)

Fig. 8. Appearance of (A) cracks and (B) transverse cracks after thermal debinding on alumina parts (Thomas-Vielma *et al* , 2007).

As can be observed in this section, the thermal and time-dependent properties of the different components of the binder play an important role in the quality of the final product. Furthermore, it also influences the choice of process parameters, as injection pressure and heating ramp, determining the cost and the productivity of the process. In this sense, understanding and improving the properties of this multi-component system or its single components have a great importance to the optimization of the PIM technology.

3.4 Possible feedstock optimization

Rheological and time-dependant properties represent powerful tools for optimizing the PIM process. The binder's rheological properties (flowability) are important to determine the optimal binder formulation or selecting a proper additive. It has been suggested in the literature that the viscosity of the binder should be less than 0.1 Pa · s in order to provide the PIM feedstock with a viscosity below 1000 Pa · s (German, 1990). Having low viscosity allows for easy molding. However, it is generally known that low viscosity materials have also low mechanical properties when solidified. Having low mechanical properties transforms into a "green part" which is prone to deformation and breakage if not handle with care, especially right after molding when the binder is still soft. For this reason a compromise has to be made between the molten rheological properties and the solid mechanical properties of binder.

Bimodality in the molecular mass distribution (Fig. 9) of binders is still an issue that has not been extensively studied. Nevertheless, recent investigations have shown its potential to bring two main benefits to the process of PIM: lower the viscosity of the feedstock in the molten state and maintaining the mechanical strength of the green part, all of this without modifying the chemistry of the binder.

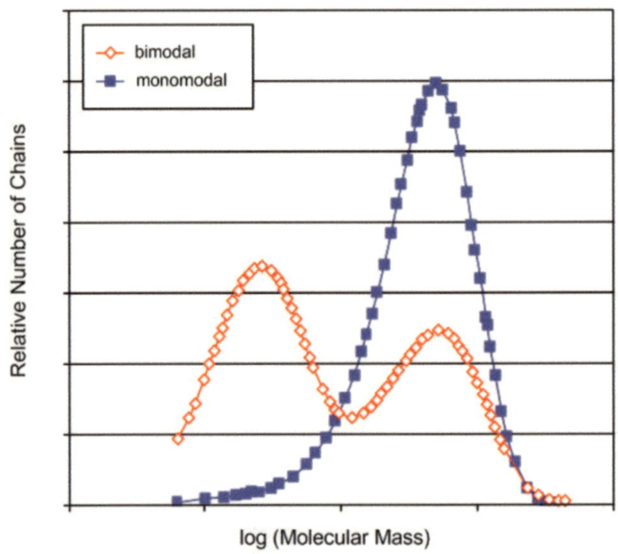

Fig. 9. Schematic representation of bimodal and monomodal molecular mass distribution in polymeric materials.

It has been observed that bimodality in the molecular mass distribution improves the shear relaxation modulus (Emri & von Bernstorff, 2006) and the shear creep compliance (Kubyshkina *et al*, 2011) of polyamide 6 in the solid state. Based on those results, Stringari *et al* (2011) studied the effect of bimodal molecular mass distribution in POM and it was shown that bimodality decreases viscosity without significantly affecting the mechanical properties (shear creep compliance) of the potential PIM binder. The mechanisms for such behavior could be explained as follows: The addition of small polymer chains through bimodality increases the flowability of the binder in the molten state, since such small molecules can act as internal lubricants for the larger molecules facilitating their movement. However, when in the solid state the smaller molecules could fit between the larger ones, creating a closely packed structure that cannot be deformed as easily, translating into improved mechanical properties.

Once good flowability of the binder has been obtained it is important to check the flowability of the feedstock material. The flowability of feedstock will not only depend on the viscosity of binder but also on the powder loading. The optimal powder loading refers to a concentration of powder for which a compound exhibits good flow properties, good dispersion and distribution, and flow stability in the shear rates applied during injection molding (10^2 - 10^5 s^{-1}). Rheological data represents an extremely useful tool to evaluate optimum powder loading. The ratio of feedstock's viscosity to binder's viscosity known as the relative viscosity is an important parameter used to determine the maximum packing fraction for a particular powder-binder compound. As the powder concentration reaches the maximum level the flow of feedstock material is restrained and a sharp increase in relative viscosity is observed (Hausnerová, 2011).

The maximum loading level in a feedstock material is dependent on the characteristics of the powders and how these particles are packed. The packing behavior of particulate materials depends largely on their particle size, shape and surface characteristics. Packing behavior has been explained by model systems with closely defined size and shape distributions. Even though real materials do not have a well-defined shape and size distribution, the principles derived from models appear to work reasonably well. Theoretical maximum packing has been studied for spherical, smooth, regular, mono-sized particles, which can readily move past another. With these particles a maximum ordered packing fraction of 0.74 has been established from geometrical principles. In order to increase the packing fraction of powders, small particles to fill in the pores in the packed structure obtained from larger particles are used and then use even smaller particles to fill in the remaining pores and so on. This approach is known as multimodal packing and much effort has been devoted to find the optimal particle size ratios that yield maximum packing fractions in the field of particulate composites (Rothon, 2003). German (1990) has already pointed out that an ideal PIM powder should combine large and small particles in a tailored size distribution. Hausnerová *et al* (1999) showed that feedstock material containing particles with a monomodal particle size distribution exhibit higher viscosity compared to feedstock with bimodal powder. The current explanation for this behavior is that the smaller particles fill the inter-particle voids created by large particles, thereby releasing previously immobilized molten binder (Hausnerová, 2011).

4. Injection molding

The second step in the PIM process is molding the feedstock into the desired shape. The most popular method is to use a reciprocating screw, horizontal, hydraulic or electric machine in which a screw stirs the feedstock inside the barrel while it is melting. After melting the feedstock the screw acts as a plunger to generate the pressure to fill the die (Stevenson, 2009). Conventional screw-type injection molding machines consist of a clamping unit, injection unit and control system. The clamping unit houses the mold which is generally comprised of two halves. When the clamping unit is closed, material can be injected into the mold, when the clamping unit is open the molded part can be removed.

The injection unit consists of a screw, a heating system and a nozzle. The screw transports the material inside a barrel, compresses it and removes any bubbles. The heating system brings the material to an appropriate temperature for easy flow. The nozzle is the conduct through which the heated feedstock is injected into the mold under pressure. The control system of a modern injection molding machine includes hardware and software where the processing conditions are set and saved to ensure the reproducibility of previously employed production cycles (Arburg, 2009).

Molding of the feedstock is comparable to the injection molding of plain thermoplastics and it has the following stages (Arburg, 2009; German & Bose, 1997):

a. The pelletized feedstock is placed in the hopper of the injection molding machine.
b. The binder in the feedstock is melted by the heating system.
c. The molten material is injected under high pressure (60 MPa or even more) into the mold cavity which is mounted in the clamping unit. The feedstock must have low enough viscosity that it can flow into the die cavity under pressure.
d. The mold remains closed while cooling channels in the die extract heat from the molten feedstock and solidify the polymer to preserve the molded shape.
e. After solidification of the binder, the nozzle of the injection die is pulled away from the mold by moving the injection unit. The clamping unit opens and the molded part is ejected by the ejector system of the machine.
f. The green part is removed from the mold. Due to the fragile nature of most green parts, the removal process is done by hand or by a robotic system in order to prevent shocks or impacts which could deform or even break the molded part.

The shaping equipment used in PIM is the same as the one used for plastic injection molding. Due to the size of molded parts, injection molding machines used for PIM are in the lower range, with clamp force typically less than 100 tons, 18 to 25 mm screws and shot size of less than 30 cm^3 (Stevenson, 2009). However, the main important difference when dealing with injection molding of any powdered part is that many of the components of the molding machine are subject to a more intense wear, particularly screws, non-return valves, cylinders and molds (Rosato & Rosato, 1995).

Injection molding machines for processing of powdery materials are optimized with wear-resistant components, through special hardening processes or utilization of alloys. For example, when dealing with stainless steel feedstock materials, hardening with carbon nitride is recommended by feedstock manufacturers. And when working with ceramic and hard metals, boride cladding or carbide hard facing are recommended. Since harder screws are more brittle, lower torque limits during startup are used to prevent screw breakage in PIM. The solid feedstock pellets cause the most abrasive wear in the feed section of the screw, thus the feedstock should be melted as early as possible in the injection cycle (Stevenson, 2009).

Screw geometry of PIM machine is adopted to lower the compression rate and extend the compression zone as compared to screws used for thermoplastics (Hausnerová, 2011). Compression ratios used in PIM tend to be in the lower range. Ratios between 1.2 and 1.8 are considered acceptable and a ratio of 1.6 is considered to be optimal for the removal of air between granules. It is also important to mention that when calculating the barrel capacity of the injection unit, the barrel rating must be scaled up to take into account the higher density of the PIM feedstock (Stevenson, 2009).

The injection molding step can create undesirable features such as gate and ejection pin marks, which must be located in non-critical locations otherwise they must be removed after fabrication. Other design limitations include gradual thickness changes, minimized wall thickness, round corners to reduce stress concentrations and risk of crack appearance, and minimum undercuts on internal bores. Whenever possible, it is important to design PIM parts with one flat surface, which allows the use of standard trays during sintering, otherwise special trays are required (Hausnerová, 2011).

5. Debinding

Before sintering, the organic binder must be removed without disrupting the molded powder particles; this process is commonly referred as debinding. Organic polymers have to be removed completely from the "green part", since carbon residues can influence the sintering process and affect the quality of the final product negatively. Moreover, binder removal is one of the most critical steps in the PIM process since defects can be produced by inadequate debinding, like bloating, blistering, surface cracking and large internal voids. It has been shown that the rate of binder removal plays a main role in the defect production due to structural changes in capillaries inside the green part (Oliveira *et al*, 2005).

The most commonly used debinding techniques include: thermal, solvent and catalytic. However there also exist some experimental techniques such as plasma debiding (dos Santos *et al*, 2004). The following sections have the aim to provide the reader with a description of the different debinding techniques showing their benefits and limitations.

5.1 Thermal debinding

Thermal debinding utilizes the mechanisms of thermal degradation of organic binders, which is based on the successive dissociation of polymers to produce light molecules that are later evaporated out of the surface of the molded part. Since the thermal degradation process is different for different polymers then thermal debinding time is greatly influenced by the type of polymer used. The binders developed in the original PIM process were a mixture of polyethylene or polypropylene, a synthetic or natural wax and stearic acid. Feedstock materials based on such binders can be removed thermally. However, it has been shown in the literature that POM and polybutyl-methacrylate (PBMA) have a much faster degradation than other polymers such as polypropylene (PP) and ethylene-vinyl-acetate (EVA) (Kankawa, 1997). It should also be noted that the choice of atmosphere under which thermal debinding is performed influences the rate of binder removal and some characteristics of the final piece such as density, carbon or oxygen content (Quinard *et al*, 2009).

In general, it can be said that thermal debinding is an inefficient process that can result in a poor etching of the piece surface if not properly controlled. Additionally, increasing the temperature too fast may produce an excessive increase of vapor pressure in the core of the molded piece leading to defects. Consequently, in order to reduce the risk of cracks or shape deformation, low heating rates are generally used, resulting in a long debinding time, ranging from 10 to 60 h (dos Santos *et al*, 2004). However, thermal debinding is still selected due to its simplicity, safety and respect for the environment as compared to solvent and catalytic binder removal (Quinard *et al*, 2009).

In order to increase the efficiency of the thermal debinding process a vacuum pump is used to continuously pull binder vapor away from the heated parts. Initially, the heating rate is typically 0.5 °C/min or less until a temperature of 100 °C is reached; this temperature is hold for approximately 4 h and later it is increased to 400 °C using a heating rate close to 1 °C/min, this temperature is maintained for 2 h. During the entire process the pieces are constantly exposed to a gas flow, in order to remove the binder vapors, which are condensed and collected in a trap. Finally when all the binder is removed the temperature is increased and the sintering process can begin. This is one of the main advantages of vacuum debinding. However, the debinding process is still around 10 hours long, since the main mechanisms of debinding is still thermal degradation (dos Santos *et al*, 2004).

Another variant of the thermal debinding process includes the use of a wicking material (porous substrate) which is in contact with the "green part" or compact. In industrial practice the compact is buried in the wicking material and therefore the binder is removed from the compact in all directions (Somasundram *et al*, 2008). A simplified model is shown in Fig. 10. The porous substrate provides a medium for capillary flow as the binder viscosity decreases due to the increase of temperature. Wicking thermal debinding is performed at temperatures where the binder melts, therefore the binder can flow out of the component into the pores of the contacting substrate. Wicking involves liquid extraction; while other thermal process requires the binder to be in gaseous state and thus the temperatures are generally lower but the process is slower due to the slower transport of liquid compared to a gas (German & Bose, 1997).

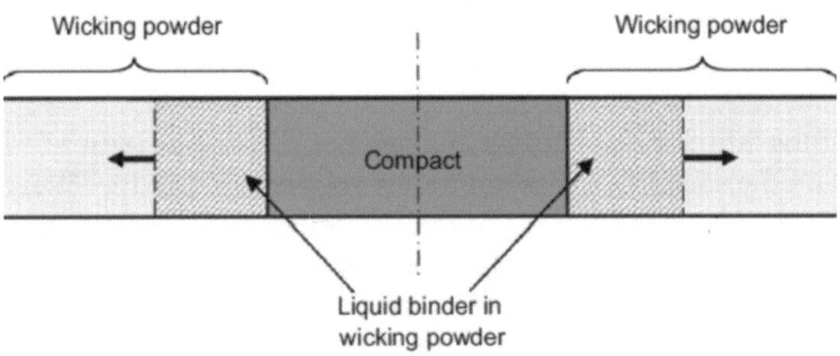

Fig. 10. Schematic cross-section of compact (green part) surrounded by wicking powder during thermal debinding (adapted from Somasundram *et al*, 2008).

5.2 Solvent debinding

Solvent debinding is done by immersing the molded part in a gaseous (Fig. 11) or liquid solvent such as ethanol, hexane, heptanes and acetone, at low temperature, typically 50 to 60 °C (Torralba *et al*, 2011). The solvent removes at least one of the binder components and produces an open porosity. The next step in solvent debinding is binder burnout to remove the backbone of the binder that provides adequate shape retention up to the onset of sintering (dos Santos *et al*, 2004; Aggarwal *et al*, 2007). Binder backbone removal is generally done thermally between 200 and 600 °C in a pre-sintering step (Tandon, 2008).

The effectiveness of solvent debinding is strongly related to the geometry of the "green part" in particular to the surface to volume ratio, since the solvent needs to penetrate the part. Other factors that influence solvent debinding include temperature and porosity

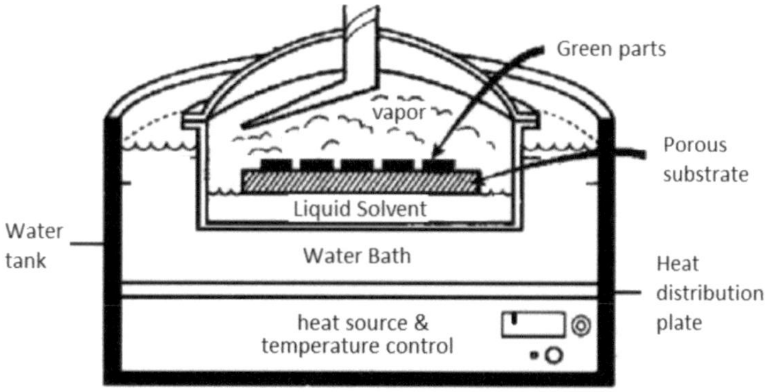

Fig. 11. Solvent vapor debing process (adapted from German & Bose, 1997).

evolution. The temperature effect is related to an increase in interaction between soluble binders and solvents as temperature increases, in other words temperature changes the solubility and diffusion coefficient of the binder (Oliveira *et al*, 2005).

There is a tendency to try to use binders that are soluble in water, since handling the aqueous solvent is much easier than handling organic solvents. Good examples of water-soluble binders include polyethylene glycols, polyethylene oxide, polyvinyl alcohols, starches and polyacrylamide. All of these polymers have monomers containing oxygen and nitrogen that are hydrophilic. The time to debind is a function of the powder material, particle size, part geometry, water temperature, water circulation and water volume relative to mass of the "green part". All of these variables are interrelated with each other and must be optimized for a particular piece. After debinding, parts are usually dried in a forced air furnace at 65 to 75 °C for at least three hours. The drying step can be included as part of the sintering program. Water can be regenerated after debinding by distillation and therefore a closed process water circuit can be used (Auzene & Roberjot, 2011).

Another type of solvent debinding is the use of a supercritical process. When applying supercritical extraction, the operative pressures of solvents such as carbon dioxide or propane are kept around 10 MPa and temperatures are less than 100 °C. The use of low temperatures results in slow diffusion of the binder to the surface of the green part, resulting in long processing times. Super critical debinding is not widely employed in commercial operations due to the long processing times and elevated cost of the necessary equipment, which require high precision control in temperature and pressure (dos Santos *et al*, 2004).

5.3 Catalytic debinding

Catalytic debinding can be used for binders that decompose into smaller molecules in the presence of a catalyst when exposed to the appropriate temperature. The most common example is a binder based on POM sometimes also referred as polyacetal. A commercial example of a feedstock with a POM based binder is Catamold® produced by BASF. By using a catalyst, the polymer at the surface of the "green part" is cracked into monomers and evaporates. As the monomers evaporate, pores are created that expose the polymer beneath the surface and the depolymerization process continues deeper into the molded part. Thus the debinding occurs from the outside inwards. Shorter debinding times are achieved with the use of the catalytic process, since the rate of diffusion of monomers is high due to the small size of their molecules (Clemens, 2009). Furthermore, the small molecules generated have a high vapor pressure, which greatly minimizes the potential for capillary condensation and allows thick part sections to be debound (Krueger, 1996).

Polyacetal-based binders depolymerizes catalytically under acidic conditions yielding formaldehyde, a direct solid to gas transition (Fig. 12). Temperature and catalyst concentration play a key role in determining the rate of debinding, as well as the particle size of the powder and geometry of the molded part. This type of debinding is performed below the melting temperature of POM, generally between 110 and 150 °C. The use of these relative low temperatures prevents the formation of a liquid phase and thus prevents deformation of the "green part" due to gravitational distortion or stress relaxation (Fu *et al*, 2005). Also the internal gas pressure is low, which minimizes the danger of crack formation and propagation. For POM, 100% nitric acid (HNO_3) is the most suitable catalyst. Even though nitric acid is a strongly oxidizing agent, its anhydrous form does not react with most of the commonly used metal powders (Krueger *et al*, 1993).

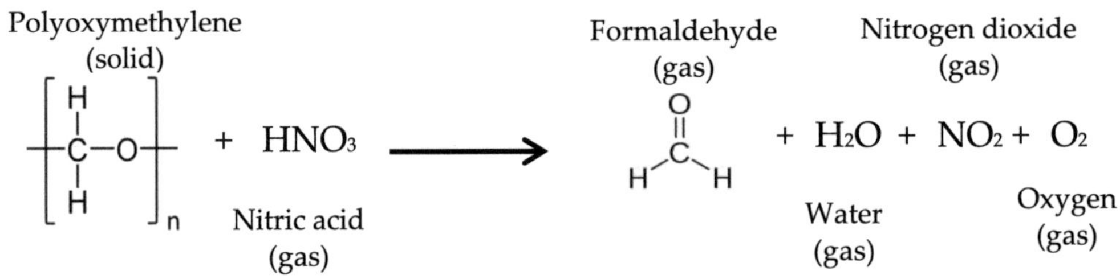

Fig. 12. Decomposition of POM in the presence of nitric acid.

The debinding time depends on the quantity of catalysts and temperature used. An increase of these two factors can shorten the time of depolymerization. However, there are some limitations as to how much temperature and catalyst content can be increased, not only due to possible damage to the molded part but also due to health concerns. As previously mentioned, POM decomposes into formaldehyde, which due to its toxicity has limitations in the allowable quantity (0.1 kg/h) and concentration (20 mg/m^3) that can be present in the working environment (Goyer *et al*, 2006). The exhaust from the debinding oven must be treated in two steps to get rid of the toxic bi-product of the depolymerization of POM, which are nitrogen dioxide (NO_2) and formaldehyde (CH_2O). First the exhaust is burned in a reducing atmosphere (no oxygen and rich in nitrogen) at a temperature of 600 °C, transforming nitric dioxide into nitrogen gas (N_2). The second step consists of burning in an oxidizing atmosphere at 800 °C to transform formaldehyde into water and carbon dioxide (Torralba *et al*, 2011). If properly treated, the exhaust fumes coming out of the debinding oven does not represent a health hazard.

It is important to mention that binders based on POM usually have a backbone polymer which is not susceptible to catalytic debinding. Such backbone polymer helps retain strength and shape stability in the "brown part". However, sintering cannot begin in the presence of this backbone polymer and thus a thermal treatment between 200 and 600 °C is applied to the part prior to the start of the sintering process (Tandon, 2008).

5.4 Comparison between debinding processes

The major differences, advantages and disadvantages between the three types of debinding techniques are summarized in Table 3.

Debinding Technique	Key Features	Advantages	Disadvantages
Thermal	Slowly heat green part to melting or degradation temperatures with a continuous sweep gas to remove binder.	One-step process, no need to handle product between debinding and sintering (unless wicking is used). Low cost installation. Applicable to a wide range of binders.	Soft binder allows warpage, poor dimensional control and relatively slow process (up to 60 h). If a wick is used problems to separate part from it.
Solvent	Green part is placed in a solvent in gaseous or liquid state to extract binder via dissolution.	Component remains rigid without chemical reactions. Lower temperatures minimize defects and distortions. Faster than thermal debinding (around 6 h)	Solvent hazard, chemical handling and environmental concern (unless water soluble binder is used). Expensive equipment if using supercritical extraction. Drying before sintering required if using liquid solvent.
Catalytic	Heat green part in atmosphere containing catalyst to depolymerize binder and sweep away monomers. Binder goes from solid to gas.	Rapid process (4 to 6h) that works well on thick and thin sections with excellent shape retention.	Possible hazards with acid catalysts and decomposition products. Exhaust products must be treated properly to prevent health and environmental hazards.

Table 3. Comparison between three major debinding techniques.

6. Sintering

The last step of the PIM process is sintering. Sintering is one of the oldest human technologies, originating in the prehistoric era with the firing of pottery. After the 1940s sintering has been studied fundamentally and scientifically leading to remarkable developments in sintering science. Nowadays, sintering can be used for the fabrication of all kinds of parts, including powder-metallurgical parts and bulk ceramic components (Kang, 2005).

Sintering is a thermal treatment that transforms metallic or ceramic powders into bulk materials with improved mechanical strength that in most cases have residual porosity. Sintering is performed at temperatures below the melting temperature of the major constituent in the metal or ceramic powder, generally within 70 to 90% of the melting point (Lame *et al*, 2003). The temperature inside the sintering furnace is high enough to start the recrystallization process of the metal or ceramic particles, but low enough so that the particles remain unmelted. At such temperatures, the particles recrystallize into each other causing them to fuse together (Boljanovic, 2010).

During sintering, solid-state atomic diffusion takes place, followed by recrystallization and grain growth. When the temperature exceeds one half to two thirds of the melting temperature of the powder material, significant atomic diffusion occurs and some chemical changes may happen on the surface of the particles, such as the vaporization of chemically bounded water. As temperature keeps increasing, thermolysis occurs which is a process that burns out the organic components such as remaining binder, dispersant, etc. During sintering there is a great deal of particle movement and mass transport. It has been identified that there are at least six different mechanisms for mass transfer involved in sintering, which are surface diffusion, evaporation-condensation, grain boundary diffusion, lattice diffusion, viscous flow, and plastic flow. These mechanisms lead to growth of necks between particles and thus increase the strength of the consolidated powders. However some mechanisms also lead to shrinkage (more than 10%) and densification. Surface diffusion is the mechanism that produces surface smoothing, particle joining and pore rounding, but not volume shrinkage. If the material has high vapor pressure, sublimation and vapor transport produce the same effects as surface diffusion. Diffusion along the grain boundaries and through the lattice produces both neck growth and volume shrinkage. Bulk viscous flow plays an important role in densification when a wetting liquid is present, while plastic deformation is important when a mechanical pressure is applied (Kang, 2007).

Fusing of the metallic particles during the sintering process has been observed using synchrotron microtomography as shown in the figures below for copper particles. It is clearly observed that during sintering the particles get closer to each other as interparticle necks grow and the porosity is reduced. From room temperature (Fig. 13A) to 1000 °C (Fig. 13B) there is no significant shrinkage but the neck formation causes small displacements and rotations of the particles that lead to a different particle packing. As a result of this rearrangement, some pores decrease in size but others increase. After sintering for 120 min at 1050 °C, the interparticle necks have grown up and it is harder to distinguish individual particles (Fig. 13C). Finally, (Fig. 13D) after the part was sintered for an additional 100 min and cooled down to room temperature, most pores have vanished but some large pores remain (Lame *et al*, 2003).

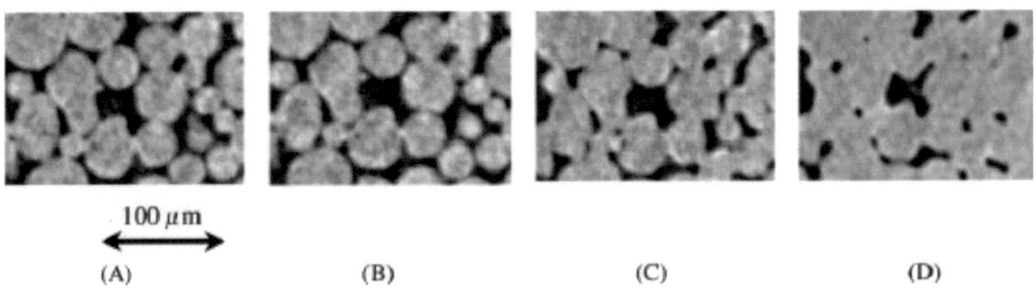

Fig. 13. Non-compacted copper particles at different stages of the sintering process, (A) at room temperature before sintering, (B) after heating to 1050 °C from room temperature in 45 min, (C) at 1050 °C after 120 min of sintering, and (D) after sintering for 220 min at 1050 °C and cooling to room temperature in 15 min (Lame *et al*, 2003).

Depending on the size of the part and the material used, the sintering time can vary. For small parts such as bushings, the average time varies from 1 to 1.5 h. For average-size ferrous parts, the sintering time can be 3 h. However tungsten parts can have a sintering time of up to 8 h. In order to harden sintered parts, a controlled cooling rate must be carried out in a separate section of the sintering furnace. Cooling is done in a protective atmosphere in order to prevent oxidation of sintered parts. Dissociated ammonia and nitrogen-based atmospheres are commonly used, however vacuum atmospheres are also used for stainless steel and tungsten parts for example. The cooling rate is critical since the mechanical properties of the sintered parts is affected by the phase transformation of the material (Boljanovic, 2010).

7. Conclusion - Closing remarks

PIM is a powerful process for the manufacturing of parts with complex geometry. It combines the design benefits of thermoplastic injection molding and the efficiency of powder metallurgy. Due to its capability is expected that PIM will growth in importance. However the current state of the art does not allow for PIM to be widely used and therefore there is a need to optimize the process to increase its efficiency and productivity, as well as the quality of the final parts. For instance, there is still room for improving the performance of binders as to obtain excellent flowability in the molten state while having high mechanical properties in the solid state. Optimization can also be brought into the powder design as to obtain the most adequate particle size distribution and size ratios for multimodal powders. Also, it has been shown that using nanoparticles brings many benefits into the PIM process, but handling procedures of these materials and their processing to avoid agglomeration is still far from optimal. In summary, PIM technology still offers a broad field of opportunities for improvement through applied research.

8. References

Ahn, S., Park, S.J., Lee, S., Atre, S.V. & German, R.M. (2009). Effect of powders and binders on material properties and molding parameters in iron and stainless steel powder injection molding process. *Powder Technology*, Vol.193, No.2, (July 2009), pp.162-169, ISSN 0032-5910.

Aggarwal, G., Smid, I., Park, S.J. & German, R.M. (2007). Development of niobium powder injection molding. Part II: Debinding and sintering. *International Journal of Refractory Metals & Hard Materials*, Vol.25, No.3, (May 2007), pp. 226-236, ISSN 0263-4368.

Arburg. (2009). Powder Injection Moulding (PIM) - Production of complex moulded parts from metal and ceramic. Arburg GmbH + Co KG, Lossburg, Germany, Available from: http://www.arburg.de/com/common/download/WEB_522785_en_GB.pdf

Auzene, D. & Roberjot, S. (2011). Investigation into water soluble binder systems for powder injection moulding. *Powder Injection Moulsding International*, Vol.5, No. 1, (March 2011), pp.54-57, ISSN 1753-1497.

Boljanovic, V. (2010). Powder metallurgy. In *Metal Shaping Processes: Casting and Molding, Particulate Processing, Deformation Processes, Metal Removal*. Industrial Press Inc., ISBN 9780831 133801, New York, USA, pp. 75-106.

Chan, T.Y. & Lin, S.L. (1995). Effects of stearic acid on the injection molding of alumina. *Journal of the American Ceramic Society*, Vol.78, No.10, (October 1995), pp. 2746-2752, ISSN 1551-2916.

Clemens, F. (2009). Thermoplastic extrusion for ceramic bodies. In *Extrusion in Ceramics-Engineering Materials and Processes,* Handle, F. (Ed.), Springer-Verlag, , ISBN 978-3540271000, Berlin, Germany, pp. 295-311.

Chuankrerkkul, N., Messer, P.F. & Davies, H.A. (2007). Powder injection moulding of cemented carbides: Feedstock preparation and injection moulding. *Journal of Metals, Materials and Minerals,* Vol.17, No.1 (May 2007), pp.53-57, ISSN 0857-6149.

Dos Santos, M.A., Neivock, M.P., Maliska, A.M., Klein, A.N. & Muzart, J.L.R. (2004). Plasma debinding and pre-sintering of injected parts. *Material Research,* Vol.7, No.3, (July/September 2004), pp.505-511, ISSN 1516-1439.

Emri, I., & von Bernstorff, B.S. (2006). The effect of molecular mass distribution on time-dependent behavior of polyamides. *Journal of Applied Mechanics,* Vol.73, No.5, (September 2006), pp.752-757, ISSN 0021-8936.

Fu, G., Loh, N.H., Tor, S.B., Tay, B.Y., Murakoshi, Y. & Maeda, R. (2005). Injection molding, debinding and sinterng of 316L stainless steel microstructures. *Applied Physics A,*Vol. 81, No.3, (May2005), pp.495-500, ISSN 0947-8396.

German, R.M. (1990). *Powder Injection Moulding.* Metals Powder Industries Federation, ISBN 0918404959, Princeton, New Jersey, USA.

German, R.M. (2008). PIM breaks the $1 bn barrier. *Metal Powder Report,* Vol.63, No.3, (March 2008), pp.8-10, ISSN 0026-0657.

German, R.M. & Bose, A. (1997). *Injection Moulding of Metals and Ceramics.* Metals Powder Industries Federation, ISBN 187895461X, Princeton, New Jersey, USA.

GIA. (2011). *Metal and Ceramic Injection Molding.* Global Industry Analysts, Inc., San Jose, USA.

Gonçalves, A.C. (2001). Metallic powder injection molding using low pressure. *Journal of Materials Processing Technology,* Vol.118, No.1-3, (December 2001) pp.193-198, ISSN 0924-0136.

Goyer, N., Bégin, D., Beaudry, C., Bouchard, M., Carrier, G., Lavoué, G., Noisel, N. & Gérin, M. (2006). *Prevention guide- Formaldehyde in the work place.* Institute de reserche Robert-Sauvé en santé et en sécurité du travail (IRSST), Montreal, Canada. Accessed: August 28, 2011, Available from: http://www.irsst.qc.ca/media/documents/PubIRSST/RG-473.pdf, ISBN 9782896310692

Hartwig, T., Veltl, G., Petzoldt, F., Kunze, H., Scholl, R. & Kieback, B. (1998). Powders for metal injection molding. *Journal of European Ceramic Society,* Vol.18, No. 9, (October 1998), pp.1211-1216, ISSN 0955-2219.

Hausnerová, B. (2011). Powder injection moulding- An alternative processing method for automotive items. In *New Trends and Developments in Automotive System Engineering,* Chiaberge M. (Ed.), InTech, ISBN 9789533075144, Rijeka, Croatia, pp.129-145.

Hausnerová, B., Saha, P. & Kubat, J. (1999). Capillary flow of hard-metal carbide powder compunds. *International Polymer Processing,* Vol.14, No. 3, (March 1999), pp.254-260, ISSN 0930-777X.

Kang, L. (2007). *Ceramic Membranes for Separation and Reaction.* John Wiley and Sons, ISBN 9780470014400, Chichester, England, 306 pages.

Kang, S.J. (2005). *Sintering: Densification, Grain Growth & Microstructure.* Elsevier Butterworth-Heinemann, ISBN 9780750663864, Oxford, U.K. 265 pages.

Kankawa, Y. (1997). Effects of polymer decomposition behavior on thermal debinding process in metal injection molding. *Materials and Manufacturing Processes,* Vol.12, No. 4, (April 1997), pp.681-690, ISSN 1042-6914.

Krueger, D., Bloemacher, M. & Weinand, D. (1993). Rapid catalytic debindng MIM feedstock, a new technology grows into a manufacturing process, In *Advances in*

Powder Metallurgy and Particulate Materials, Metal Powder Industries Federation, Princeton, New Jersey, USA, Vol. 5. No.2, pp. 165-180, ISSN 1065-5824.

Krueger, D.C. (1996). Process for improving the debinding rate of ceramic and metal injection molded products. US Patent 5531958.

Krug, S., Evans, J.R.G. & ter Matt, J.H.H. (2000). Residual stresses and cracking in large ceramic injection mouldings subjected to different solidification schedules. *Journal of the European Ceramic Society,* Vol.20, No.14-15, (December 2000), pp.2535-2541, ISSN 0955-2219.

Krug, S., Evans, J.R.G. & ter Maat, J.H.H. (2001). Transient effects during catalytic binder removal in ceramic injection moulding. *Journal of the European Ceramic Society,* Vol.21, No.12, (October 2001), pp.2275-2283, ISSN 0955-2219.

Krug, S., Evans, J.R.G. & ter Maat, J.H.H. (2002). Differential sintering in ceramic injection moulding: particle orientation effects. *Journal of the European Ceramic Society,* Vol.22, No. 2, (February 2002), pp.173-181, ISSN 0955-2219.

Kubyshkina, G., Zupančič, B., Stukelj, M., Grošelj, D., Marion, L. & Emri, I. (2011). The influence of different sterilization techniques on the time-dependent behavior of polyamides. *Journal of Biomaterials and Nanobitechnology* (accepted for publication), ISSN 2158-7027.

Lame, O., Bellet, D., Di Michiel, M. & Bouvard, D. (2003). In situ microtomography investigation of metal powder compacts during sintering. *Nuclear Instruments and Methods in Physics Research B,* Vol.200, (January 2003), pp.287-294, ISSN 0168-583X.

Liu, Z.Y., Loh, N.H., Tor, S.B. & Khor, K.A. (2003). Characterization of powder injection molding feedstock. *Materials Characterization,* Vol.49, No. 4, (November 2002), pp.313-320, ISSN 1044-5803.

Mathew, B.A. & Mastromatteo, R. (2003). MIM Focus- Metal injection moulding for automotive applications. *Metal Powder Report,* Vol.58, No.3, (March 2002), pp.32-35, ISSN 0026-0657.

Merz, L., Rath, S., Piotter, V., Ruprecht, R., Ritzhaupt-Kleissl, J. & Haussel, J. (2002). Feedstock development for micro powder injection molding. *Microsystem Technologies,* Vol.8, No. 2-3, (March 2002), pp.129-132, ISSN 0946-7076.

Neikov, O.D., Naboychenko, S., Gopienko, V.G. & Frishberg, I.V. (2009). *Handbook of Non-Ferrous Metal Powders: Technologies and applications,* Elsevier, ISBN 9781856174220, Amsterdam, The Netherlands, 634 pages.

Oliveira, R.V.B., Soldi, V., Fredel, M.C. & Pires, A.T.N. (2005). Ceramic injection molding: influence of specimen dimensions and temperature on solven debinding kinetics. *Journal of Materials Processing Technology,* Vol.160, No.2, (March 2005), pp.213-220, ISSN 0924-0136.

Pioter, V., Bauer, W., Benzler, T. & Emde, A. (2001). Injection molding of components for microsystems. *Microsystem Technologies,* Vol.7, No. 3, (March 2001), pp.99-102, ISSN 0946-7076.

Quinard, C., Barriere, T., Gelin, J.C. (2009). Development and property identification of 316L stainless steel feedstock for PIM and µPIM. *Powder Technology,* Vol.190, No.1-2, (March 2009), pp.123-128, ISSN 0032-5910.

Rahaman, M.N. (2003). *Ceramic processing and sintering,* Marcel Dekker, ISBN 9780824709884, Basel, Switzerland, pp. 49-122.

Rosato, D.V. & Rosato, D.V. (1996). Specialized injection molding process. In *Injection Molding Handbook,* 2nd edition. Chapman & Hall GmbH, ISBN 9780792386193, Weinheim, Germany, pp. 1005-1058.

Rothon, R.N. (2003). Particulate-Filled Polymer Composites, 2nd edition. Smithers Rapra Technology Limited, ISBN 9781859573822, Shropshire, United Kingdom, pp. 29-33.

Ruprecht, R., Gietzelt, T., Müller, K., Piotter, V. & Haußelt, J. (2002). Injection molding of microstructured components from plastics, metals and ceramics. *Microsystem Technologies*, Vol.8, No.4-5, (May 2002), pp.351-358, ISSN 0946-7076.

Schrader, G.F., Elshennawy, A.K. & Doyle, L.E. (2000). Powder Metalurgy. In *Manufacturing Process and Materials*. Society of Manufacturing Engineers, ISBN 0872635171, Dearborn, Michigan, USA, pp. 199-209.

Shearwood, C., Fu, Y.Q., Yu, L. & Khor, K.A. (2005). Spark plasma sintering of TiNi nano-powder. *Scripta Materialia*, Vol.52, No.6, (March 2005), pp.455-460, ISSN 1359-6462.

Somasundram, I.M., Cendrowicz, A., Wilson, D.I. & Johns, M.L. (2008). Phenomenological study and modelling of wick debinding. *Chemical Engineering Science*, Vol.63, No.14, (July 2008), pp.3802-3809, ISSN 0009-2509.

Stevenson, J.F. (2009). Powder Metal Injection Molding. In *Injection Molding – Technology and Fundamentals*, Kamal, M.R., Isayev, A.I., & Liu, S.J. (Ed), Cal Hanser Verlag, ISBN 978156904343, Munich, Germany, pp. 309-340.

Stringari, G., Zupančič, B., Kubyshkina, G., von Bernstorff, B. & Emri, I. (2011). Time-dependant properties of bimodal POM – Application in powder injection molding. *Powder Technology*, Vol.208, No. 3, (April 2011), pp.590-595, ISSN 0032-5910.

Tandon, R. (2008). Metal injection moulding. In *Encyclopedia of Materials: Science and Technology*, Buschow, K.H.J., Cahn, R.W., Flemings, M.C., Ilscher, B., Kramer, E.J., Mahajan, S. & Veyssiere, P. (Ed.), Elsevier Science Ltd, ISBN 9780080431529, Amsterdam, The Netherlands, pp.5439-5442.

Tay, B.Y., Loh, N.H., Tor, S.B., Ng, F.L., Fu, G. & Lu, X.H. (2009). Characterisation of micro gears produced by micro powder injection moulding. *Powder Technology*, Vol.188, No. 3, (January 2009), pp.179-182, ISSN 0032-5910.

Thomas-Vielma, P., Cervera, A., Levenfeld, B. & Várez, A. (2008). Production of alumina parts by powder injection molding with a binder system based on high density polyethylene. *Journal of the European Ceramic Society*,Vol.28, No. 4, (October 2007), pp.763-771, ISSN 0955-2219.

Thornagel, M. (2010). PIM 2010- Simulating flow can help avoid mould mistakes. *Metal Powder Report*, Vol.65, No.3, (March-April 2010), pp. 26-29, ISSN 0026-0657.

Torralba, J.M., Hidalgo, J. & Jimenez-Morales, A. (2011). Powder injection moulding: Processing of small parts of complex shape, *Proceedings of ICIT&MPT 2011 8th International Conference on Industrial Tools and Material Processing Technologies*, pp. 53-58, ISBN 9789616692021, Ljubljana, Slovenia, October 2-5, 2011.

Tseng, J.W. & Hsu, C.K. (1999). Cracking defect and porosity evolution during thermal debinding in ceramic injection moldings. *Ceramics International*, Vol.25, No. 5, (July 1999), pp.461-466, ISSN 0272-8842.

Yang, W.W., Yang, K.Y. & Hon, M.H. (2002). Effects of PEG molecular weights on rheological behavior of alumina injection molding feedstocks. *Materials Chemistry and Physics*, Vol.78, No. 2, (February 2002), pp.416-424, ISSN 0254-0584.

Zauner, R. (2006). Micro powder injection moulding. *Microelectronics Engineering*, Vol.83, No. 4-9, (April - September 2006), pp.1442-1444, ISSN 0167-9317.

4

Wick Debinding – An Effective Way of Solving Problems in the Debinding Process of Powder Injection Molding

Lovro Gorjan
Jožef Stefan Institute,
Research and Development Center for Ignition Systems and Electronics d.o.o.,
Slovenia

1. Introduction

Powder injection molding (PIM) has been shown itself to be a successful shaping technique for producing complex-shaped ceramic, metal or cermet parts. The process starts with preparing a high solid loading suspension, where ceramic or metal powder is mixed with a thermoplastic material. At high temperature the suspension is fluid and can be injected into molds by applying a pressure. Inside the mold the suspension takes the shape of the mold and then cools below the melting point of the thermoplastic material and solidifies into a green body. After the molding cycle the green body consists of solid particles held together by the thermoplasic phase, which serves as a binder.

The challenging and time-consuming operation in the powder-injection molding process is removing the binder from the green bodies prior to the sintering, without causing any deformation or cracks. The debinding process is difficult because green bodies contain relatively large amount of poorly volatile binder in the solid state, i.e. below the melting point. Debinding is usually achieved by slowly heating the green bodies, causing the binder to decompose and vaporize. This is the thermal debinding process. The difficulties are especially severe in low-pressure injection molding, since in this case the binder does not contain a backbone polymer that would hold the particles firmly in place during the debinding. Low-pressure injection molding (LPIM) is a variant of injection molding where relatively low pressures are used, typically less than 0.7 MPa, and the pressure is applied using compressed air instead of a screw (like in the more common high-pressure variant). The liquid medium in the feedstock is a low-melting-point wax, which is crucial for the low viscosity of the molten feedstock. The advantages of LPIM, in comparison with other ceramic injection techniques, include the lower cost of the molds, less die wear and less expensive and simpler equipment for the injection molding (Zorzi et al., 2003; Cetinel et al., 2010; Loebbecke et al., 2009; Gorjan et al., 2010). The method has also been shown to be appropriate for the shaping of microcomponents (Cetinel et al., 2010; Bauer & Knitter, 2002; Wang et al., 2008).

However, an effective way of reducing the formation of defects in the process of binder removal exists. That is, to introduce an additional debinding step – debinding in a wicking embedment (Curry, 1975; German, 1987; Wei, 1989; Liu, 1999; Bao & Evans, 1991; German,

1990) or wick-debinding. A wicking agent can be in the form of a porous solid substrate plate or in the form of a loose powder or granulate. The granular form offers a gentle physical support for samples, regardless of their shape, and thus prevents certain flaws, such as distortion and cracking. The capillary extraction is uniform over the entire surface of the green body, which ensures that debinded parts also have, as much as possible, a uniform structure after the wick-debinding. A solid plate does not offer so many benefits; however it has one advantage over the granular form of wicking agent, i.e., there are fewer practical problems when handling the compacts after the debinding. The wick-debinded parts do not have to be cleaned and are simply transferred to the sintering furnace.

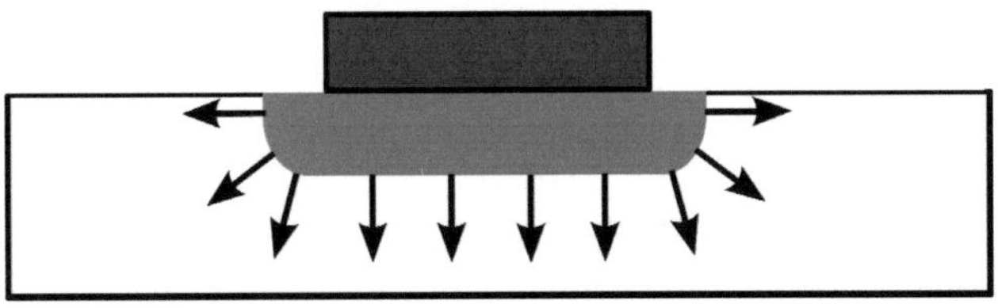

Fig. 1. Wick-debinding on a porous plate. The molten binder is extracted from the green body into the porous supporting plate.

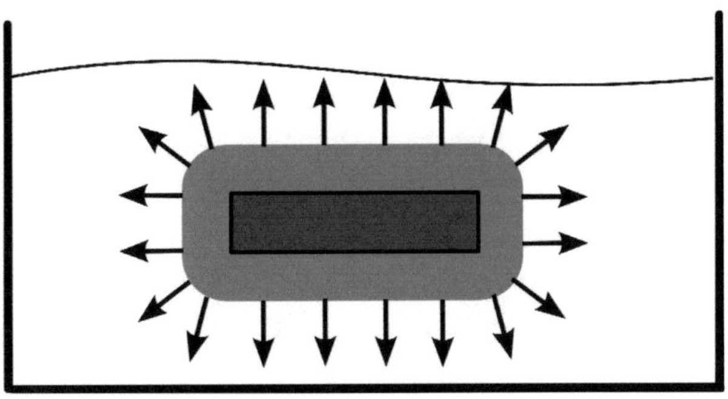

Fig. 2. Wick-debinding in a embedment of porous powder or granulate. The molten binder is extracted in all directions from the green body.

The wicking embedment can be utilized with great success in either the high- or the low-pressure injection molding. However, its use is more beneficial in the low-pressure variant, where the debinding is a more delicate operation.

2. Fundamentals

The basic principle of wick-debinding is the phenomenon called capillarity, which is a spontaneous flow of liquid into small pores. This effect occurs because of the attractive forces between the liquid and the solid surface of the pores and the surface tension of the liquid. The attraction of the liquid to the surface causes the adhesion of the liquid and the solid, which results in the liquid wetting the surface. The wetting is characterized by a wetting angle, which depends on the interactions between the liquid phase, the solid phase and the atmosphere. The smaller the wetting angle the better is the wetting and the liquid easily spreads over the surface.

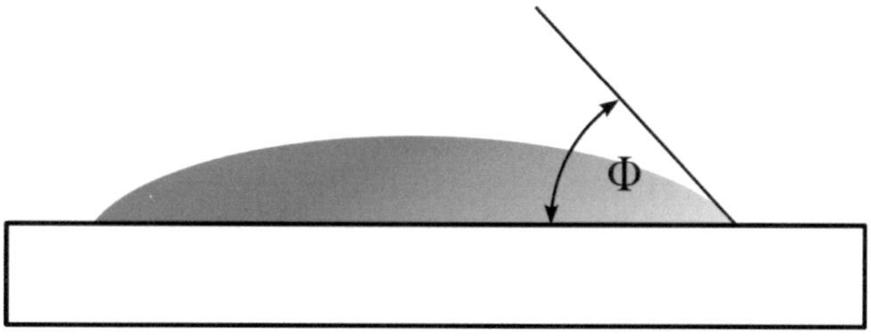

Fig. 3. Sketch of a droplet of liquid on a solid surface showcasing the wetting phenomenon, characterized by the wetting angle (Φ). When a liquid wets a surface it spreads over it.

An interesting phenomenon occurs when the liquid is inside a small pore. When the liquid wets the surface of a small pore at a certain angle (Φ), the surface becomes concavely curved as is sketched in Fig 4. Any curved liquid surface causes a pressure difference across the interface ($\Delta P_c = P_V - P_L$) between the liquid and the surrounding atmosphere.

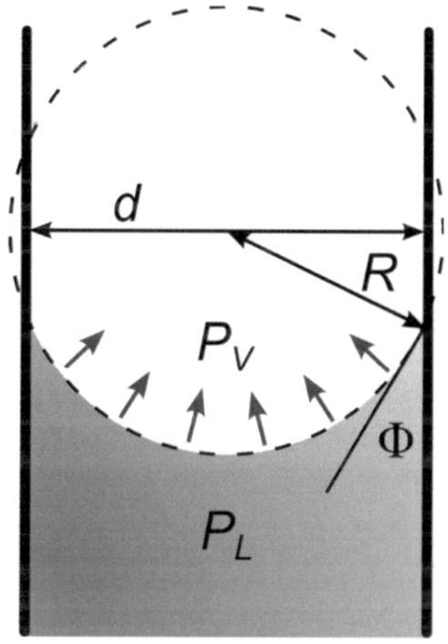

Fig. 4. The liquid, that wets the surface, inside a small, cyllindrical pore forms a concave spherical surface that causes a pressure difference between the liquid phase.

The equilibrium pressure difference is described by the Laplace-Young equation:

$$\Delta P_c = P_L - P_V = \gamma\left(\frac{1}{R_1} + \frac{1}{R_2}\right) \tag{1}$$

where ΔP_c [Pa] is the pressure difference between the liquid phase and the air phase, γ [N/m] is the surface tension, and R_1 and R_2 are the principal radii of curvature. As the capillary surface is concave towards the atmosphere, the liquid pressure is lower than that of the atmosphere, possibly reaching negative values, which is called a tensile stress inside the liquid (Bouzid et al., 2011).

In the case of a small, cylindrically shaped, pore channel the surface of the liquid is symmetrical and $R_1 = R_2 = R$. On small scales gravity is not strong enough to significantly influence the shape of the liquid surface so the surface has a spherical shape. If the wetting angle is considered the curvature in the small, tube-shaped, pore channel can be reasoned from Fig. 4:

$$R = \frac{d}{2\cos(\Phi)} \tag{2}$$

Combining equations (1) and (2) we obtain a correlation between the capillary pressure, the wetting angle and the pore diameter:

$$\Delta P_c = \frac{4\gamma\cos(\Phi)}{d} \tag{3}$$

From equation 3 it is clear, that the capillary pressure is inversely proportional to the pore diameter. Because the capillary pressure is larger for smaller pores, the liquid is forced to move from the larger to the smaller pores. So in the equilibrium state the liquid would fill the smallest pores of the system. The main idea of wick-debinding is to get a green body, heated to the temperature where the binder is molten, in contact with a material that has finer pores than the pores of the green body. Capillarity would then cause the binder to move from the green body into the material in the contact.

The wetting angle must be quite small for practical use. If the surface is not wetted by the liquid ($\Phi > 90°$) then $\cos(\Phi)$ has a negative value, which means that the capillary pressure would be opposite and the liquid would not enter the porous media.

In any case, in a real system the porous media consist of pores of different sizes and shapes. Even for a green body made of packed monosized spherical particles there are voids of different sizes and shapes. However, real powders are composed of particles that are different in size and shape, which leads to an even more complex pore structure and a wider size distribution of pores. A labyrinth of interconnected voids is present in the green body and also in the wicking agent. Because of the complexity of real systems, the equation (3) is difficult to use directly. However, in real systems it has been experimentally observed that the liquid enters a porous body with a front (Bao & Evans, 1991; Somasundram, 2008) and a single value of characteristic capillary pressure at the front can be successfully used.

Another important thing to consider in the debinding is the kinetics of the process. It is important, from a practical point of view, that the process is reasonably fast. The kinetics of wick-debinding, besides capillary pressure, also depends on the resistance to flow through the porous media. Each individual channel has a certain resistance - a viscous drag that limits the velocity at which the liquid is flowing through.

The motion of liquid substances is generally described by the Navier-Stoker equations, which arise from applying Newton's second law to fluid motion. However, these equations are too complicated for practical use in describing debinding phenomena since the shape of the liquid surface would present boundary conditions that are too complex. However, with the development of computer software for liquid mechanics and because of the constant increase in computer power it is possible that accurate simulations of debinding will be developed in the future. Nevertheless, a simplified theoretical approach in dealing with the phenomena of debinding has produced satisfactory results.

The flow through an idealized single, long, circular, pore channel is described by the Hagen-Poiseuille equation (4), which is also an exact solution of the Navier-Stokes equations with certain assumptions, such as steady state, axisymmetric flow with no radial and swirl components of velocity.

$$q = \frac{d^2 \Delta P}{32 \mu L} \tag{4}$$

where q [m/s] is the flux or flow per area, ΔP [Pa] is the pressure difference between the ends of the pore channel, μ [Pa s] is the viscosity and L [m] is the length of the pore channel.

The smaller the pore, the larger the viscous drag. This generally means that small pores present a high resistance to flow. Again, like in the case of using equation (3), the Hagen-Pouseille equation is due to the extremely complex shapes of pore channels in real systems, inappropriate for calculations, but nevertheless it demonstrates that despite the high capillary pressures, liquid transport through small pores can be slow. However, regardless of the complexity of pore channels, the flow of a liquid through porous material can be successfully described by a simple equation called the Darcy's law:

$$q = -\frac{K \nabla P}{\eta} \tag{5}$$

where q [m^3/ (m^2 s)] is the volumetric flux, K [m^2] is the parameter called permeability, η [Pa s] is the viscosity and BP [Pa] is the pressure gradient.

The law was formulated in the 19th century by the French engineer Henry Darcy based on the results of water flow through sand (Richardson & Harker, 2002). It is a constitutive equation with a similar meaning for fluid flow as Ohm's law for the electricity and Fourier's law for the conductive heat transfer. Darcy's law has been experimentally confirmed on many different material combinations and is considered well proven. It has also been derived from the Navier-Stokes equations.

The permeability (K) is a characteristic parameter of a porous substance that depends on the size, shape and interconnectedness of the individual pore channels and on the fractional porosity. The complex shape of pore channels makes a permeability difficult to calculate or predict from basic principles.

Many empirical equations have been used to determine the permeability from basic powder-compact properties, such as particle diameter (d), specific surface (S) and fractional porosity (E). Some of them are listed below (Bao & Evans, 1991; German, 1987):

$$K = \frac{E^3}{5S^2(1-E)^2} \tag{6}$$

$$K = \frac{E^4 d^2}{90(1-E)^2} \tag{7}$$

$$K = 190\left(\frac{E^2}{S}\right)^{2.42} \tag{8}$$

These permeability correlations have been tuned for a forced flow through the porous material, i.e. , the flow of liquid that is pushed through the material by applying an external pressure. However, the permeability can be significantly different in the case of capillary extraction, where the liquid is sucked out of green body by capillary forces.

The wicking agent must extract the liquefied binder from the green body, which is itself a porous body. If the molten binder is removed from the green compact then a new surface must be formed in the interior of that compact. This new surface, which initially appears in larger pores, causes a capillary pressure in the opposite direction and resists extraction. A competition for the binder emerges between the two porous media. Only the smaller pores of the wicking agent might have a capillary suction that is strong enough to exceed the capillary pressure of the green body. The liquid then travels into the wicking agent only through these pores. In contrast, if the liquid were to be forced by the external pressure through the wicking agent it would travel mostly through larger pores, which present a smaller resistance.

The measured permeability, or that calculated from equations 6-8, could be significantly higher than in the case of capillary extraction. This means, that conventional methods of measuring permeability, such as measuring the forced gas flow through a sample of a porous material, cannot be used to determine the permeability for the capillary-extraction phenomenon. The mismatch between the forced flow and the capillary-extraction permeability is especially large in the granular form of the wicking agent. A characteristic case for porous material in the form of large granules with a fine porosity is schematically presented in Fig 5. If the fluid is forced through such a material the permeability would appear much larger than if this granulate was extracting the liquid from another porous material, for which a strong capillary pressure is required.

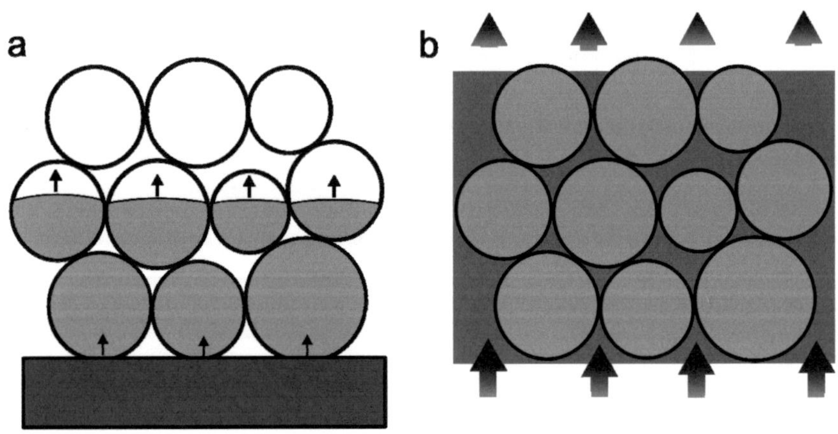

Fig. 5. In the capillary extraction the liquid flows only through smaller pores inside the granulae, wheras in the case of liquid flow forced by external pressure the majority of flow would be between the granulae.

Besides the kinetics of the capillary-extraction process it is crucial that the powder compact of green body retains its shape after the process has been completed and no flaws are introduced. One of the most critical moments in the process is the point when the binder melts. At this stage the compact becomes quite weak and soft. This is especially critical in the case of low-pressure injection molding, where only one component binder i.e., paraffin wax, is used. It is because of a characteristic known as the yield stress that the green body retains its shape. The suspension behaves like a rigid body below the yield stress and flows like a liquid above the yield stress. The yield stress is mostly governed by the particle size, shape, solid content and the inter-particle forces.

However, a large yield stress is undesirable for the molding step, since it results in a low moldability of the suspension (German, 1990). Suspensions with a high yield stress must have high powder content, which also increases viscosity, which is again undesirable in the molding step (German, 2003).

Fortunately, the yield stress of the molded green body can be significantly larger than the yield stress of the suspension before molding. This is because the particles rearrange during molding and solidification into a denser configuration - a consequence of the shrinkage of the binder after solidification. An increased attractive inter-particle interaction occurs in the denser form (Dakskobler & Kosmač, 2009). Ideally the process does not reverse during re-melting. If the inter-particle forces are high enough, then the particle arrangement will not change; instead the expanding binder will be exuded from the body, while the particle arrangement remains intact. A series of photographs of a LPIM sample, taken with an optical microscope during heating from below to above the melting temperature of the binder, is shown in Fig. 6.

A molten paraffin binder exudes out of the green body. This happens without any disruption of the powder skeleton.

The extent of the exudation effect depends on the amount of low-melting point wax in the body. During the melting the volume of the wax increases by 15%. The effect is most pronounced in the low-pressure injection molding where the amount of wax is large - around 60 vol%. In high-pressure molding the amount of wax is around 10% - 30%.The large amount of wax is an important factor that explains why in LPIM extreme difficulties are encountered when a wicking agent is not used. In the HPIM process the yield strength during melting does not pose that big a problem, because the additional high-melting point polymer ensures that the particles are held in position. It has been shown that the yield stress of molded parts can also increase with the storage time after the molding (Novak et al., 2000; Cetinel et al., 2010).

Water from a humid environment can penetrate the green body and interfere with the bonds between surfactant molecules and the surface of the particles. The strength of the inter-particle forces increases, which leads to a significant increase in the yield strength. This effect can be made even faster, if the molded bodies are soaked in water (Novak et al., 2000).

Wicking embedment offers another benefit. It guarantiees a gentle physical support for the parts. If the debinding takes place on a hard substrate there is danger that certain flaws will occur, as schematically presented in the sketch in Fig. 7. The suspended parts of the green

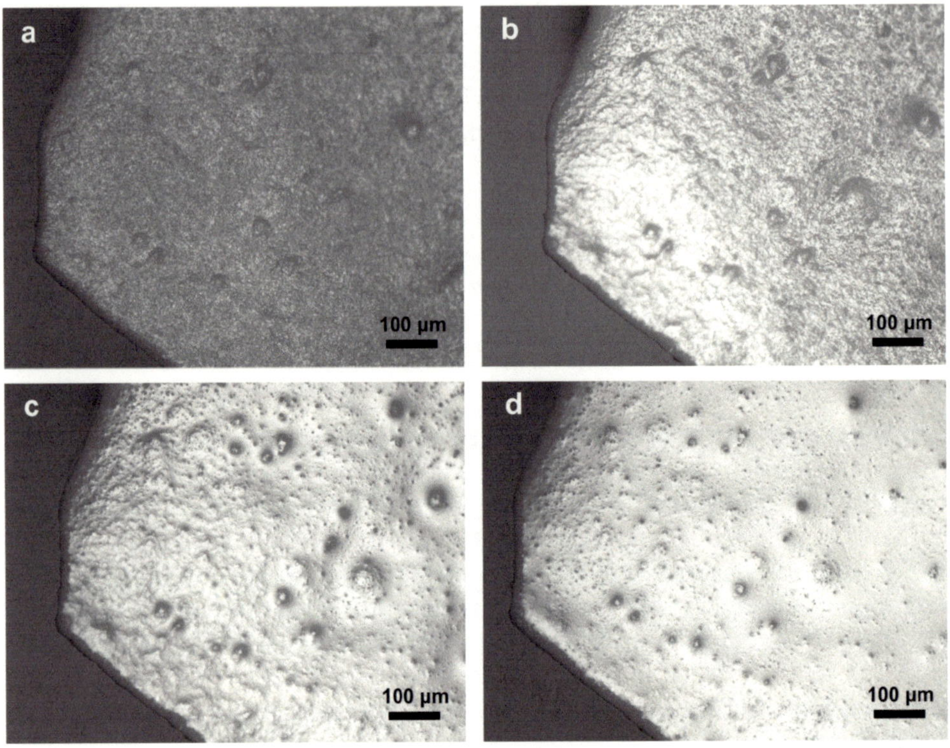

Fig. 6. Exudation of the paraffin-wax binder during heating above the melting point of the paraffin, as observed with optical microscope. The photograph a) shows the state before the melting, b) shows the first molten paraffin exuding from the green body, c) shows the situation 1 minute after the b) and d) shows state 10 min after the b).

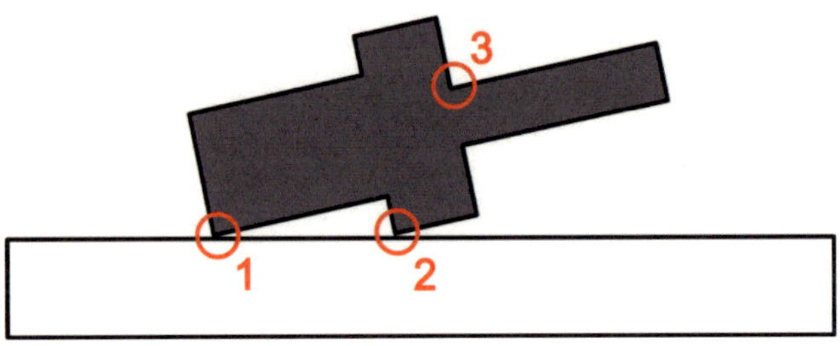

Fig. 7. Green bodies with a complex shape can pose difficulties if they are debinded on a solid substrate. Small areas on which the green body rests on the substrate (1 and 2) can be deformed due to the large compressive stress. Suspended parts of the body can bend due to gravity or even crack at the point where the tensile stress is the highest (3). The wicking embedment can successfully reduce these flaws, since the support pressure is well spread over the green body's surface.

body could bend or crack and point pressure areas where the green body rests on the solid substrate can deform.

3. Overview of theoretical work

Because of the complexity of the capillary system in the porous green body and the wicking agent the accurate and general theoretical model is difficult to obtain. Since the systems can be quite different, the extraction can also show different behaviour. The existing theoretical models predict different behaviors during the debinding and many even contradict each other. The basis of all models is Darcy's law and some form of capillary-pressure description. One of the first to theoretically describe the process of wick debinding for injection-molded samples was German (German, 1987), who in 1987 proposed a model, where he assumed that the binder is extracted from a molded compact as a continuous body in liquid form, leaving behind a binder-free region.

A partially debinded compact should, therefore, have a characteristic binder distribution with a binder-saturated region near the contact with the wicking powder and a region with no binder on the other side. A sharp border between these regions should be present - a sign of the trailing front of the molten binder. This model is simple and has frequently been used as a basis for research in wick debinding. Monte-Carlo simulations of binder removal based on German's assumptions have also been conducted (Shih & Houring, 2001; Lin & Houring, 2005). These simulations focused on binder penetration in the wicking embedment and examined the case where pieces are not completely surrounded by the embedment.

However, German's model has been criticized, on the basis of experimental data. Contradicting this model, many researchers observed that the binder is uniformly distributed inside the body at all stages of the debinding process (Liu, 1999; Bao & Evans, 1991; Vetter et al., 1994; Kim et al., 1999; Somasundram, 2008). There is also the question of how the air can enter behind the trailing front into the binder-free region if the molded pieces are completely surrounded by the wick (Somasundram, 2008). Furthermore, the debinding rate does not correspond well with some experiments (Vetter et al., 1994). It has also been observed that the permeability in a wick embedment can have important effects and can be a limiting factor, rather than the flow through a sample, as was suggested in German's model [Vetter et al., 1994; Somasundram, 2008]. With more precise examinations of the binder-removal rate it has been confirmed that wick-debinding must take place via more than a single mechanism.

One clearly observable effect, for example, is a rapid decrease in the binder content at the beginning of the process. This has been attributed to the pressure arising from the thermal expansion of the binder [Somasundram, 2008, Gorjan et al., 2010]. Before the debinding process, molded parts contain binder in the solid state, then during the melting a large, and relatively sudden, expansion of the binder occurs. For example, the density of the paraffin drops by around 15% during melting (Gorjan et al., 2010).

With further studies of the kinetics of capillary extraction it has been found that the molten binder inside the body exists in different states, a differentiation based on the position inside the body. It can behave as a 'mobile binder' located in the larger voids between the powder particles, which can flow due to the pressure gradient caused by the wicking embedment and as an 'immobile binder' located on the surfaces of the particles and inside the smaller voids, which cannot be moved due to the capillary suction – it is too strongly bonded to the powder and trapped inside smaller pores (Gorjan et al., 2010).

There can also be shrinkage during the debinding, which is inversely related to the ceramic volume fraction, with less shrinkage in green bodies with a high solid loading. Very little or no shrinkage occurs at a volume fraction of around 64% [Wright et al., 1990, Gorjan et al., 2010]. In order to avoid a large shrinkage a green body must be made with high a green density. A high green density is also beneficial for the sintering process; however, a high solid loading is detrimental for the molding step. It is always necessary to make a certain compromise.

Capillary extraction effectively removes only a part of the binder, because there is always a certain amount of the binder "trapped" inside the finest pores of the green body. This residual binder must be removed in the form of a gaseous phase. In the case of oxide ceramics the removal of the remaining binder can be achieved by controllably burning the binder. If the temperature during wick-debinding is raised above approximately 200°C then an organic binder starts to decompose due to oxidation reactions.

All of the binder can be removed if the temperature is increased above approximately 600°C, where even carbon burns. However, when all of the binder is removed from the body, the body becomes extremely brittle and weak. In this state it would be impossible to remove it from the embedment and clean it without causing serious damage. One solution is to further heat the system to the temperature where first stage of sintering starts. Pre-sintered or 'biscuit sintered' parts can then be safely removed from the embedment and since they contain no binder they can also be sintered without problems. However, practical problems can accompany this procedure. For example, if alumina parts are debinded a high temperature is required for the pre-sintering and at this temperature the wicking agent also starts to lose the fine porosity and can stick strongly to the surface of ceramic parts.

Another, economically even more acceptable option is to heat the samples after the capillary extraction to the temperature where the organic binder starts to decompose and then hold the parts at this temperature. It has been observed that at this temperature some amount of paraffin wax cures into a hard, brown-colored, non-volatile resin, which remains in the parts and is stable at the dwell temperature of around 200°C. This curing effect drastically improves the strength of the samples, which increases with the dwell time (Gorjan et al., 2011).

Parts processed in this way can be made sufficiently strong for handling without any risk of damage. They are also appropriate for green machining, like cutting, boring and grinding. Since they contain a very small amount of the binder, rapid heating inside the sintering furnace can be applied and the curing of the binder does not influence the strength of the sintered ceramic parts.

A procedure of debinding, where the benefits of wick-debinding are fully used, has also been developed, while the main drawback, i.e., additional cleaning and handling operations are avoided. According to the patent the debinding and sintering can take place in a single furnace (Gorjan & Dakskobler, 2011). This can be achieved by using a high purity-carbon granulate, which completely burns after its role as the wicking agent is completed.

4. Practice

In a practical use the wick-debinding process can offer significant benefits. Faster and safer debinding can be achieved in comparison with debinding a without the wicking agent. One of the most important factors in the debinding practice is to avoid the introduction of defects in green bodies. Potential defects include the loss of a compact's shape through distortion, warping, cracking and also the undesirable strong adhesion of the wicking powder on the surface of the debinded parts.

For example, in a low-pressure injection-molding, shaping technique it is almost impossible to debind samples without using a wicking agent. In HPIM the use of wick-debinding can be avoided, since the green body retains its strength after the wax has been melted due to the presence of polymer, which binds the particles together. Also in the case of HPIM, the wick-debinding reduces the possibility of flaws.

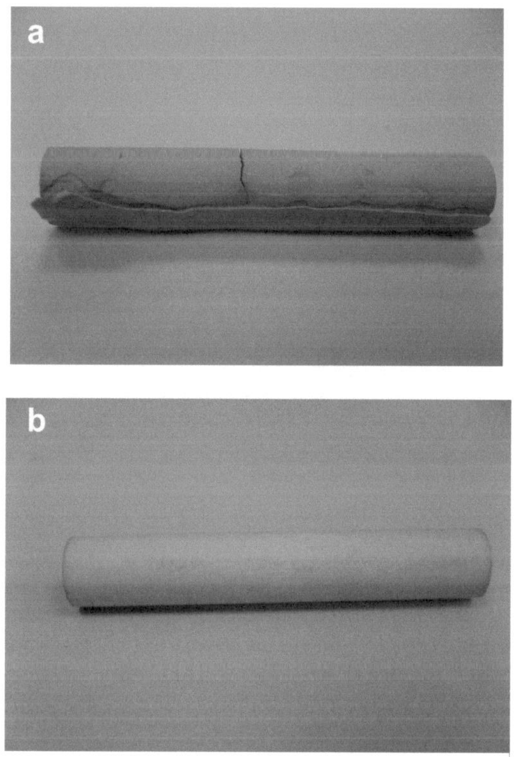

Fig. 8. Wick-debinding can significantly reduce the formation of flaws. Photograph a) shows the low-pressure injection molded sample, debinded without wicking embedment, while the photograph b) shows the sample which was debinded in the embedment of highly porous alumina wicking agent.

A major practical problem of wick debinding is the danger of causing defects when the parts are removed from the wick embedment and cleaned afterwards. Because the debinded parts can be quite fragile, a gentle and manually intensive operation is required. If the debinded compacts are strong, then a more robust handling such as sieving can be applied. During this handling the breaking of parts can occur.

In the practice a wicking embedment must also satisfy some additional considerations besides having good capillary-extraction characteristics. It is the most practical if it is in the form of granules with a size of 50–200 μm. This size of granules ensures uniform contact with a green body and has, at the same time, good flowability. This flowability is crucial for easy handling. Smaller pores are powders tend to form dust, which is undesirable. Also, the granules are easier to clean from the surfaces of the parts after the debinding process. Each individual granule contains a very fine porosity, which is crucial for a highly efficient capillary extraction.

The correct temperature regime must be used in order to achieve debinding. A slow heating rate must be applied in order to give the wicking agent time for extraction. Typical debinding cycles last from 20 hours to several days.

The adhered wicking agent causes problems, because it would lead to a rough surface after the sintering. Therefore, it should be thoroughly cleaned from the debinded parts.

Fig. 9. Alumina wicking agent in the granular form. Photograph a), taken with optical microscop, shows granules of the wicking agent. Photograph b), taken with scanning electrone microscope show the surface of one granule. It can be seen, that the granula contains very fine porosity, which is a condition for efficient capillary extraction.

Fig. 10. Comparison of the properly debinded part (left) with the deformed part (right), which was deformed and had granulate wicking agent adhered to the surface. The defect was caused when the part was embedded into a too hot wicking agent, which had not been cooled enough after a thermal regeneration.

Fig. 11. Wick-debinded parts are loaded on a tray for the sintering process. Successfully wick-debinded samples contain an open porosity and are ready for a fast sintering cycle, in which complete burnout of the residual binder takes place.

After single or multiple uses the wicking agent accumulates a certain amount of organic phase – binder degradation products. This phase decreases the porosity of the wicking agent and thus its capillary-extraction ability. However, it can be regenerated by heating it to around 600°C, where all the organics burn. In practice, a wicking granulate with different amounts of residual organic phase can be used for debinding different parts. Small parts are debinded in the embedment, which is rich in organics, whereas the large parts are debinded using freshly regenerated granulate with a maximum capillary-extraction ability. As a result, the embedment can thus be consequently used for ever smaller parts.

5. Conclusion

Removing the organic binder from the powder-injection-molded parts with the use of highly a porous granular embedment has been shown to be an effective method. It offers many benefits, such as shorter debinding time due to capillary extraction. Also, it guarantees a gentle physical support for the parts and therefore reduces certain flaws, such as distortion and cracking. Wick-debinding also has an important drawback, such as the practical problems of cleaning the debinded bodies. These drawbacks are the reason, that wick debinding is avoided if possible. In the case of high pressure injection molding it is possible to avoid using the wick embedment because of the use of high melting point polymeric binders in addition to the low melting point wax.

However in the case of low-pressure-injection molding, where the debinding process is even more delicate the use of wick-debinding has a firm place. Furthemore, future improvements in wick-debinding and the developments of novel procedures can make this highly effective way of removing the binder from injection-molded parts easier to apply and more popularize in the industry.

6. References

Bao Y.; Evans J. R. G. (1991). Kinetics of Capillary Extraction of organic Vehicle from Ceramic Bodies. Part I: Flow in Porous Media. *Journal of European Ceramic. Society*, Vol. 8, No. 2, (March 1991), pp. 81 - 93, ISSN 0955-2219

Bauer W.; Knitter R. (2002). Development of a rapid prototyping process chain for the production of ceramic microcomponents. *Journal of Materials Science*, Vol. 37, No. 15 (August 2002), pp. 3127 - 3140, ISSN 00222461

Bouzid M.; Mercury L., Lassin A.; Matray J. M.; Azaroaual M. (2011). In-pore tensile stress by drying-induced capillary bridges inside porous material, *Journal of Colloidal and Interface Science*. Vol. 355, No. 2, (March 2011), pp. 494 - 502, ISSN 0021-9797

Cetinel F. A.; Bauer W.; Muller M.; Knitter R.; Hausselt J. (2010). Influence of dispersant, storage time and temperature on the rheological properties of zirconia-paraffin feedstock for LPIM, *Journal of European Ceramic. Society*, Vol. 30, No. 6, (Januar 2010), pp. 1391 - 1400, ISSN 0955-2219

Curry J. D. (1977). US patent number 4011291: Apparatus and method of manufacture of articles containing controlled amounts of binder, filed September 1975, published March 1977

Dakskobler A.; Kosmač T. (2009). Rheological properties of re-melted paraffin-wax suspensions used for LPIM, *Journal of European Ceramic. Society*, Vol. 29, No. 10, (July 2009), pp. 1831-1836, ISSN 0955-2219

German R. M. (1987). Theory of thermal debinding, *International Journal of Powder Mettalurgy*. Vol. 23, No. 4, (October 1987), pp. 237 - 245, ISSN 0888-7462

German R. M. (1990). Debinding Practice, In: *Powder Injection Molding*, german R. M., pp. 281 - 316, Metal Powder Industries Federation, ISBN 0-918404-95-9, New Jersey, 1990

German R. M. (2003). Overview of Process Variations, In: *Powder Injection Molding - Design and Applications*, German R. M., pp. 9 - 20, Innovative Material Solutions, Inc., ISBN 0972764208, State College, Pennsylvania

Gorjan L.; Dakskobler A.; Kosmač T. (2010). Partial wick-debinding of low-pressure powder-injection-moulded ceramic parts. *Journal of European Ceramic. Society*, Vol. 30, No. 15 (Januar 2010), pp. 3013 -3021, ISSN 0955-2219

Gorjan L.; Dakskobler A. (2011). *Postopek toplotne obdelave oblikovancev s sintranjem : patentna prijava 201100196*. Ljubljana: Urad RS za intelektualno lastnino, (2011)

Gorjan L.; Dakskobler A.; Kosmač T. (2011). Strength Evolution of Injection-Molded Ceramic Parts During Wick-Debinding. *Journal fo American Ceramic Society*, doi: 10.1111/j.1551-2916.2011.04872.x, ISSN 0002-7820

Kim S. W.; Lee H. W.; Song H. (1999). Effect of minor binder on capillary structure evolution during wicking, *Ceramics Internetional*, Vol. 25, No. 7, (September 1999), pp. 671 - 676, ISSN 02728842

Lin T. L.; Houring L. W. (2005). Investigation of wick debinding in metal injection molding: numerical simulations by the random walk approach and experiments, *Advanced Powder Technology.*, Vol. 16, No. 5 (September 2005), pp. 495 - 515, ISSN 0921-8831

Liu D. M. (1999). Control of yield stress in low-pressure ceramic injection moldings, *Ceramics International*, Vol. 25, No. 7 (September 1999), pp. 587 - 592, ISSN 02728842

Loebbecke B.; Knitter R.; Hausselt J. (2009). *Rheological* properties of alumina feedstocks for the low-pressure injection moulding process, *Journal of European Ceramic. Society* ,Vol. 29, No. 9 (June 2009) pp. 1595-1602, ISSN 0955-2219

Novak S.; Dakskobler A.; V. Rubitsch. (2000). The Effect of water on the behaviour of alumina-paraffin suspensions for low-pressure injection mouldnig (LPIM), *Journal of European Ceramic. Society*, Vol. 20, (November 2000), pp. 2175 - 2181, ISSN 0955-2219

Richardson J. F.; Parker J. H. (2002). Flow of Fluids through Granular Beds and Packed Columns, In: *Coulson &Richardson's Chemical engineering, vol. 2: Particle technology & separation processes*, Richardson J. F., Parker J. H., pp. 191 - 234, 2002, Butterworth-Heinemann, ISBN 0 7506 4445 1, Oxford

Somasundram I. M.; Cendrowitz A.; Wilson D. I.; Johns M. L. (2008) Phenomenological study and modelling of wick debinding, *Chemical Engineering Science*, Vol. 63, No. 14 (July 2001), pp. 3802 - 3809, ISSN 0009-2509

Somasundram I. M.; Cendrowitz A.; Johns M. L.; Prajapati B.; Wilson D. I. (2010). 2-D simulation of wick debinding for ceramic parts in close proximity, *Chemical Engineering Science*, Vol. 65, No. 22 (November 2010), pp. 5990-6000, ISSN 0009-2509

Shih M. S.; Houring L. W. (2001). Numerical simulation of capillary-induced flow in a powder-embedded porous matrix, *Advanced Powder Technology*, Vol. 12, No. 4 (December 2001), pp. 457-480, ISSN 09218831

Vetter R.; Sanders M. J.; Majewska-Glabus I.; Zhuang L. Z.; Duszczyk J. (1994). Wick-Debinding in Powder Injection Molding, *International Journal of Powder Metallurgy*, 1994, Vol. 30, No. 1 (Januar 1994), pp. 115-124, ISSN 0888-7462

Wang J.; Liu G.; Xiong Y.; Huang X.; Tian Y. (2008). Fabrication of ceramic microcomponents and microreactor for the steam reforming of ethanol, *Microsystem Technologies*, Vol. 14, No. 9 (October 2008), pp. 1245-1249, ISSN 09467076

Wei T. S.; German R. M., US Patent number 5028367: Two-stage fast debinding of injection molding powder campacts, filed December 1989, published July 1991

Wright J. K.; Edirisinghe M. J.; Zhang J. G.; Evans J. R. G. (1990). Particle Packing in Ceramic Injection Molding, *Journal of American Ceramic Society*, Vol. 73, No. 91 (September 1990), pp. 2653-58, ISSN 0002-7820

Zorzi J. E.; Perottoni C. A.; Da Jornada A. H. (2003). A new partially isostatic method for fast debinding of low-pressure injection molded ceramic parts, *Matterials Letters*, Vol. 57, No. 24 (August 2003), pp. 3784 – 3788, ISSN 0167-577X

5

Micro Metal Powder Injection Molding

Kazuaki Nishiyabu
Kinki University,
Japan

1. Introduction

Powder injection molding (PIM), which encompasses metal powder injection molding (MIM) and ceramic powder injection molding (CIM) is a net-shape process for the manufacturing of high volume and high precision components for use in a variety of industries. The micro-miniaturization of dimension and structures in MIM is facing with various technical problems, such as incomplete filling to narrow cavity, failure in demolding of fragile green compacts, and deformation in debinding and sintering process. Therefore micro MIM (μ-MIM) process is a more sophisticated process for tiny metal components and micro structured parts. This chapter introduces a general flow of MIM process, the material properties of the feedstock and focuses on the unique phenomena in the micro injection molding and the filling behaviour. A flow simulation of micro gear and micro dumbbell tensile specimen will be carried out and the flow pattern by short shot test and internal pressure measured will be compared to the simulation results. The production method of micro sacrificial plastic mold insert MIM (μ-SPiMIM) process has been proposed to solve drastically the specific problems involving the miniaturization of MIM parts. The sacrificial plastic mold (SP-mold) is prepared by injection-molding polymethylmethacrylate (PMMA) polymer into Ni-electroform. Micro-sized stainless steel 316L powder feedstock is injection-molded into the SP-mold which consists of micro multi-pillar structures. The effects of metal particle size and processing conditions on the quality of molded and sintered parts are evaluated. For the higher quality of μ-SPiMIM process, the feedstock composed of nano-sized Cu powder and oxymethylene-based binder is adequately prepared and molded into PMMA films with fine line-scan structures which are prepared by nano-imprint lithography (NIL) technique. From the evaluation results on the effects of particle size of metal powder and processing conditions toward the high precision of sintered parts, it is concluded that the μ-SPiMIM process has a great potential to produce precisely the complex metallic parts with fine micro-structures.

1.1 What is metal powder injection molding (MIM)?

Metal powder injection molding (MIM) is a manufacturing method that combines traditional powder metallurgy (P/M) with plastic injection molding as shown in Fig.1. Over the past decade it has established itself as a competitive manufacturing process for small precision components that would be costly to produce by alternative methods. It can be used to produce comparatively small parts with complex shapes from almost all types of materials such as metals, ceramics, inter-metallic compounds, and composites (German, 1984). Recently MIM

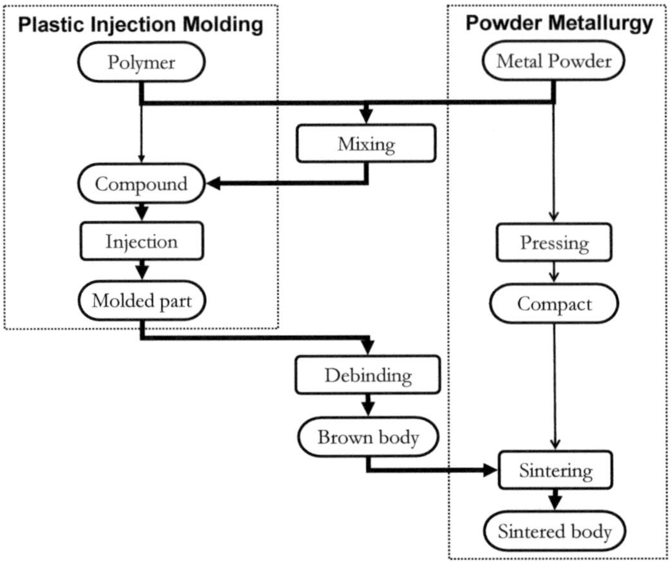

Fig. 1. Flow of metal powder injection molding (MIM) process indicated by bold line.

has been studied not only for hard metals, but also for materials such as titanium, copper and aluminium (German and Bose, 1997). Unlike in the case of P/M, MIM requires mixing metal powders with a large amount of polymeric binder. Afterwards the organic constituents are removed in a debinding step using solvent extraction or pyrolysis. The brown body is held in the molded form by only metal powder after debinding. The mixing and debinding of polymer with metal powder is a very unique and an original process of MIM. Thus it is very effective for manufacturing higher functional metal parts to apply this unique process.

1.2 Micro-miniaturization techniques by MIM

Fig.2 shows some typical examples of commercial MIM products, such as connector, impeller, cam, planetary-gear set and micro-reactor. In these conventional MIM products, as the size become smaller, the dimensional accuracy becomes higher as it falls within a few tenth micrometers. As the size of MIM products decreases much further, the dimensional accuracy is hoping to ensure within a few micrometers. In practical productions, however, it cannot be achieved easily. The production method of these tiny metallic parts which have micro-size and micro-structure are called μ-MIM, which is capable of manufacturing the micro-structured parts such as micro-pillars (Löhe and Haußelt, 2005).

1.3 Technical problems and solutions to miniaturize MIM parts

The μ-MIM process is very useful for the manufacturing of micro-sized and micro-structured metallic parts, but it is facing with various technical problems in each process. For example, it is difficult to fill feedstock completely into a narrow cavity and to demold fragile green compacts from a metallic mold in injection molding process. A careful handling is also required in the debinding and sintering processes. There are many other technical problems such as measuring of the density, the shape and mechanical properties of

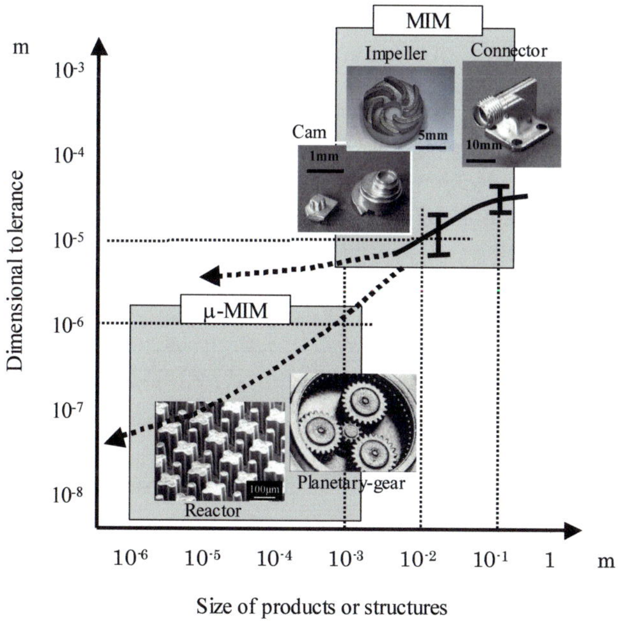

Fig. 2. Dimensional tolerance versus size of MIM Products.

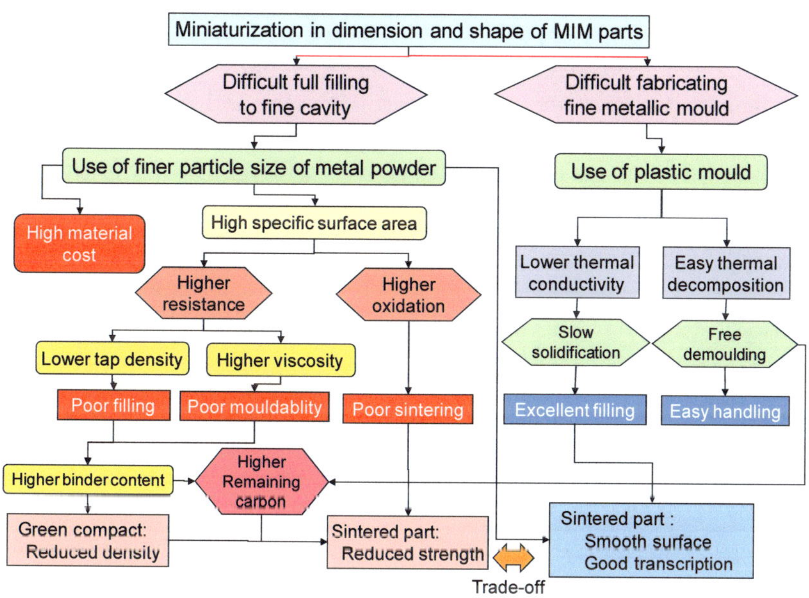

Fig. 3. Technical problems and its solutions to miniaturize MIM parts.

sintered parts. In Fig.3, the solutions for filling of viscous feedstock into the fine mold and fabricating of fine mold are shown in addition to the effectiveness and the disadvantages. The use of finer powder is essential to fill the feedstock into several tenth and a few micron-sized cavity. However, nano-sized powder has extremely high specific surface area, thus the tap density is very low and the viscosity of the feedstock increases remarkably. It is also susceptible to oxidation and relatively high production cost. Therefore these properties result in adverse affect for the production and the utilization. The fluidity of the feedstock is improved generally by increasing the binder content, but it makes lower the quality and the mechanical property of the sintered parts because the density of green compact become lower and the remaining carbon content increases. On the other hand, it is not easy to fabricate precisely a metallic mold, but a plastic is more superior for processability of the fine mold. In addition, a plastic mold is much superior for filling of the feedstock because its low thermal conductivity which delays solidification of melted materials. Moreover it is not necessary to demold a green compact from a plastic mold which can be decomposed thermally in debinding process. As a result, if there is a solution to the problem in sintering and injection of finer metal powder into plastic mold, a sintered part with smooth surface and good transcription can expect to be manufactured by MIM process. Therefore the trade-off problem needs to be solved by any innovative technologies.

2. Metal powder injection molding for small components

2.1 Material properties of MIM feedstock

The metallic powders used for MIM process are plain and low alloy steels, high speed steels, stainless steels, superalloys, intermetallics, magnetic alloys, hardmetals, and titanium and so on (Osada et al, 2007). Among them, stainless steels are commonly used. Fig.4 shows distributions of particle diameter of 316L stainless steel powder (10μm and 2μm in mean diameter, Epson Atmix Co., Ltd., PF-20J, PF-2F) produced by water-atomization method. Fine powders sintered more readily than coarser ones, but there is a number of limiting factors. The metallic powders are compounded with wax and polymeric binder by high-pressure kneader and are granulated by plunger-type extruder. The least possible amount of binder should be used, but an appropriate volume fraction of binder to powder exists. In industrial practice, the ratio varies from about 0.5 to 0.7.

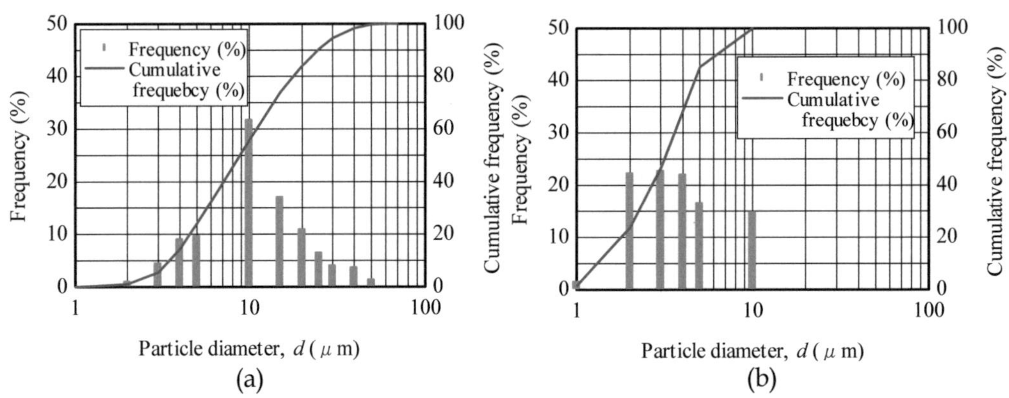

Fig. 4. Distributions of particle size of stainless steel powders used for MIM feedstock; (a) 316L powder (D_{50}=10μm); (b) 316L powder (D_{50}=2μm).

Fig.5 shows the melt viscosity of the feedstock with various fractions of 316L stainless steel powders with different mean particle sizes, which was measured by a capillary graph. The binder used for the feedstock was polyacetal polymer and paraffin wax. The volume

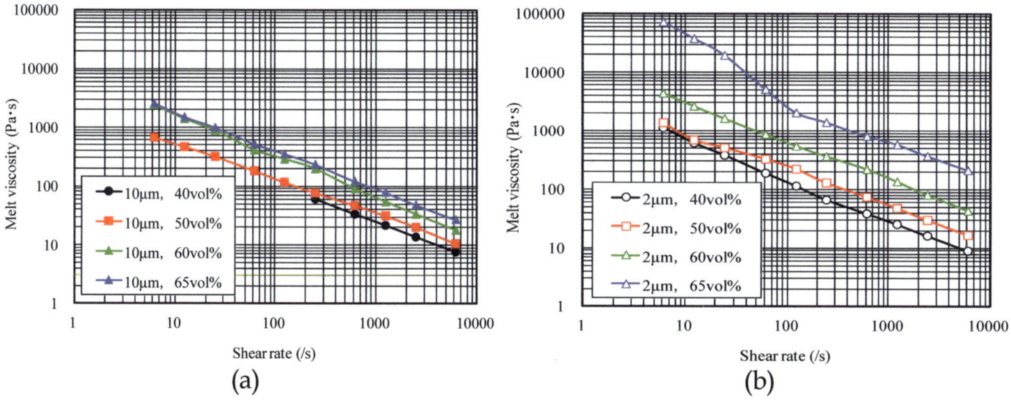

Fig. 5. Melt viscosities of MIM feedstock with various fractional metal powders; (a) 316L powder (D_{50}=10μm) feedstock; (b) 316L powder (D_{50}=2μm) feedstock.

fraction of the metal powder was varied from 40 to 65%. It is obvious that the melt viscosity of the feedstock increases remarkably as volume fraction of metal powder increases, and fine powder makes the melt viscosity of feedstock much higher than the coarse one.

2.2 Effects of capacity of mixing machine and injection molding machine

The feedstock used for MIM is generally prepared by mixing metal powder and binders with a twin screw extruder and kneader. A highly-homogenized feedstock is significant to manufacture the high quality of sintered parts, but it is not easy due to a big difference in specific gravity between metal powder and binders. The homogeneity of the feedstock was evaluated with the coefficient of variation (CV). Fig.6 shows the CV values of binder weight contained in a pellet of feedstock which was prepared with various mixing volumes such as 100cc, 1000cc and 40,000cc in a lot. The experimental result shows that the variation of binder content can reduce significantly with decreasing of the mixing volume.

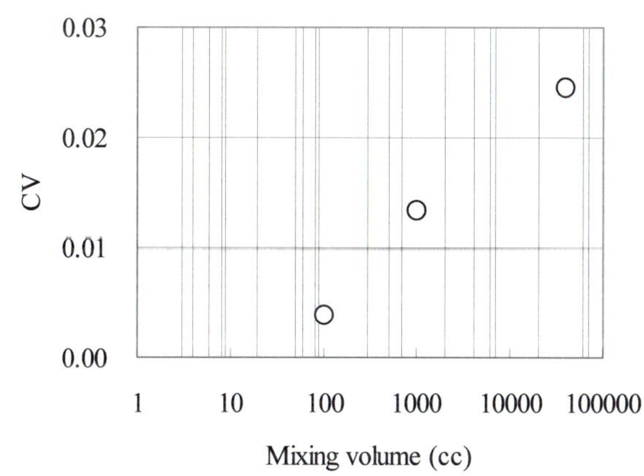

Fig. 6. Variation of binder weight in a pellet prepared with various mixing volumes.

Fig.7. shows the distributions of weight of green compacts prepared by two types of injection molding machines with varied capacities. Micro injection molding machine

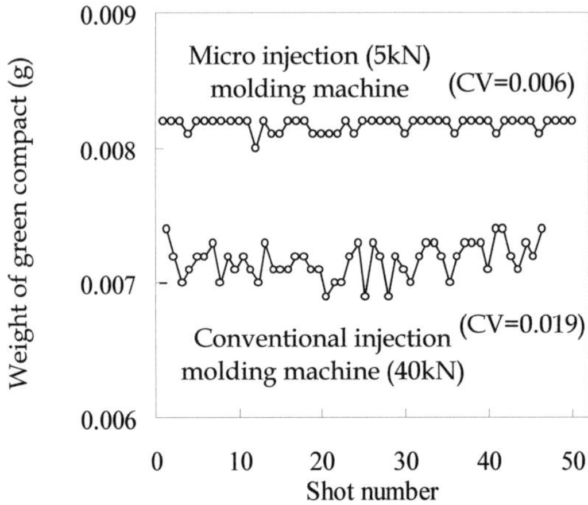

Fig. 7. Distributions of weight of green compacts prepared by injection molding machines with varied capacities.

(Battenfeld GmbH, Microsystem50) has 50kN clamping force and is smaller than conventional one with 400kN (Nissei Plastic Industrial Co., Ltd., PN40-2H). The diameter of injection plunger in micro injection molding machines is 3mm which is much smaller than that of 19mm in conventional one. Minimum volume for fabricating of micro injection molding machine is 120mm^3 which is much smaller than the conventional one. It is found that small capacity injection molding machine can be reduced the variation in weight of green compacts as compared to conventional one.

2.3 Filling behavior and flow simulation

2.3.1 Micro gear

In recent years, some advanced micro-manufacturing processes and the micro-sized gears made of metals and some advanced ceramics were demonstrated (Löhe and Haußelt, 2005). The micro-planetary gear motors made of Ni-Fe and Ni-based bulk metallic glasses were developed by X-ray lithography, electro-deposition and injection molding method (Ishida et al., 1995). However, micro-sized gears made of general-purpose durable materials are demanded for miniaturization and reliability improvement of products, and also the manufacturing is aiming to achieve a high economical efficiency for industrial needs. Authors have studied the tribological properties of micro-gear manufactured by MIM process and were evaluated quantitatively, thus the wear mechanisms were clarified (Kameo et al., 2006) and the accuracy of the ultra-compact planet gear was also evaluated by measuring the variation in dimensions of the gear teeth with digital image analysis (Nishiyabu et al., 2008). Fig.8 shows the figures of the micro-planetary gear composed of three types of gearwheels manufactured by μ-MIM process and the dimension of the planet gear (module: m=0.07mm, number of teeth: z=24). The materials used for producing the ultra-compact gears are stainless steel 17-4PH water-atomized powder (D_{50}=2μm) and oxymethylene-based binders. The volume fraction of powder in the feedstock is 60%. The feedstock was injection-molded using high-speed injection molding machine (FANUC Ltd., S-2000i 50A). The green compacts were debound at 600°C for 2hrs in N_2 gas, and sintered at 1150°C for 2hrs in Ar gas. Also the sintered parts were age-hardened at 480°C for 1hrs. From

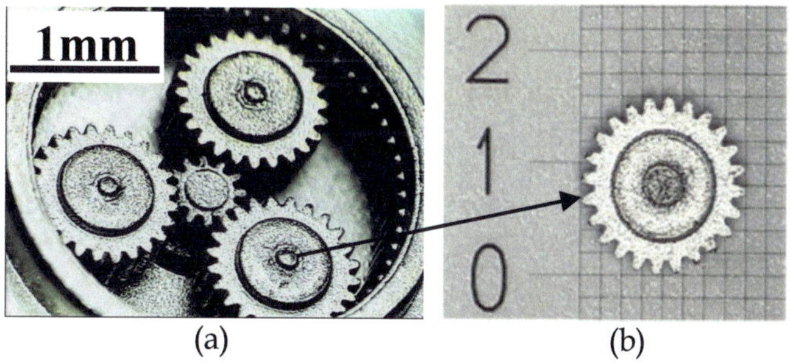

Fig. 8. Micro-planetary-gear manufactured by MIM process; (a) Planetary-gear set; (b) Planet gear.

the evaluation results, it was clarified that the accuracy class of the ultra-compact planet gear have not come up to that of ground precise gear which is equivalent to around five. However it is reviewed that the accuracy class of ultra-compact gear is around seven which is an acceptable level for general-purpose applications from a practical point of view.

Computational fluid dynamics (CFD) analysis can provide valuable information to mold designers and manufacturers (Nishiyabu et al., 2008). Flow simulation of micro-planetary-gear manufactured by MIM process was done using Moldex™ software. The finite element model shown in Fig.9 is used. In total, the cavity meshes and mold base meshes included approximately 2.3million elements. Whilst the material for the analysis has the material properties of MIM feedstock, it is not currently possible to accurately model all of the complex flow characteristics of MIM feedstock, such as layer slip, compressibility and jetting. The material in the analysis has similar flow characteristics to a heavy plastic. Most commercial injection-molding CFD programs cannot accurately create micro-parts model using standard machine settings as they have unsuitable values for shot weights and flow rates etc. From the flow simulation results shown in Fig.10, it is clarified that the filling time is approximately 1.8ms for spur-runner part but only 0.2ms for cavity one. Also the filling of gear teeth is at the end of flow, and is hard to apply high pressure into fine wall of teeth. Extra care must be taken with runner shapes and gate locations in micro-analysis due to the exceptionally small and thin cavity shapes.

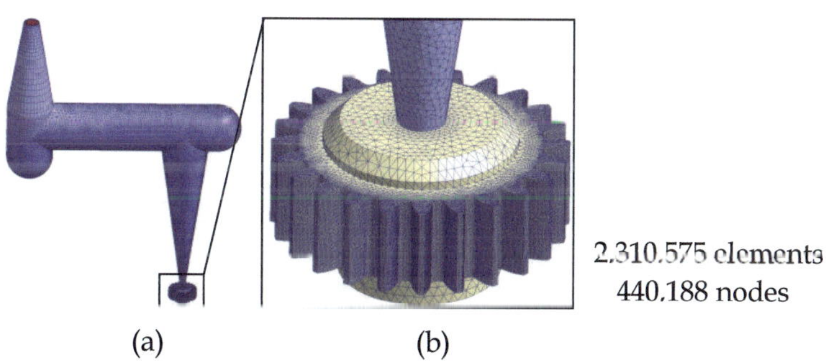

Fig. 9. Micro-planetary-gear manufactured by MIM process; (a) Runner and spur part; (b) Micro-gear part.

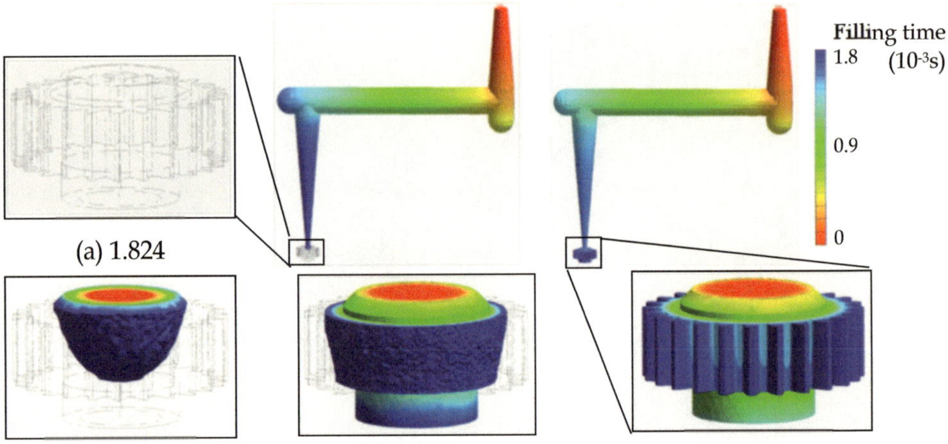

(b) 1.896 ms (b) 1.987ms (c) 2.033ms

Fig. 10. Filling state of micro-planetary-gear with spur-runner part.

2.3.2 Micro-dumbbell tensile specimen

As the size of MIM parts decrease, the test specimen is also necessary to downsize appropriately for actual evaluation. Authors (Nishiyabu et al., 2005) used micro-dumbbell tensile specimen with 0.1mm wide in narrowest portion as shown in Fig.11. The reason why the dumbbell shape with large volumes of clamps at the ends was attached is to investigate the performance of filling and the change of internal pressure when it was molded with an in-line screw type injection molding machine. The jigs for clamping are also manufactured by MIM, and the specimen can be observed by scanning electron microscope (SEM). The motion of deformation and damages occurred on the surface is recorded.

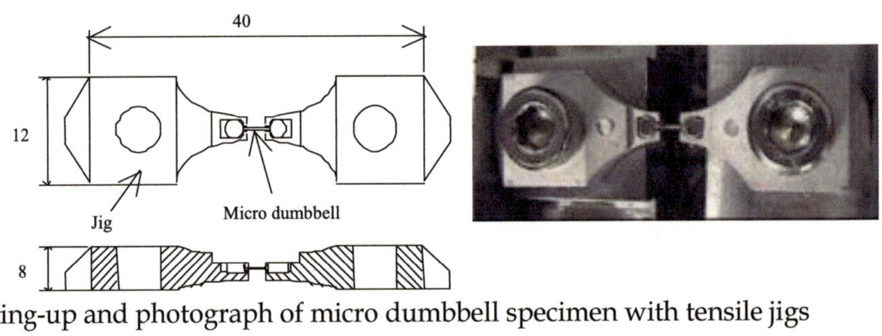

(a) Setting-up and photograph of micro dumbbell specimen with tensile jigs

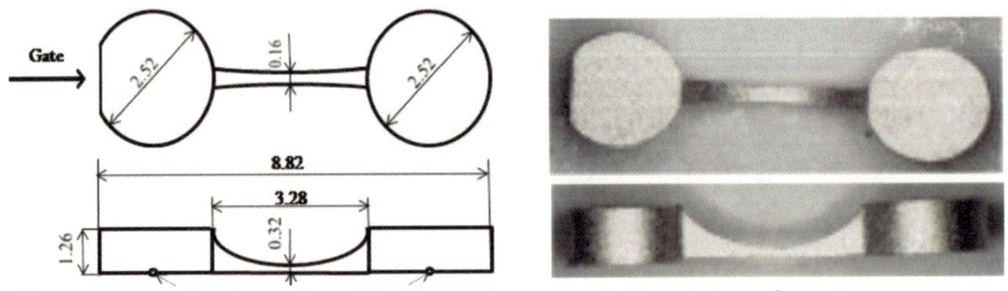

(b) Detail geometry of micro dumbbell specimen and the sintered part

Fig. 11. Geometry of micro dumbbell specimens.

The conditions of injection molding for micro dumbbell specimens were examined with various injection pressures and speeds, but the other conditions such as holding and back pressure, injection and holding time, metered value, molten material temperature and mold temperature were constantly-applied. The results of the experiments using D_{50}=10μm and D_{50}=2μm 316L stainless powder feedstock in constancy of metal powder fraction of 50vol.% are shown in Fig.12(a) and (b), respectively. These diagrams show the filling behaviour. In case of 10μm powder feedstock, the maximum injection pressure was limited at 150MPa because of the high melt viscosity, while no limitation existed for 2μm powder feedstock. In case of 2μm powder feedstock, the short shot phenomenon was observed in the wide range of injection pressure and speed. In either case, the short shot was likely to occur at lower injection speed, while flash is significant at higher injection speed, because the melt viscosity of feedstock depends on shearing rate as shown in Fig.5. With this preliminary examination, it is concluded that the suitable injection pressure and speed are 20-70MPa and 300-400mm/s for 10μm powder feedstock while 70MPa and 400mm/s for 2μm one.

The finite element model of micro-dumbbell with spur-runner parts as shown in Fig.13 was used for flow simulation. The cavity pressure profiles obtained by numerical analysis and

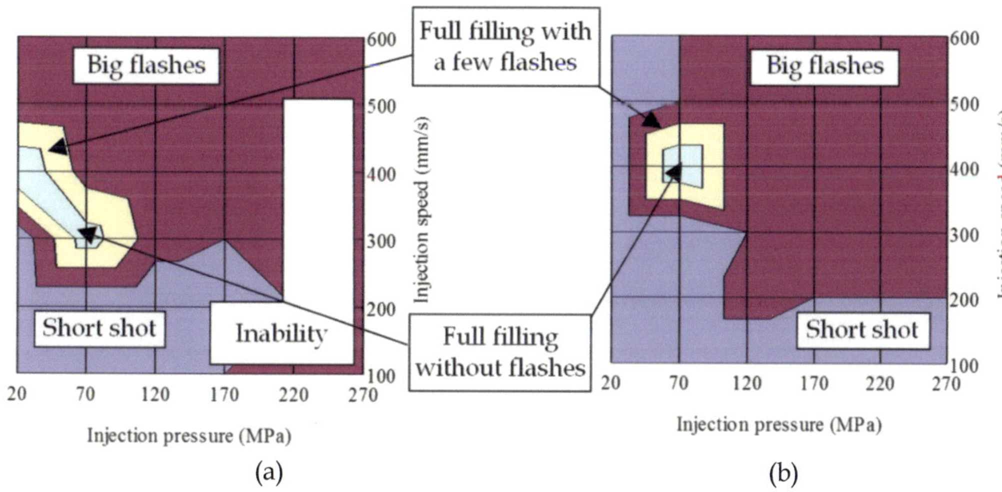

Fig. 12. Dependence of injection molding conditions on filling behaviour in micro dumbbell specimens molded using varied particle size feedstock; (a) D_{50}=10μm powder feedstock; (b) D_{50}=2μm powder feedstock.

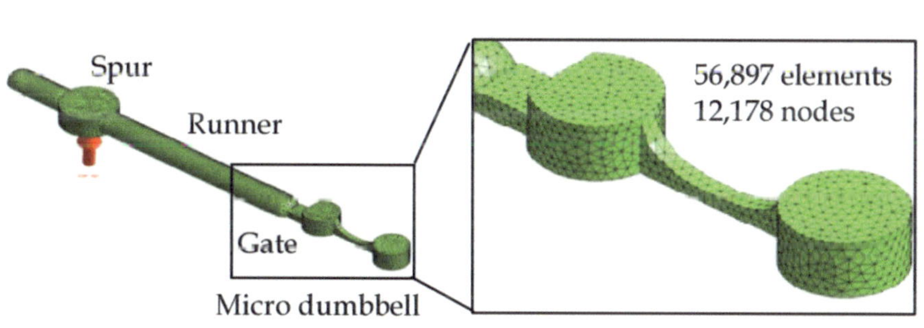

Fig. 13. FE model for micro-dumbbell with spur-runner parts.

experiment are shown in Fig.14. The far gate side is applied much lower pressure than the near gate. The results agree qualitatively, however it has large differences quantitatively between the simulation and the experiment.

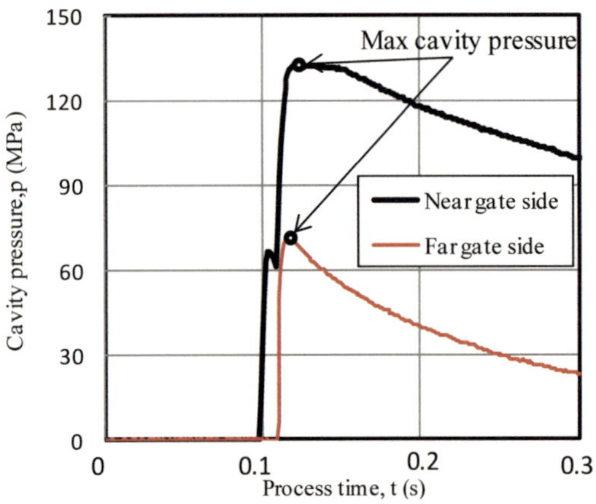

(a) Simulation

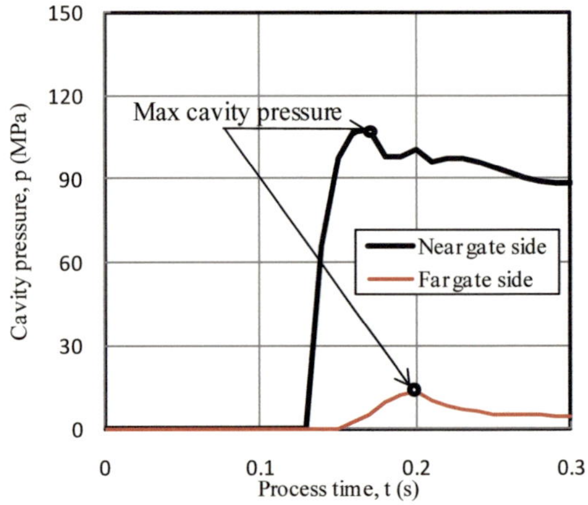

(b) Experiment

Fig. 14. Cavity pressure versus process time.

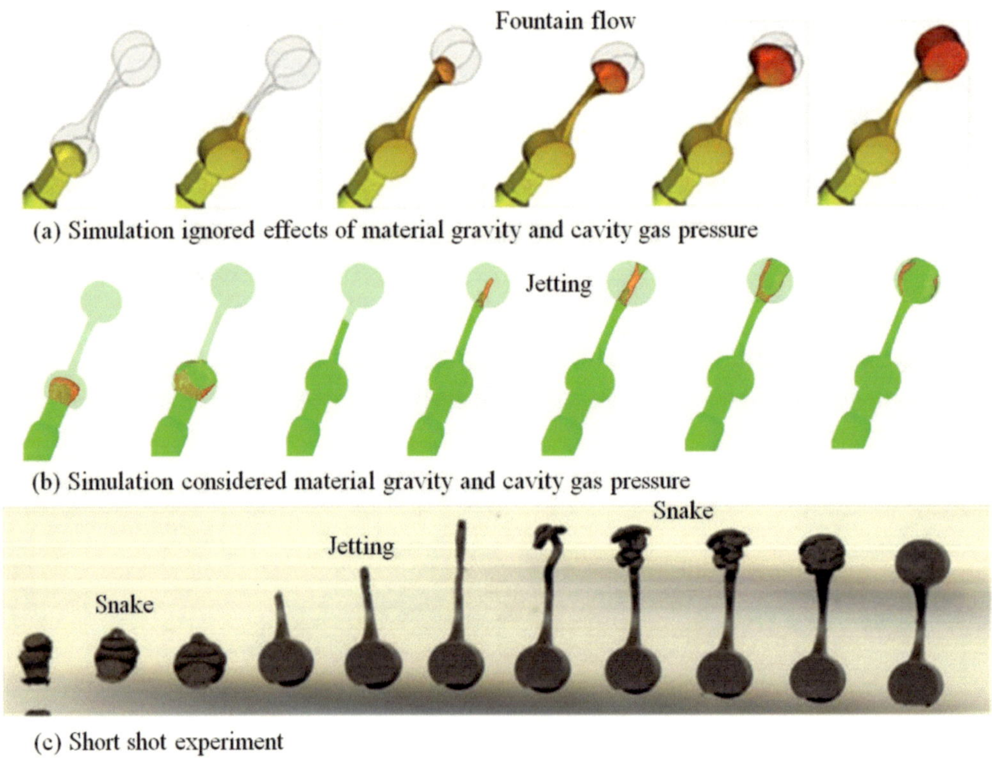

Fig. 15. Comparison of filling between simulation and short shot results.

The result in Fig.15 shows that one of the reasons to be considered is difference in filling between simulation and short shot. When the flow simulation considered the effects of material gravity and cavity gas pressure (Fig.5(b)), the results could be simulated as a jetting phenomenon and it could not be identified the unstable filling state such as snake phenomenon (Fig.5(c)). Thus it is considered that the pressure is resulted to reduce significantly in the experiment.

3. Metal powder injection molding for micro pillar-structured parts

3.1 Concept of sacrificial plastic mold insert MIM (SPiMIM) process

Fig.16 shows the flow of sacrificial plastic mold insert MIM (SPiMIM) process which is basically divided into three steps; 1) manufacturing of SP-mold, 2) injection molding of MIM feedstock into SP-mold insert, and de-molding the green compact and SP-mold insert part as one component, handling to the debinding-sintering process, and finally 3) debinding to eliminate the SP-mold and polymeric binder followed by sintering process. Therefore, the μ-SPiMIM process has great potentials to improve the filling, de-molding and handling, and to produce the tiny parts with 3 dimensional complex shapes and fine structures. The SP-molds used in this process can be manufactured by several types of methods such as injection molding, machining, rapid-proto typing, hot-embossing and lithography and so on (Nishiyabu et al., 2007).

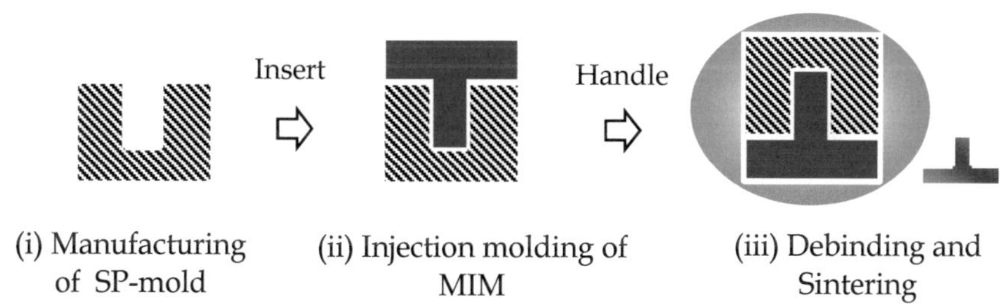

Fig. 16. Flow of micro sacrificial plastic mold insert MIM (SPiMIM) process.

Fig.17 shows some examples of SPiMIM products such as an optical fibre connector, a zigzag spring and micro impeller. These shapes cannot be manufactured easily because they has hollow, under-cuts and external screw and very narrow portions. The core and mold insert are made of polymethylmethacrylate (PMMA) polymer by conventional plastic injection molding for optical fibre connector and a zigzag spring, while micro rapid plot typing is used to make the closed mold to manufacture a micro-impeller.

3.2 Differences in filing behavior of metallic mold vs. plastic mold

The differences of mold materials on the filling behaviour were investigated by flow analysis using commercial plastic injection molding simulation software (Moldflow™, MPI ver.4.0). The main difference in property of mold materials is thermal conductivity, which of steels (46.2W/mK) is 230 times higher than plastics (0.2W/mK). From the analytical results in zigzag spring specimen as shown in Fig.18, it was confirmed that filling of feedstock could be accomplished with lower injection pressure by using plastic mold. This is

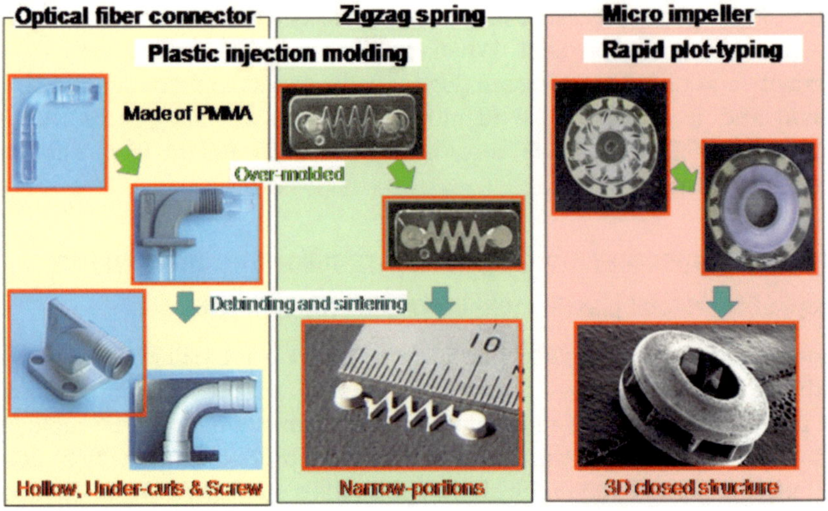

Fig. 17. Examples of metal parts manufactured by SPiMIM process.

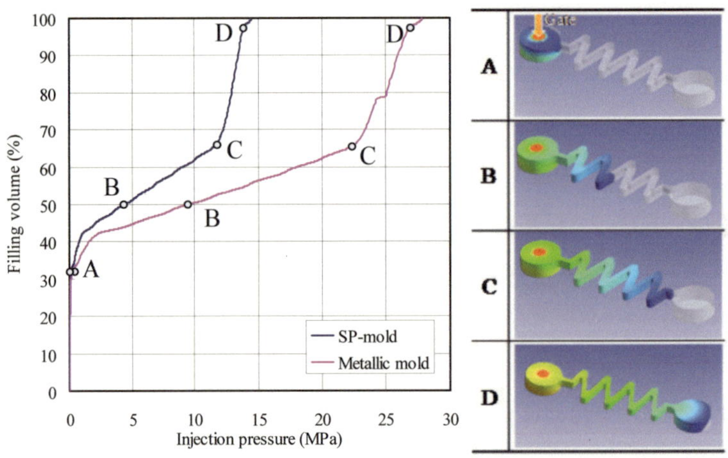

Fig. 18. Effects of mold material on filling behaviour in zigzag spring specimen.

considered because the plastic mold cools the feedstock much slower than metallic one. This is due to the low thermal conductivity of plastic and heat generated by friction between viscous flowing feedstock and plastic mold. In consequence of high viscosity of the feedstock, there is no rapid increase in plastic mold. Therefore, it can be expected to easily fill the viscous feedstock in narrow cavity of micro parts. In particular, with injection molding of micro pillar-structured parts, it is effective to use plastic mold.

3.3 Effectiveness of plastic in mold fabrication for μ-MIM

The products shown in section 3.1 cannot easily be manufactured by conventional MIM process, but they are not so tiny parts. Therefore, the SPiMIM method was applied into μ-MIM process by using finer tooling production method as below. Fig.19 shows the situation of MIM process overwritten in a road map on micro-nano processing developed by Japan

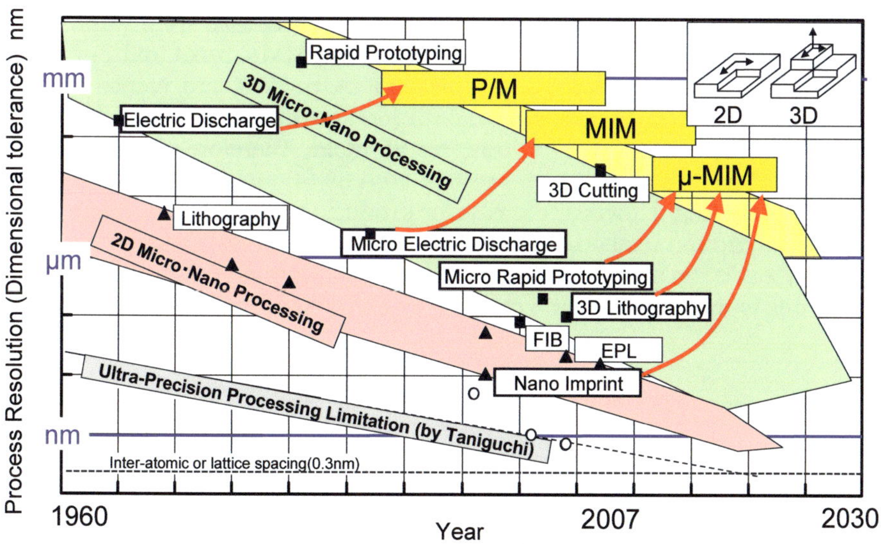

Fig. 19. Situation of MIM process in road map on micro-nano processing.

Society of Mechanical Engineers (JSME), which was originally surveyed by Prof. Hata. It is obvious from this road map that the process resolution is becoming progressively more accurate from micro-scale to nano-scale with the passing of the year by two or three dimensional micro-nano processing such as several lithography techniques, nano-imprint, micro electric discharge and micro-rapid prototyping and so on. By using these three dimensional micro-nano processing's to make the molds, the dimensional tolerance of P/M and MIM is becoming higher and it is even more so high for μ-MIM. However, most of these ultra-precision processing's is capable of applying for polymers, not for metals. Therefore, it is suggested that the use of plastic mold in MIM process is effective for the manufacturing of fine mold in addition to providing a high performance in filling and easy handling.

3.4 Combination of LIGA and MIM process

In producing of the higher precision parts, it is necessary to use the molds that are more precise than conventional molds manufactured by cutting and electric discharging. LIGA ((Lithographie , Galvanoformung , Abformung) is the abbreviation of German words for lithography, electroforming and molding, which is one of the micro-processing techniques applied for semiconductor production such as Micro-Electro-Mechanical-System (MEMS) or Micro System Technology (MST). LIGA is mainly consisted of 3 processes; the first is irradiation process to transcript the mask shape to resin, namely resist, the second is developing process to remove unnecessary part of the resist, and to make negative structure of the desired body, the third is molding process by electro-forming metal structure in the resist. The main features of LIGA process are to create the fine profiles with dimensional accuracy in nanometre order, and micro-structures with high aspect ratio. However, the materials applicable to LIGA are limited, and the process is not suitable for mass-production because of its high cost. The manufacturing method of micro sacrificial plastic mold insert metal injection molding (μ-SPiMIM) combined LIGA process, namely LIGA/μ-SPiMIM process has been proposed to solve specific problems involving the miniaturization of MIM. As shown in Fig.20, the μ-SPiMIM method inserted ultra-fine molds which were fabricated by LIGA process, namely LIGA/μ-SPiMIM process, was proposed (Nishiyabu et al, 2007). Two types of SP-molds with fine structures such as 1) PMMA resist and 2) PMMA mold injected into Ni-electroform which is a typical LIGA process, are named as "One-step transcription method" and "Two-steps transcription method", respectively. In one-step transcription method, PMMA resist is used for SP-mold. Therefore, this is very good for transcription, because original shaped resist is used as SP-mold, and out-gassing is very easy in injection molding. However the cost for producing PMMA resist is extremely high. In two-steps transcription method, Ni-electroform is used only for injection molding of SP-mold, it is preferable for mass-production with cost efficient, but the quality reduction of transcription due to the twice injection molding operations is a main issue.

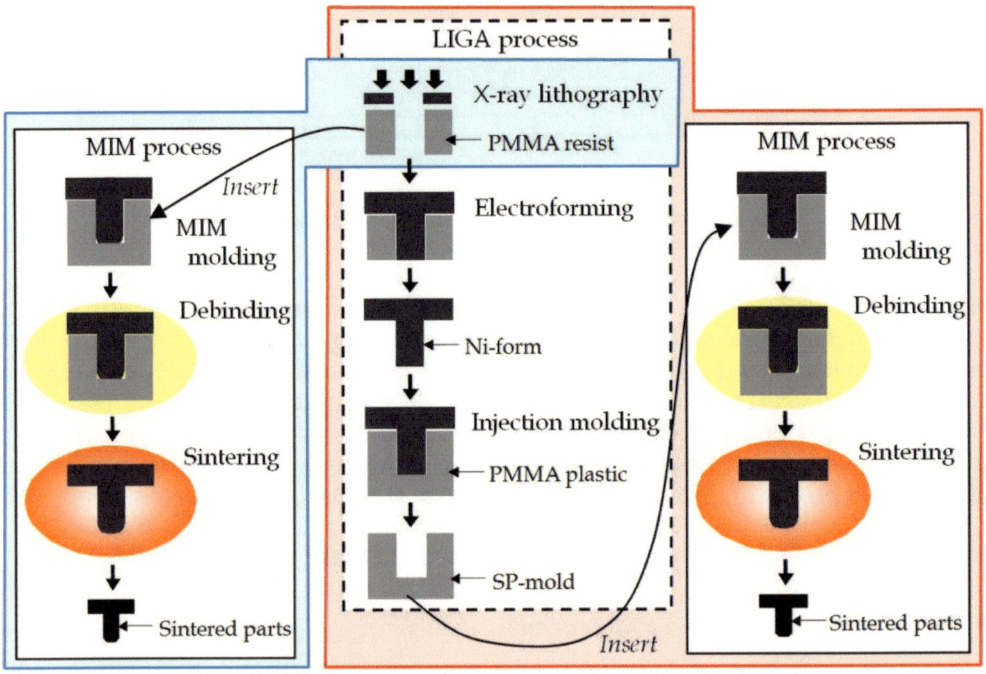

Fig. 20. Flow of two types of transcription methods in LIGA based μ-SPiMIM Processes.

3.5 Manufacturing by resist/μ-SPiMIM process

To demonstrate the possibility of Resist/μ-SPiMIM process, authors tried to manufacture multi-pillar structures assuming a micro-fluidic device, such as micro-reactor and micro-mixer as shown in Fig.21 (Nishiyabu et al, 2007). The resist film shown in Fig.21(a) was used as SP-mold. The dimension of a cylinder is 200μm in height and 50μm in diameter. The sintered part with 30-40% shrunk dimension after sintering was obtained as shown in Fig.21(b). It is also shown in Fig.21(c) that micro fluidic chip made of stainless steel 316L can be fabricated with relative ease by Resist/μ-SPiMIM process. Though the aspect ratios of these structures are not so high, it is very difficult to mold such a fine structure with conventional injection molding using metallic mold. Due to the fragility of green compact and fine structures with high aspect ratio, the de-molding is extremely difficult, because the surface area of multi-pillar structures increase exponentially with increasing in number of

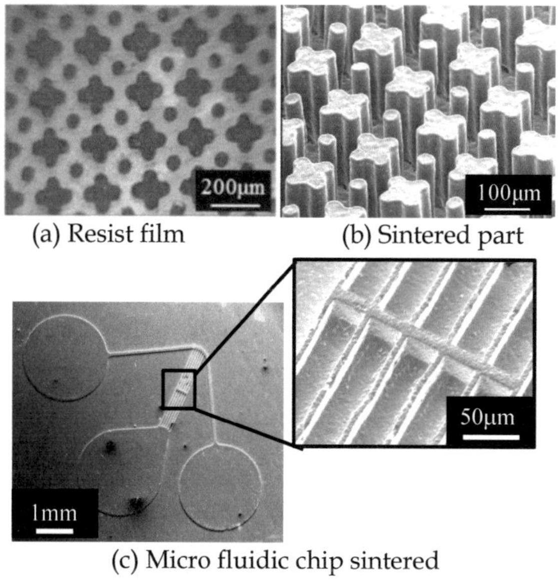

Fig. 21. Resist film and sintered parts manufactured by Resist/µ-SPiMIM process.

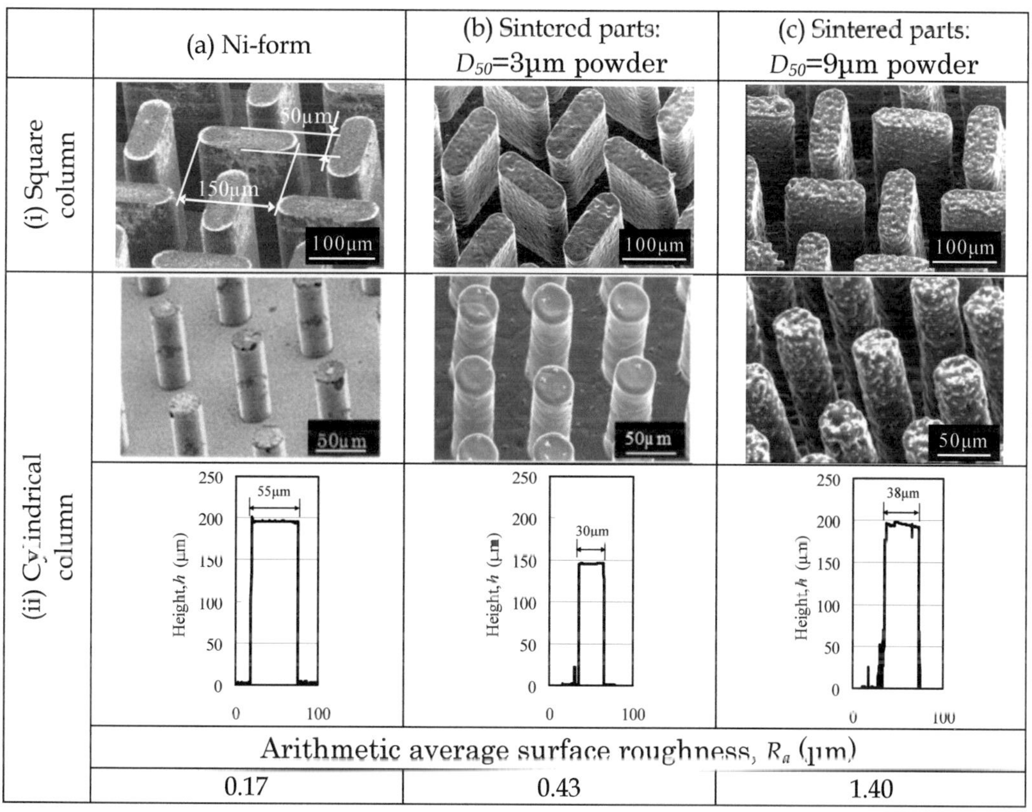

Fig. 22. SEM images, profiles and roughness of Ni-forms and sintered parts manufactured by Resist/µ-SPiMIM process.

pillars. However, the de-molding is not required in μ-SPiMIM process, therefore this process possesses great advantages in producing the tiny parts with micro-structures. Fig.22 shows SEM images, profiles and roughness of Ni-forms and sintered parts manufactured by Resist/μ-SPiMIM process using varied particle diameter of stainless steel 316 powder. The shape accuracy and surface roughness of sintered parts manufactured using finer powder is considerably improved compared to coarse ones and it is close to Ni-forms. It is also obvious to be able to manufacture metallic parts smaller than SP-mold.

3.6 Defects in molding in LIGA/μ-SPiMIM process

LIGA/μ-SPiMIM process requires two steps transcription, which is 1) plastic injection molding for SP-mold, and 2) metal powder injection molding for green compact. Thus, it is very important for obtaining accurate green compact to manufacture precisely SP-mold. However, the defects such as the development of weld line and short filling often occurred around pillars of molded parts in plastic injection molding. In order to reduce those defects to the scale of having no influences after sintering, the molding conditions for manufacturing the SP-mold should be optimized and also the effects of metal particle size and processing conditions on the shrinkage, transcription and surface roughness of sintered parts should be investigated. The specimen for SP-mold with numerous micro-pillars is shown in Fig.23. The size of square column is 150μm on a side and 200μm in height. In case of improper molding conditions, defects such as weld line and short filling caused by injection molding of SP-mold as shown in Fig.24(a). As the result, the sintered parts with unwanted shape such as convex line and rounded edge as shown in Fig.24(b) were obtained.

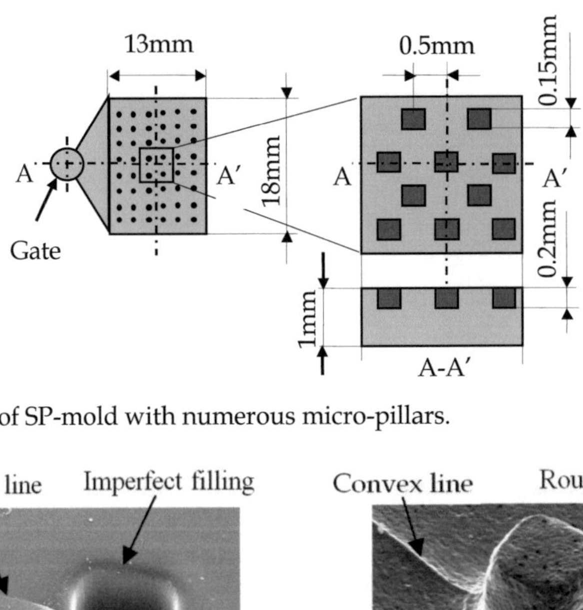

Fig. 23. Geometry of SP-mold with numerous micro-pillars.

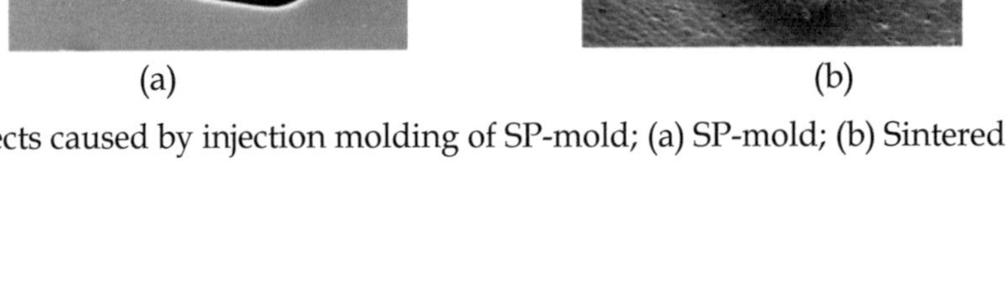

Fig. 24. Defects caused by injection molding of SP-mold; (a) SP-mold; (b) Sintered part.

The effects of injection molding condition on defects were investigated. Fig.25 shows the depth of weld line and width of short filling developed in SP-molds fabricated under various injection speeds and mold temperatures, but the material temperature and holding pressure were held constant. It is obvious from these graphs that the both sizes of weld line and short filling decreased as injection speed and mold temperature increased. As the results of the manufacturing of SP-molds under the maximum injection speed and highest mold temperature, the size of the defects was reduced as shown in Fig.26. The size of weld

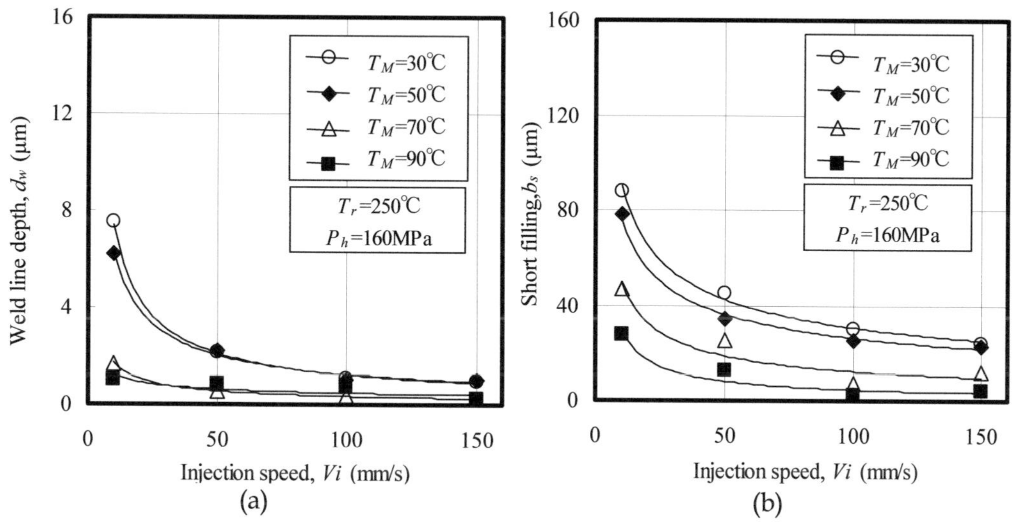

Fig. 25. Effects of injection speed and mold temperature on molding quality of SP-mold; (a) Depth of weld line; (b) Width of short filling.

	(a) Origin condition	(b) Optimized condition
Molding condition	v_i=10mm/s, T_M=60°C T_r=195°C, P_h=30MPa	v_i=150mm/s, T_M=90°C T_r=250°C, P_h=160MPa
SEM image	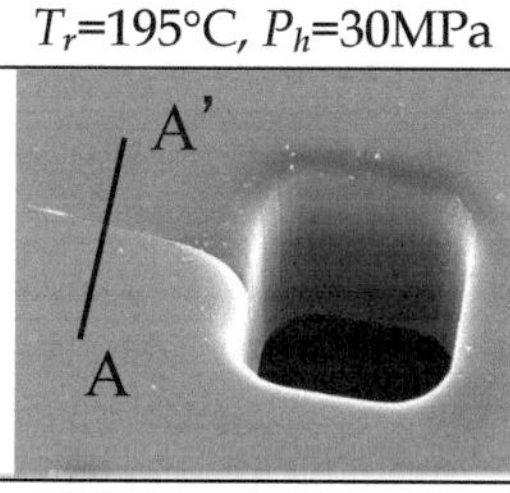	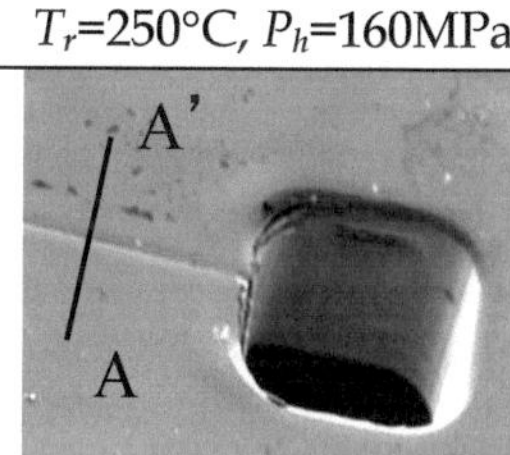
A-A′ Profile	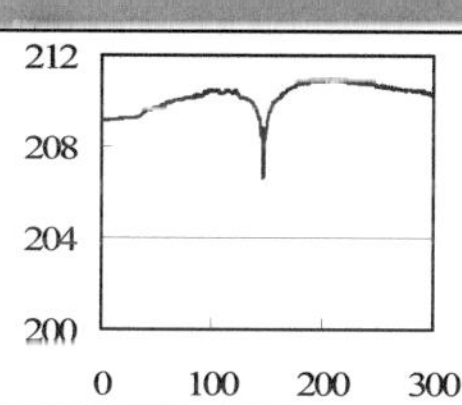	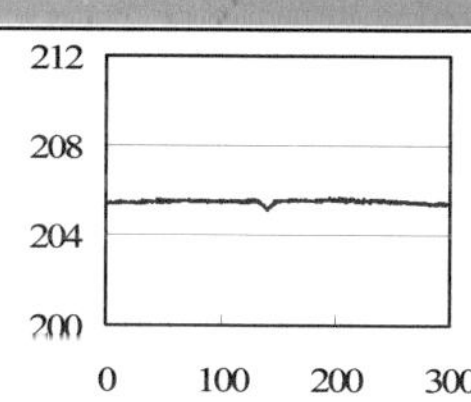
Size of weld line	4μm deep, 90μm wide	0.4μm deep, 9μm wide

Fig. 26. SEM images and profiles of SP-molds prepared before and after optimization of molding condition.

line is 0.4μm deep and 9μm wide, which the size is equivalent to disappear after sintering. Micro-pillar structured parts were manufactured by LIGA/μ-SPiMIM process as shown in Fig.27. The SP-mold (Fig.27(b)) was prepared by injection molding PMMA polymer into Ni-form (Fig.27(a)). Although the defects that occurred during the injection molding such as weld line and rounded edge, SP-mold can be manufactured with low cost and high cycle time. MIM feedstock prepared varied particle diameter of stainless steel 316L powder were injection-molded in the SP-mold, the sintered parts (Fig.27(c)(d)) were obtained after debinding and sintering. As similar to sintered parts manufactured by Resist/μ-SPiMIM process, micro-parts with much higher quality in shape and surface could be obtained by using fine powder better than coarse one.

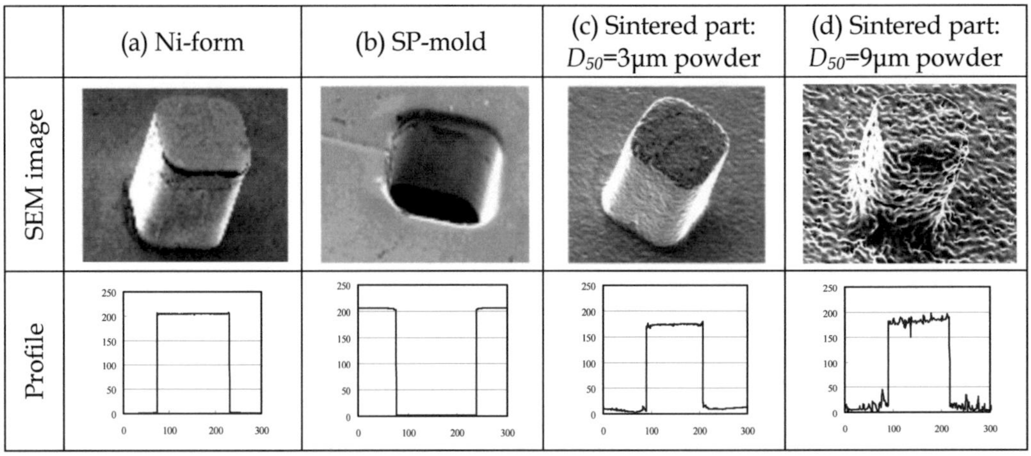

Fig. 27. SEM images and profiles of Ni-form, SP-mold and sintered parts manufactured by LIGA/μ-SPiMIM process.

4. Use of nano-scale powder in micro sacrificial plastic mold insert MIM

4.1 Material properties of MIM feedstock using nano-scale powder

Further improvements on the quality of μ-MIM products are required in practical productions. In general the use of finer metal powders is one of solutions to improve the dimensional accuracy and surface roughness of μ-MIM products. It can be also expected to inhibit the grain growth by reducing the sintering temperature. On the other hand, sintering inhabitation by higher oxidation, lower packed density, higher viscosity and higher cost are some of drawbacks in MIM production. MIM feedstock is prepared using various particle sized metal powders from micro-scale to nano-scale and the effects of particle size on material properties of MIM feedstock were investigated.

The metal powders used for the experiments are five types of pure Cu powders as shown in Fig.28. The reasons for pure Cu powder is used as the research material are that Cu has a high thermal conductivity, thus it is expected to be used for microscopic structures with high-specific surface area but it is not easy to mass-produce the metallic parts with such a structure, besides Cu can be reduced easily by H_2 gas because of the high Gibb's energy. The micro-sized Cu powder is a conventional material manufactured commercially by a water-atomization method (d_{50}=8.2μm, d_{50}=20μm) or wet-electrolytic method (d_{50}=1.5μm,

d_{50}=0.30μm). On the other hand, the nano-sized Cu powder is an ultra-fine material produced by a radio-frequency thermal plasma method (d_{BET}=0.13μm). The primary particles are completely different in size and shape from micro-sized Cu powder. As nano-sized powder has a large specific surface, melt viscosity of the feedstock increases more significantly. Therefore it is a key technology for deriving the effectiveness of nano-scale powders to select the component of binder and its fraction of MIM feedstock. Multi-component binder composed of polyacetal polymer and paraffin wax is used. The binder content is predicted by referring 35vol.% of the practical optimum binder content for Cu powder (d_{50}=8.2μm) which are using in conventional MIM production. Tap densities of various sized Cu powders are shown in Fig.29(a). Tap density of Cu powder increases exponentially as the particle size of Cu powder increases. As for the binder content, it decreases when the particle size increase as shown in Fig.29(a). As the results, a finer

	(a) d_{BET}=0.13μm	(b) d_{50}=0.3μm	(c) d_{50}=1.5μm	(d) d_{50}=8.2μm	(e) d_{50}=20μm
Producti-on method	Radio frequency plasma method	Wet reduction method	Water atomization method		
SEM image	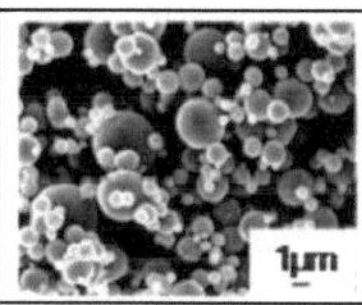	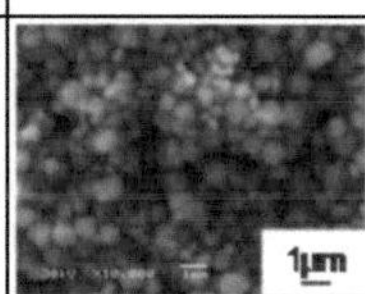		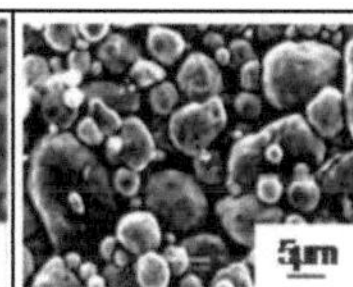	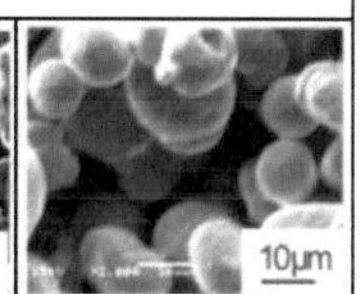
Tap density	0.66g/cm³	2.10 g/cm³	3.50 g/cm³	4.80 g/cm³	4.81 g/cm³
Specific surface	5.1 cm²/g	1.6 cm²/g	0.65 cm²/g	0.101 cm²/g	0.125 cm²/g

Fig. 28. Properties of Cu powders with various particle sizes.

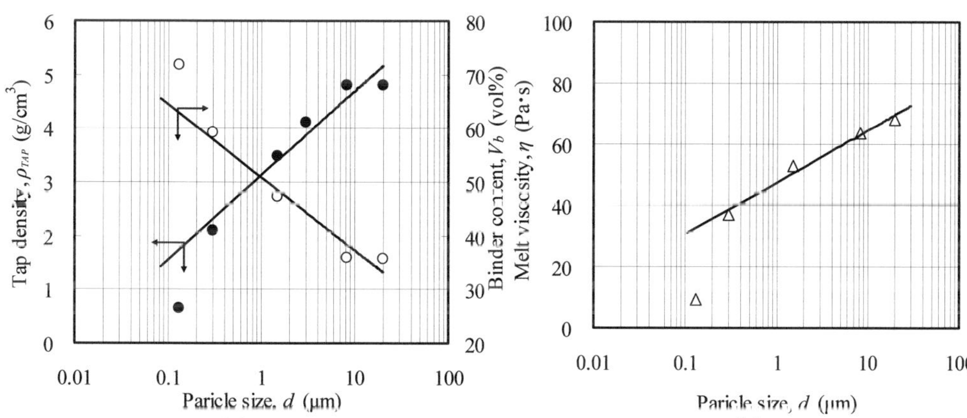

(a) (b)

Fig. 29. Properties of MIM feedstock with various particle sizes of metal powder; (a)Tap density and binder content; (b) Melt viscosity.

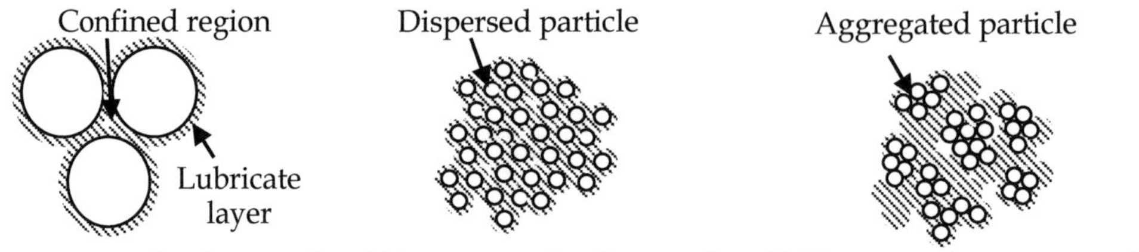

(a) Micro powder dispersed (b) Nano powder dispersed (c) Nano powder aggregated

Fig. 30. Configuration interaction of binder in MIM feedstock with varied particle.

powder has a larger specific surface, the melt viscosity of the feedstock increases more significantly as shown in Fig.29(b). When the binder content is predicted from the equation based on the space rate estimated from tap density, and the calculation of the binder content which the spaces are filled up, were made sometimes it shows that the binder content is too high and the melt viscosity becomes much lower. This reason is assumed to be due to remarkable agglomeration of nano-sized particles as shown in Fig.30. Thus the melt viscosity of feedstock was tried to keep constant by changing binder content using the equation based on the space rate estimated from tap density.

4.2 Molding machine for micro sacrificial plastic mold insert MIM

It is often desirable to conduct a material parametric study using a very small amount of nano-sized powder. The use of relatively large amount of feedstock is required for a conventional injection molding machine. Therefore, a direct mixing-injection molding machine as illustrated in Fig.31 has been developed. This machine is small enough to be placed on top of a table, and it enables the mixing of metal powder and binders followed by injection molding, therefore it can achieve molding without pelletizing. The capacity of the furnace is 0.05cm^3 in volume which is equivalent to the general size of a single feedstock pellet. A mixing is completed by a rotation of the plunger with 3mm in diameter. The mixing condition is basically controlled by furnace temperature, rotation speed and mixing time. The procedure of injection molding on this machine which is metal powder and binders are previously metered with balance and homogeneously mixed into the furnace. The feedstock is fully injected into the cavity at high speed when the plunger is pressed by compressed air with 0.1-0.85MPa. Therefore, this operation can achieve an accurate control on the volume of feedstock for each shot. In case of μ-SPiMIM process, resist film and NIL film is inserted into the mold cavity as shown in Fig.31(d) and (e), respectively. The green compact and SP-mold are ejected as one component.

4.3 Effects of particle size in resist sacrificial plastic mold insert MIM

In Resist/μ-SPiMIM process, the resist film made of PMMA polymer with numerous micro-holes was used as SP-mold. The feedstock was prepared using Cu powders with various particle sizes and the green compacts could be prepared with a high efficiency in experiment using a small amount of feedstock by a small molding machine introduced in previous section. The sintered parts with micro-pillar structure were obtained after

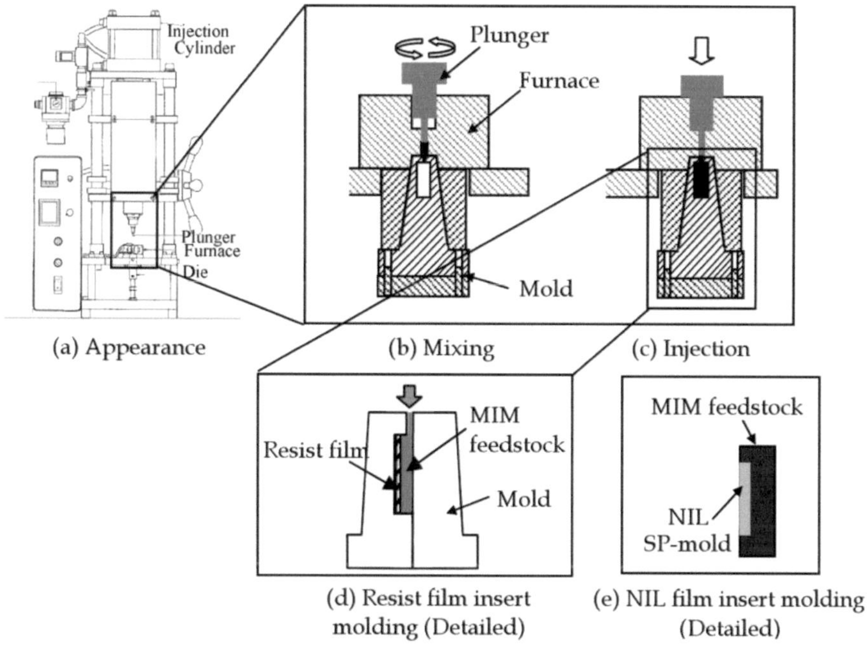

Fig. 31. Scheme of small direct mixing-injection molding machine.

debinding and sintering process. Fig.32 shows the SEM images of green compacts and sintered parts produced by various sized Cu powders. As the particle size of Cu powder used decreases, the filling of the green compacts in molding was improved except for d_{50}=0.13μm powder specimen. The decreasing of particle size results in a marked improvements of surface roughness, transcription and dimensional variation of sintered parts. Fig.33 shows SEM images of green compact and sintered part manufactured by using d_{50}=0.3μm Cu powder. For the entire body of green compact, the feedstock was filled fully in fine pillar structure, but a slight deformation is visible in sintered part. This is considered to be due to large amount of binder, and further study on the decision of binder content in particle agglomeration is needed for the quality improvement.

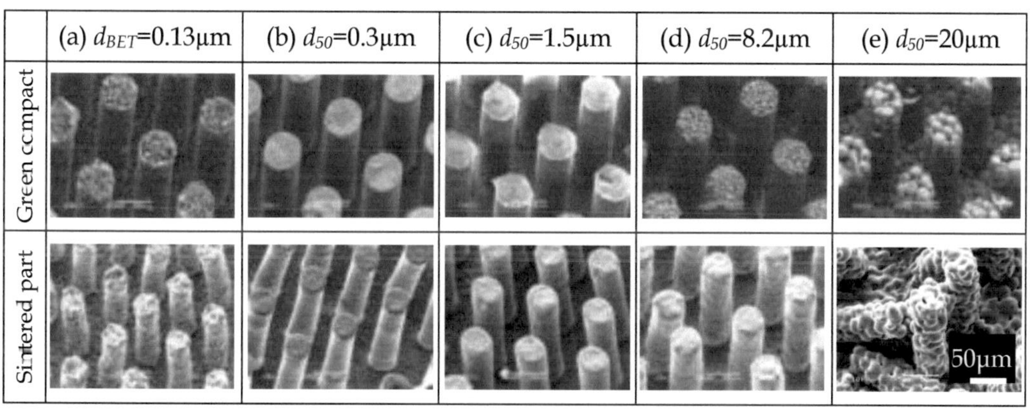

Fig. 32. SEM images of green compact and sintered part manufactured by using Cu powders with various particle sizes.

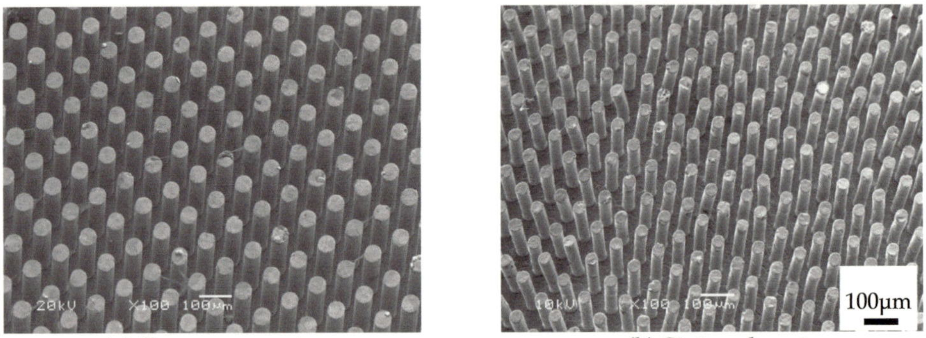

(a) Green compact (b) Sintered part

Fig. 33. SEM images of green compact and sintered part manufactured by using d_{50}=0.3μm Cu powder.

4.4 Nano-imprint lithography (NIL) sacrificial plastic mold insert MIM

Fig.34 shows the flow of NIL/μ-SPiMIM process. Thermal NIL technique is an application for the hot embossing technique but it can achieve higher resolution than conventional ones. The main features of NIL process are to create fine profiles with dimensional accuracy in nanometre order, and microstructures with several micrometers. However, the materials applicable to NIL process are limited only to polymers and glasses. Thus the combination process named as NIL/μ-SPiMIM which was used NIL process for manufacturing SP-mold in μ-SPiMIM process was proposed. As shown in Fig.34(i), thermal NIL process mainly consists of three steps; 1) Heating to soften PMMA film more than the glass-transition

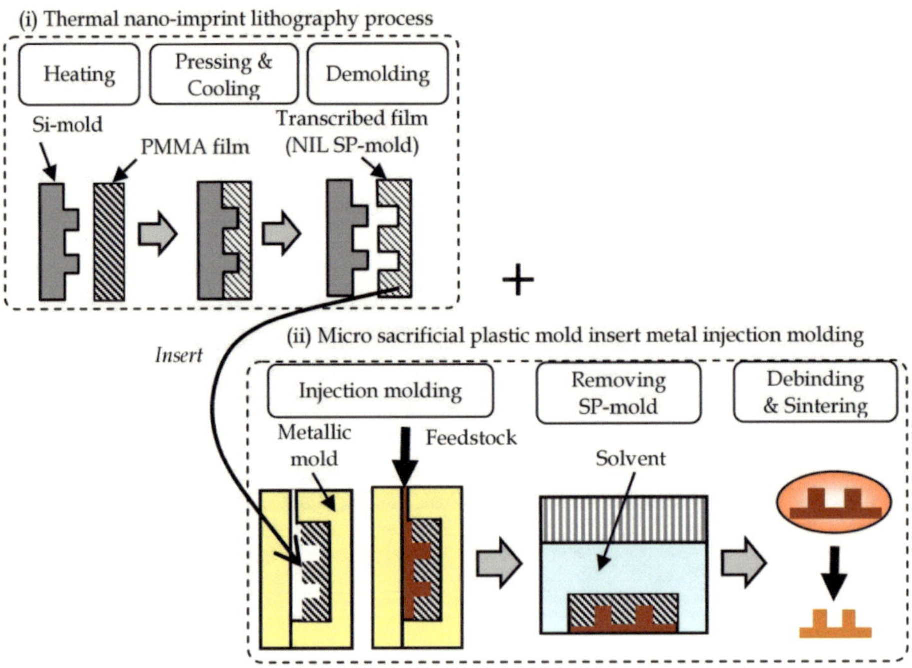

Fig. 34. Flow of NIL/μ-SPiMIM process.

temperature, 2) Pressing and cooling to transcript Si-mold shape to PMMA film and 3) De-molding the film from Si-mold. Subsequently the μ-SPiMIM process is proceeded by three steps as shown in Fig.34(ii); 1) Injection molding of MIM feedstock into the SP-mold, and ejecting the green compacts and SP-mold as one component, 2) Removing the SP-mold and 3) Debinding polymeric binders followed by sintering.

The metal powder used for the experiments is nano-sized Cu powder which is manufactured commercially by a wet-reduction method (d_{50}=0.7μm, tap density: 3.17g/cm^3, specific surface area: 1.69m^2/g). The solid loading is 50vol.% and melt viscosity of the feedstock is attained to 43.7 Pa-s. The feedstock composed of nano-sized Cu powder and oxymethylene-based binder was adequately prepared and molded into NIL film made of PMMA polymer with fine line-scan structures (5μm or 10μm in width and 10μm in height), and the molded parts (with a single length of 4mm) were sintered in a reductive gas atmosphere followed by solvent debinding of the films. The debinding and sintering condition was optimized by investigating the effects of sintering temperature and atmosphere gas on density, shrinkage, composition and profile accuracy with thermo-gravimetric analysis, carbon and oxygen analysis and SEM observation.

Fig.35 shows the SEM images of green compacts, and sintered parts processed at 573K and 973K. In the green compacts with both L/S=5μm and 10μm, the feedstock has be seemingly filled into the micro channels completely, but the polymer binder builds up at the upper corners of microscopic structures. The sintered parts processed at 573K have many

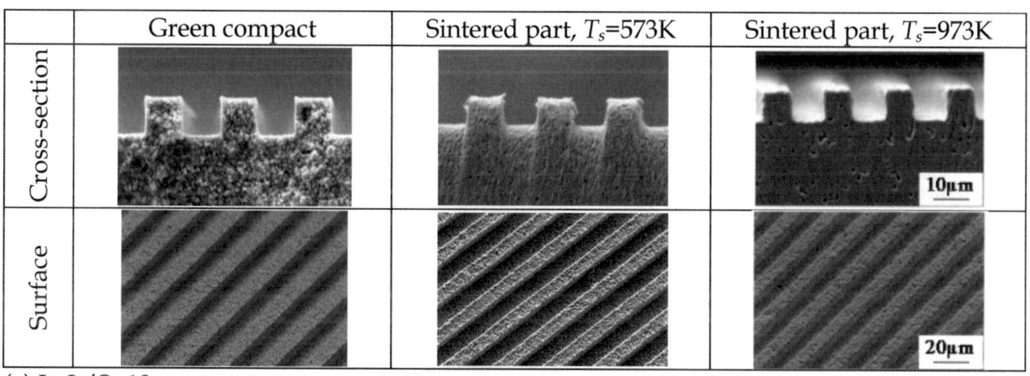

(a) In L/S=10μm structure

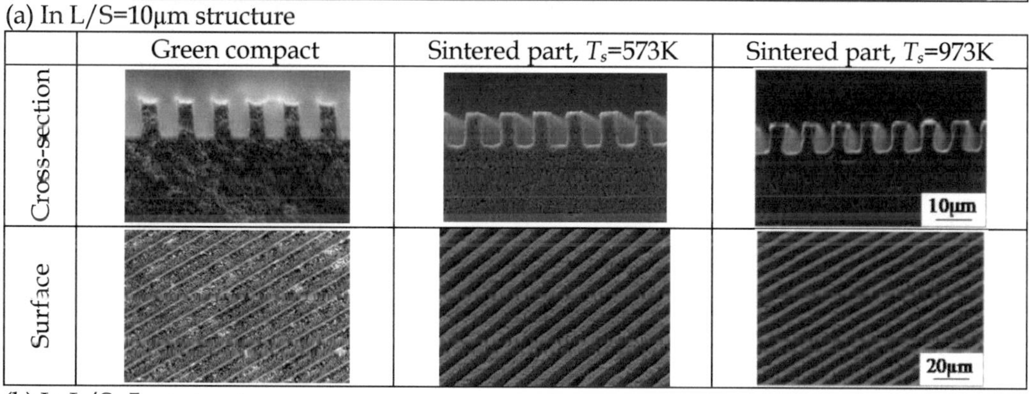

(b) In L/S=5μm structure

Fig. 35. SEM images of green compacts and sintered parts.

sub-micron pores, because both debinding and oxidizing have not been completed at the low temperature. Then the microscopic structures are kept in accurate shape. In general the sintered parts processed at higher temperature shrink more, and the corner of microscopic structures becomes dull correspondingly. However the sintered parts processed at 973K keep the edge sharpness under optimized debinding-sintering conditions. The green compacts have slight concave portions on the top face of microscopic structure, but it was attained to fill the feedstock into SP-mold with sufficiently high transcription. On the other hand, the sintered parts processed at 973K shrank 20% in both height and width, and became round at the top and bottom of corner portions. As the sintering temperature is raised, the shrinkage ratio increased remarkably up to 873K and further increased gradually. It was also seen that the shrinkage ratios of L/S=5μm structures are larger than that of L/S=10μm ones and the whole bodies.

The processability of a variety of μ-SPiMIM processes as above-described is summarized in Fig.36 in compared with the other precision processing and machining methods. A conventional rapid plot typing is difficult to manufacture SP-mold with micro-scale structure and high dimensional accuracy. Then RP/μ-SPiMIM method is not superior compared to micro machining such as micro-cutting, micro-EDM and micro-casting on the size of products, but it has a great potential to manufacture complex shaped parts such as micro-impeller shown in Fig.17. Further micro-miniaturization and surface quality improvement of rapid plot typing are required for μ-SPiMIM process, micro rapid plot-typing using stereo-lithography and 3D-printing technology is a prospective method for manufacturing of fine structured SP-mold. On the other hand, LIGA/μ-SPiMIM and Resist/μ-SPiMIM are very hopeful combined methods for manufacturing metallic micro-structured parts with high aspect ratios. The size possible for manufacturing is ranged from hundreds micrometers to tens micrometers. The problems are included high manufacturing cost and shape limitation of SP-mold. Furthermore, NIL techniques has a possibility for

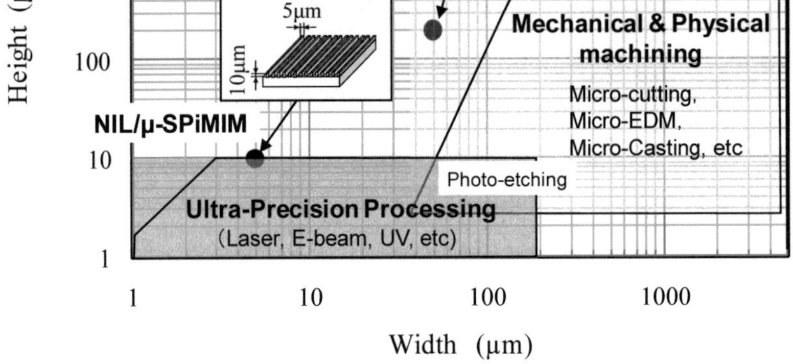

Fig. 36. Processability of a variety of μ-SPiMIM processes.

manufacturing a single digit micrometer-sized structured part, but technical problems on quality such as high-densification and sintering by using nano-scale metal powders should be cleared in NIL/μ-SPiMIM process at this moment. In addition, evaluation methods and designing of micro devices are hoped with effectiveness of these micro-structure and a variety of metallic own properties. These μ-SPiMIM processes have a unique advantage on fabricating as a one component with macro-scale and micro-scale structures made from a variety of materials. Also the micro-scale open and closed porous structures can be formed in sintering parts. This is not accomplished by semiconductor processes and depositions methods. Therefore it is hoping to use advanced applications such as micro reactor and micro-patterned electrodes with catalyst activity for fuel cell and battery manufacturing or micro-sensors for medical devices.

5. Conclusion

In this chapter, a general characteristic of MIM process on materials and conditions for manufacturing of small metallic parts with high quality was described utilizing actual data, and the complex flow characteristics of MIM were introduced on two kinds of small components, such as micro gear and micro dumbbell specimen. Then the technical problems to be solved for micro-miniaturizing of MIM parts were addressed, and the effectiveness of sacrificial plastic mold in micro-MIM process was shown by citing some example productions of micro-structured parts. A variety of methods for fabricating of the sacrificial plastic mold such as rapid proto-typing, plastic injection-molding, LIGA process and nano-imprint lithography process, were introduced by showing the investigation results on the effects of metal particle size and processing conditions. The use of nano-sized metal powder was applied in micro MIM process inserted sacrificial plastic molds made of resist or nano-imprint lithography, the results that the decreasing of particle size improved the surface roughness and shape-transcription of sintered parts were shown obviously. In collusion, micro sacrificial plastic mold insert metal injection molding, named as μ-SPiMIM method has a great potential to solve technical problems occurring in the μ-MIM process, it can be produced precisely the 3 dimensional complex metallic metal components with single-digit micrometer structures.

6. Acknowledgment

Author deeply appreciated for research foundation supports and understandings to President Dr. Shigeo Tanaka from Taisei Kogyo Co., Ltd., and great efforts of experimental works to many former graduated students from Osaka Prefectural College of Technology.

7. References

German, R.M. (1984), "Powder Metallurgy Science", Metal Powder Industry, ISBN 978-0918404602

German, R.M. and Bose, A. (1997), "Injection Molding of Metals and Ceramics", Metal Powder Industry, ISBN 978-1878954619

Löhe, D. and Haußelt, J. (2005), Microengineering of Metals and Ceramics: Part I and Part II. *Advanced Micro & Nanosystem,* Vol.4, WILEY-VCH Verlag GmbH & Co. KGaA, pp. 89, 127, 258, 276, 306, 354, 441, 464, ISBN 978-3527323784

Ishida, M., Takeda, H., Nishiyama, N., Shimizu, Y., Kita, K., Saotome, Y. and Inoue, A. (2005), "Characterization of Super-precision Microgear made of Ni-based Metallic Glass", *Journal of Metastable and Nanocrystalline Materials*, 24-25, pp. 543-546

Kameo, K., Nishiyabu, K., Friedrich, K. and Tanimoto, T. (2006), "Sliding Wear Behavior of Stainless Steel Parts Made by Metal Injection Molding (MIM)", *Wear*, 260, pp. 674-686

Nishiyabu, K., Kakishita, K., Osada, T., Matuzaki, S. and Tanaka, S. (2005), "Micro Evaluation Method for Quality Improvement of Micro Metal Injection Molding by Direct Mixing-Injection Molding Machine", *Proceeding of EuroPM2005*, 182-1-6

Nishiyabu, K., Kanoko, Y. and Tanaka, S. (2007), Innovations in Micro Metal Injection Molding Process by Lost Form Technology, *Materials Science Forum*, Trans Tech Publications Ltd., Vol. 534-536, pp. 369-372

Nishiyabu, K., Kakishita, K. and Tanaka, S. (2007), "Micro Metal Injection Molding Using Hybrid Micro/Nano Powders", *Materials Science Forum*, Trans Tech Publications Ltd., Vols. 534-536, pp. 381-384

Nishiyabu, K., Andrews, I. and Tanaka, S. (2008), "Accuracy Evaluation of Ultra-compact Gears Manufactured by The MicroMIM Process", *Powder Injection Molding International*, Vol.2, No.4, pp. 60-63

Osada, T. Miura, H., Ymagami, T., Nishiyabu, K., and Tanaka, S. (2007), "Gas Nitriding Mechanism in Titanium Powder Injection Molded Products", *Materials Science Forum*, Trans Tech Publications Ltd., Vols.534-536, pp. 361-364

6

Ceramic Injection Molding

Zdravko Stanimirović and Ivanka Stanimirović
IRITEL A.D.,
Republic of Serbia

1. Introduction

Powder injection molding (PIM), which encompasses metal injection molding (MIM) and ceramic injection molding (CIM), is a net-shaping process which enables large scale production of complex-shaped components for use in a diverse range of industries. It combines plastic injection molding techniques and performance attributes of ceramic and metal powders. Ceramic injection molding (CIM) uses ceramic powders such as alumina, zirconia, titania, ferrite powders, etc. It was introduced in 1940's, but for the next thirty years it was of minor interest to ceramic industry. In 1970's and 1980's CIM provided cost-effective fabrication method for mass production of ceramic parts for automotive industry. Today more than 300 companies practise PIM. Most of them practise MIM technology (>70%). Small percentage (5%) produce metals, ceramics and carbide components and about 25% practice CIM technology. This positive tendency can be attributed to unique properties of ceramic materials. They have excellent mechanical properties and low specific weight. Also, they are suitable for applications under extreme conditions (high temperatures, corrosive atmospheres, abrasive conditions, high loads at high temperatures). This combination makes them interesting for a wide variety of applications.

The ceramic injection molding process consists of four basic steps: feedstock preparation, injection molding, debinding process and sintering (Fig. 1).When powder technologies are in question, the key step in production process is choosing the adequate ceramic powder. Specific surface area, particle size, size distribution, particle shape and purity of the powder influence properties of the feedstock. Typical particle sizes in CIM are 1-2μm, but also much finer particles down to submicron or nano region are being used in advanced CIM. CIM uses a feedstock of composite granulate. A high concentration of ceramic powder is mixed with a thermoplastic binder system to form moderate viscosity feedstock - homogenous powder-binder mix that is free of agglomerates, has optimum ceramic/binder content and still maintains sufficient fluidity (Rak, 1999).

The feedstock is molded using injection molding equipment similar to that used for polymer injection molding. Injection molding involves concurrent heating and pressurization of the feedstock. It requires close monitoring in order to minimize molding defects. As a result, a green body is obtained (Fig. 2). After molding, the binder is extracted from the green body. Debinding usually takes place in two steps. Immersion is the first step. Soluble component of the binder is removed and system of pore channels develops to allow removal of the remaining component. The second step is thermal debinding and the insoluble component is being removed by thermal decomposition thus resulting in brown body (Fig. 2).

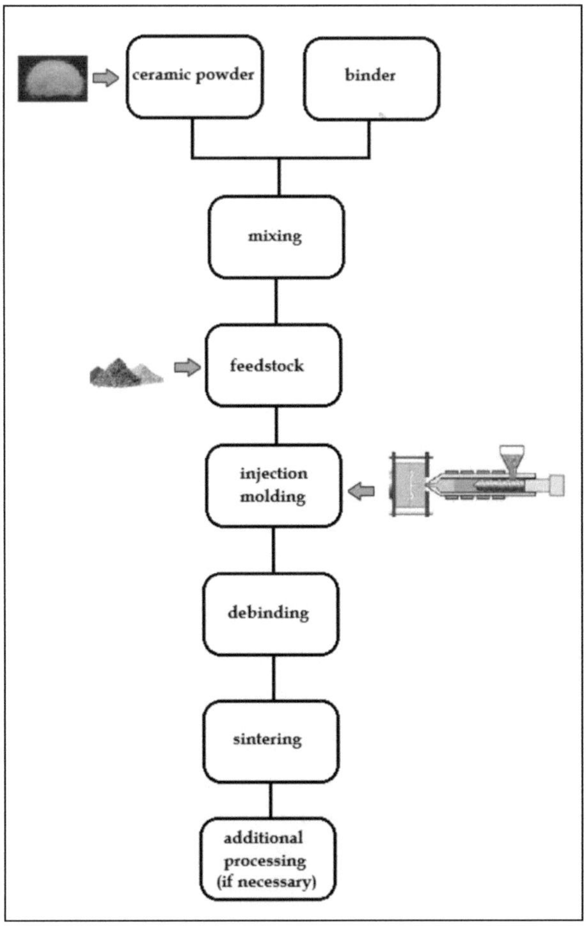

Fig. 1. The ceramic injection molding process.

Fig. 2. CIM component: green body, brown body and sintered part, respectively.

After debinding process, the sintering process takes place. Sintering parameters depend on the type and electronic properties of the ceramic powder used and, as a result, CIM components are obtained. Quality control of ceramic components in the green, brown and sintered state is commonly carried out by visual inspection and weighting. In that way surface cracks, impurities, voids, pores, distortions, incomplete parts and skin marks can be detected. Measuring the density of sintered components is another indispensable method for characterisation of CIM parts. Additional processing after sintering is optional depending on the type of component and specific application and in standard applications is seldom required.

In the recent past extensive international research and development activities were performed. Ferrite ceramics, piezoelectric ceramics and alumina were recognized as the most commonly used ceramic powders in production of wide range of CIM components for various applications and for that reason they are described in this chapter. Also, as an insight in future trends of CIM technology development, advanced CIM technologies are presented.

2. Ferrite ceramics for CIM

Ferrites are ceramic materials based on iron-oxide. They exhibit soft magnetism and therefore are being used in a variety of applications such as antennae, transformer cores, microwave waveguides, etc. There are three main types of ferrites: Mn-Zn ferrite, Ni-Zn ferrite and Mg-Zn ferrite. Ferrites have several advantages when compared to other materials: temperature and stability, high resistivity, wide frequency range and low loss combined with high permeability. Disadvantages are low saturation flux density and low tensile strength. Differences between soft ferrites and other magnetic materials are presented in Table 1 (Z. Stanimirović & I. Stanimirović, 2010).

There are several techniques available to forming ferrite specimens: grinding, extrusion, pressing and injection molding. Most ferrites are commercially produced by a dry pressing process. The powder flows into a die cavity and upper and lower punches at about 10 tons per surface square inch are being applied. Since the pressing is being done in vertical direction, resulting specimen geometries are limited to simple geometric shapes. Grinding is the most economical forming technique to produce non standard ferrite cores. It requires no tooling since cores are ground from isostatically formed sintered bars. Extrusion is an ideal technique for forming long rods and bars.

Parameter	Ferrites	Magnetic Alloys	Iron Powder
Initial permeability range	5-15000	5000-300000	5-150
Curie temperature range [°C]	100-500	500	600
Loss factor [10^{-6}] 10kHz 100kHz 160kHz	 5 10 25	 8 80 4000	 25 30 100
Resistivity [Ωm]	$10\text{-}10^{8}$	10^{-5}	10^{4}
Saturation flux density range [mT]	300-500	800-2400	1000-1200

Table 1. Differences between soft ferrites and other magnetic materials.

However, in recent years ceramic injection molding technique (Rodrigez et al., 2003; Zlatkov et al., 2008) has been applied as an alternative forming process. Injection molded ferrite parts can be produced from very simple forms to quite complex shapes. Further processing is rarely required, but if necessary, this can be achieved using conventional tools. Parts produced through this process can have very intricate shapes and tight tolerances. Injection molded ferrite components have properties similar to conventionally produced parts (Skolyszewska et al., 2003; Zlatkov et al., 2010).

Manganese zinc ferrite is a magnetically soft material suitable for use as magnetic cores in low frequency range (1kHz-1MHz). For soft ferrite magnetic core production uniaxial powder pressing technique is usually used. However, CIM as an extremely flexible technology enabled production of Mn-Zn ferrites with characteristics comparable with commercial samples prepared by conventional methods. There are two reasons for CIM investigations of Mn-Zn ferrites: the shape complexity and the better permeability.

The starting material used for the production of Mn-Zn ferrite samples was commercially available Mn-Zn ferrite powder shown in Fig. 3(a). Prior to injection molding, powder was processed in a conventional manner. Mn-Zn ferrite powder in combination with binder (combination of polypropylene, microcrystalline wax and stearic acid) was used for feedstock production. Feedstock contained 10.5% of binder (9% wax, 1% wax with lower melting temperature, 1% of stabilizer) and 68% of Mn-Zn ceramic ferrite powder. Photograph of Mn-Zn ferrite feedstock is shown in Fig. 3(b).

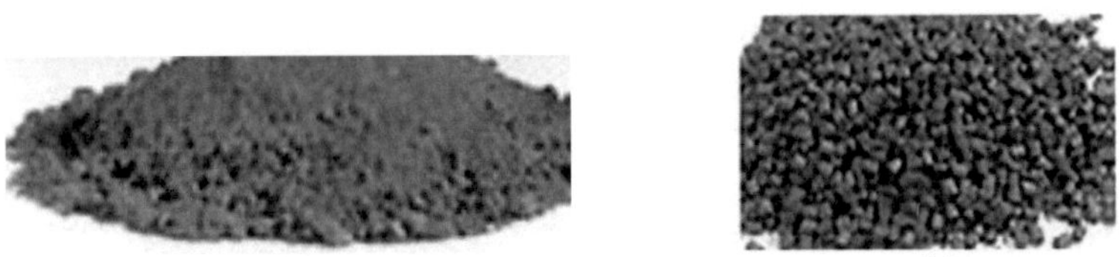

(a) (b)

Fig. 3. Mn-Zn ferrite: powder (a) and feedstock (b).

Injection molding was performed using molds in shapes required to form ring and disc shaped specimens. The injection molding process was carried out in Battenfeld HM 250/60-B4 machine and main injection molding parametres are given in Table 2. Debinding of green parts after injection molding was performed in two steps: solvent and thermal debinding. Thermal debinding was performed during initial priod of sintering (150-800°C heating period) and the green ferrite samples were sintered at 1280-1320°C/1-4h in nitrogen atmosphere.

Parameter	Setup
Injection temperature (°C)	120-160
Mold temperature (°C)	30-45
Injection speed (ccm/s)	3-20
Injection pressure (bar)	300-800
Cooling time (s)	10
Sample ejection pressure (bar)	20-40

Table 2. Main injection molding parameters.

Photographs and dimensions of injection molded ring and disc shaped CIM Mn-Zn specimens are given in Fig. 4 and Fig. 5. The main properties of injection molded sintered Mn-Zn ferrite ring shaped specimens are given in Table 3 and the comparative properties of injection molded sintered Mn- Zn ferrite ring and disc shaped specimens are given in Table 4.

Fig. 4. Disc shaped CIM Mn-Zn ferrite samples (diameter d=16mm, thickness t=5mm).

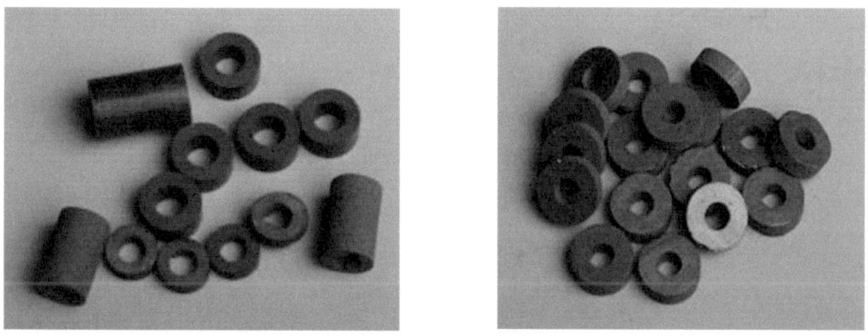

Fig. 5. Ring shaped CIM Mn-Zn ferrite samples (outer diameter d_o=15mm, inner diameter d_i=6mm, thickness t=5mm).

T_{sint}/t [°C/h]	Grain size [μm]	Relative density [% TD]	Initial relative permeability		Loss factor [10^{-6}]	
1320/2	5.83	89.2	1518	1305	13.21	13.9
1320/2	4.80	-	1238	1360	17.43	11.71
1320/4	7.71	-	1463	1463	23.52	21.60
1280/2	4.20	87.1	922.5	922.5	24.15	23.33
1250/4	4.50	90.4	717.7	-	37.18	-
1320/2	4.33	90.1	1609	1456	9.0	9.18
1320/2	n.a.	85.1	733.4	716.3	50.94	42.06

Table 3. The main properties of injection molded sintered ring shaped Mn-Zn ferrite specimens.

	Ring shaped specimens	Disc shaped specimens
Initial relative permeability	750±25%	900±25%
Operating frequency range [MHz]	0.1-1	0,01-0,5
Relative loss factor	$8 \cdot 10^{-6}$-$30 \cdot 10^{-6}$	$5 \cdot 10^{-6}$-$25 \cdot 10^{-6}$
Magnetic Induction [mT]	390	390

Table 4. Comparative properties of injection molded sintered Mn-Zn ferrite ring and disc shaped specimens.

Experimental work demonstrated that Mn-Zn ferrite ceramics can be prepared using injection molding technique but the process is not trivial. For example, a special attention must be paid to initial filling of the mold. Due to uneven shrinkage rate during solidification, creation of stresses within the body of the sample may occur resulting in nucleation of voids or cracks. Also, if air pockets remain within the body of the specimen this may lead to poor properties of the realized component.

Problems may also occur during the burn out of the binder. Burn-out can be performed in air but this may cause excessive specimen shrinkage and surface layer exfoliation due to oxidative reaction. For that reason, nitrogen atmosphere was used. Sintering and densification of the specimen during sintering is a key factor that determines magnetic properties of the sample. Highest temperature used (1320 °C) yielded the highest sample densities and as illustration SEM microstructure of CIM Mn-Zn ring shaped ferrite specimen sintered at 1320°C/2h is shown in Fig. 6.

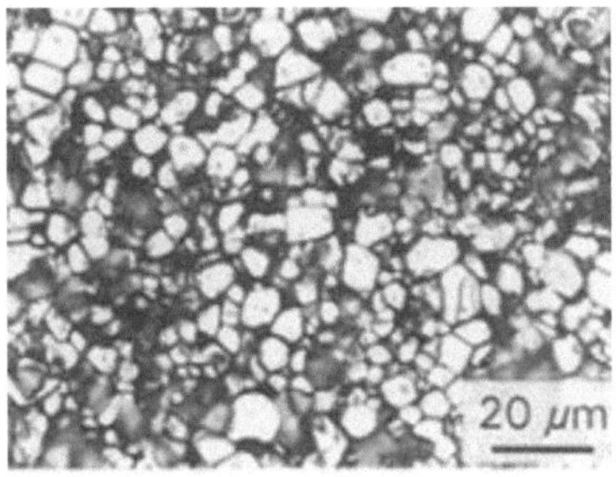

Fig. 6. SEM microstructure of CIM Mn-Zn ring shaped ferrite specimen sintered at 1320°C/2h.

The grain sizes of injection molded specimens are usually smaller than grain sizes of conventionally produced Mn-Zn ferrites. Injection molding process leads to similar specimen densities when compared to conventional methods but grain sizes are up to 50% smaller. Increase of the sintering time from 2h to 4h results in grain growth normally expected for Mn-Zn samples. Smaller grain sizes lead to more grain boundaries and therefore more pinning centers resulting in lower permeability. Grain growth during prolonged sintering cycle leads to greater densities and increased permeability but at temperatures higher than 1180°C (Pigram & Freer, 1994) zinc volatilization occurs leading to zinc loss. For that reason, temperature of 1150°C (Pigram & Freer, 1994) is recommended for adequate grain growth without zinc volatilization. Therefore, further work to optimize the processing conditions is desirable. The goal is to increase the grain size and initial permeability of realized specimens.

Experimental work on Mn-Zn ferrite has shown that ferrite ceramics can be prepared using injection molding technique although the process is not trivial and optimization of both tools and processing conditions is essential. Mn-Zn sintering obtained by CIM technology is sensitive process compared to the conventional pressing technology, but the obtained results are satisfactory. Realized Mn-Zn ferrite specimens have high green-state strength, ideal for production of delicate and complex shapes. Also, specimens exhibit satisfactory structural integrity and magnetic properties, as well as densities similar to conventionally produced material.

3. Piezoelectric ceramics for CIM

Piezoelectric ceramics is one of the functional materials which have unique electrical properties with broadening range of applications. Lead zirconate titanate ($PbZrTiO_3$) and barium titanate ($BaTiO_3$) are ceramic materials that have found widespread use - especially lead zirconate titanate that is being widely used in sensors, transducers, microactuators, multilayer capacitors and micro-electromechanical systems (MEMS). These materials are known for their superior piezoelectric and ferroelectric properties. When a mechanical force is applied, piezoelectric materials generate electrical voltage. Conversely, when an electric field is applied, these materials induce mechanical stress or strains. These effects are known as direct piezoelectric effect and converse piezoelectric effect, respectively.

Conventional powder metallurgy method is a commonly used method to produce piezoelectric materials. It starts with powder preparation. The powder is pressed to required shapes and size, and green shapes are processed into mechanically strong and dense ceramics. Machining process is being used for achieving desired shapes of the components. Elecroding and poling are the final sterps of the process. When complex shapes are in question, cutting and machining of piezoelectric ceramics are time consuming. There are also cost considerations because of the cost of the die and the waste material.

The most of the published papers have dealt with fabrication and electrical properties of piezoelectric ceramics produced using conventional powder metallurgy method (Fig. 7). However, a little work has been carried out on the fabrication and characterisation of piezoelectric ceramics prepared by ceramic powder injection molding method (CIM) (Luo et al., 2006; Wang et al., 1999, Zlatkov et al., 2008). In order to synthesise piezoelectric ceramics

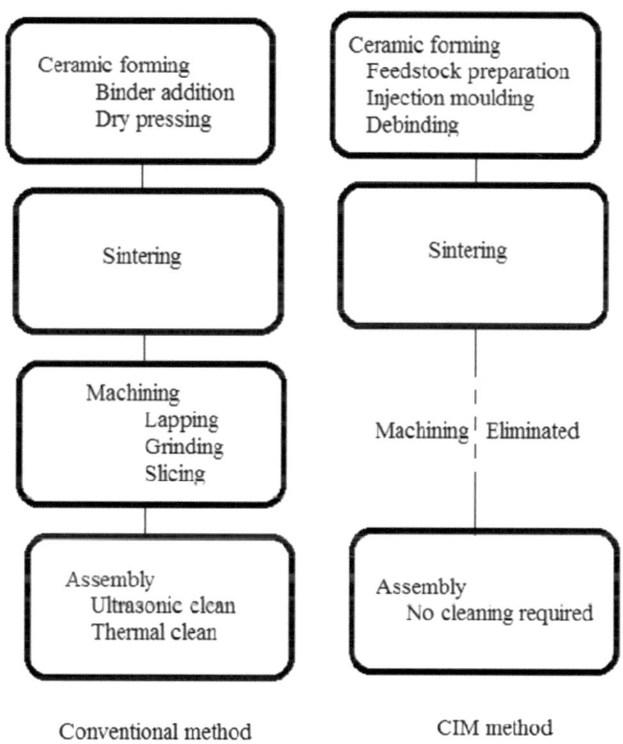

Fig. 7. Piezoelectric ceramics production: conventional powder metallurgy method vs. CIM method (I. Stanimirović & Z. Stanimirović, 2010).

using CIM process, four basic steps should be performed: feedstock preparation, ceramic injection molding, debinding and sintering. Components produced by CIM are expected to have more complex shapes and more homogeneous microstructure than components produced by conventional metallurgy method. Also, reduced machining and recycling use of feedstock are significantly reducing fabrication costs.

In order to explore the feasibility to synthesise piezoelectric ceramics by CIM, two series of test samples were realized (I. Stanimirović & Z. Stanimirović, 2010). Commercially available $BaTiO_3$ and $PbZrTiO_3$ powders were used. Basic properties of used powders are given in Table 5. Also, photographs of $BaTiO_3$ and $PbZrTiO_3$ powders are given in Fig. 8, as well as the micrograph of $PbZrTiO_3$ powder particles Fig. 9(a).

Ceramic powder:	$BaTiO_3$	$PbZrTiO_3$
Density [g/cm^3]	6,2	7,8
Dielectric constant $\varepsilon_{33}/\varepsilon_0$	1700	2200
Isolation resistance [Ωm]	10^{11}	10^{11}
Electromechanical coupling coefficient k_{33}	0,50	0,69
Piezoelectric coefficient d_{33} [C/N]	$120 \cdot 10^{-12}$	$450 \cdot 10^{-12}$
Curie temperature T_C [°C]	118	350
Specific surface area [m^2/g]	2,6	2,7

Table 5. Basic properties of $BaTiO_3$ and $PbZrTiO_3$ powders.

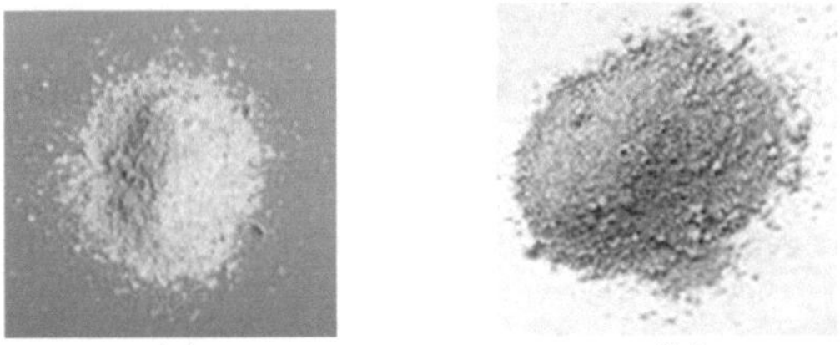

(a) (b)

Fig. 8. $BaTiO_3$ powder (a) and $PbZrTiO_3$ powder (b).

(a) (b)

Fig. 9. Scanning electron micrograph of $PbZrTiO_3$ powder particles (a) and $PbZrTiO_3$ feedstock (b).

These starting materials in combination with binder were used for feedstock production. Each feedstock contained 10.5% of binder (9% wax, 1% wax with lower melting temperature, 1% of stabilizer). Photograph of $PbZrTiO_3$ feedstock is shown in Fig. 9(b).

The feedstock was then heated to a sufficient temperature - such that it melted and injected into a mold cavity where it cooled and formed desired shape. The injection molding process was carried out in Battenfeld HM 250/60-B4 machine and the main parameters of injection molding corresponded to ones listed in Table 2. Dimensions of green bodies were 20mm×10mm×2mm. In accordance with the binder system, debinding procedure was performed. A two-stage debinding technique was applied. The solvent debinding stage was followed by thermal debinding stage. The main debinding parameters are given in Table 6. The slow heating rate prevented defects such as micro-cracks, slumping and blistering of the parts to be induced during debinding process.

In a water bath:	24h in a destilled water
In chamber drying device with fan:	4h at 80°C
	80°C - 145°C, rising degree [20°C/h]
	145°C - 155°C, rising degree [0.5°C/h]
	155°C - 160°C, rising degree [0.2°C/h]
	At 160 °C holding time 4h
	160°C - 170°C, rising degree [2-5°C/h]
	170°C - 220°C, rising degree [10°C/h]
	220°C - 300°C, rising degree [20°C/h]
	At 300 °C holding time 2h

Table 6. Two-stage debinding procedure.

After the debinding process, the debinded parts were sintered in an air atmosphere. In order to minimize lead loss from $PbZrTiO_3$ bodies that occur at about 800°C, these samples were sintered in presence of a lead source. Basic information about the sintering process is given in Table 7.

Butch kiln	ELEKTRON 1500
Sagger	Alumina 98%
Sagger cover	Alumina 98%
Bottom plates	ZrO_2
Setting and safety powder	ZrO_2 and Pb_3O_4
Manual filling	20 psc. green bodies/sagger
Sintering conditions	Air atmosphere
	21°C -600°C, rising degree [100°C/h]
	$PbZrTiO_3$: 600°C-1250°C, $BaTiO_3$: 600°C-1260°C, rising degree [150°C/h]
	$PbZrTiO_3$: 1250 °C, $BaTiO_3$: 1260 °C, holding time 2h
	Cooling 1250°C /1260°C–21°C: Natural

Table 7. Sintering process.

Dimensions of sintered samples were 16.67mm×8.43mm×1.52mm ($PbZrTiO_3$ samples) and 16.6mm×8.42mm×1.5mm ($BaTiO_3$ samples). After the sintering process, $PbZrTiO_3$ samples were silver plated using screen printing process and fired in an air atmosphere at 750°C/10min. All samples were then polarized and functional and electrical measurements were performed (Table 8). Photographs of CIM $PbZrTiO_3$ and $BaTiO_3$ samples are given in Fig. 10 and Fig. 11.

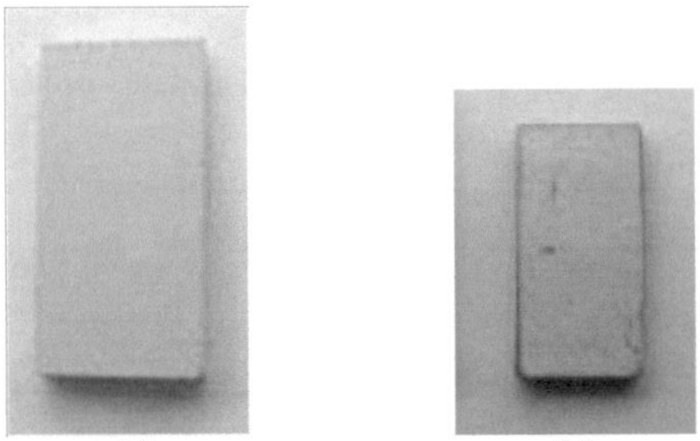

(a) (b)

Fig. 10. CIM $PbZrTiO_3$ samples: green (a) and sintered and silver plated (b).

Fig. 11. CIM $BaTiO_3$ samples.

	$BaTiO_3$	$PbZrTiO_3$
Dielectric constant $\varepsilon_{33}/\varepsilon_0$	2000±30	1700±30
Loss tangent $tg\delta_{max}$	0.02	0.02
Electromechanical coupling coefficient k_{33}	0.65	0.50
Piezoelectric coefficient d_{33} [C/N]	$550\pm50 \cdot 10^{-12}$	$120\pm50 \cdot 10^{-12}$

Table 8. Basic properties of $BaTiO_3$ and $PbZrTiO_3$ samples.

In order to evaluate feasibility of producing piezoelectric ceramics by conventional metallurgy method and CIM, we compared results for CIM $PbZrTiO_3$ samples obtained from our study (Table 9) with those found in literature (Gu et al., 2008). Table 9 lists dielectric constants, loss tangents, electromechanical coupling coefficients and piezoelectric coefficients for samples obtained by those two methods.

The dielectric constant is a measure of charge stored on an electrode material brought to a given voltage. It strongly depends on sintering temperature for both CIM and conventional metallurgy method. When conventional method is concerned, dielectric constant increases with sintering temperature due to the increase in $PbZrTiO_3$ grains. For CIM samples, dielectric constant increases with sintering temperature to 1250°C and then decreases with sintering temperature because of the deceased density due to lead loss. Therefore observed difference in dielectric constants obtained by two methods can be attributed to microstructure development and change in grain size with variation of sintering temperature.

	CIM method	Conventional powder metallurgy method (Gu et al., 2008)
Dielectric constant $\varepsilon_{33}/\varepsilon_0$	1700±30	1210
Loss tangent $tg\delta_{max}$	0.02	1.0
Electromechanical coupling coefficient k_{33}	0.50	0.27
Piezoelectric coefficient d_{33} [10^{-12}C/N]	120±50	97

Table 9. Comparative properties of $PbZrTiO_3$ samples.

Electromechanical coupling coefficient and piezoelectric coefficient are slightly lower than those from reference, while the loss tangent is higher for samples realized using CIM method than for samples obtained by conventional metallurgy method. They strongly depend on processing parameters, especially sintering temperatures, and by further adjustment of various processing parameters, CIM technology can provide $PbZrTiO_3$ components with application-specific properties similar to those provided by conventionally produced components.

Obtained results have shown that piezoelectric ceramics can be successfully produced by CIM method. Sintering temperature was found to play important role in physical, mechanical and electrical properties since it affects sample density and porosity. The obtained sample properties were comparable to those found in literature. It is important to note that in comparison with conventional powder metallurgy, CIM samples have more homogenous microstructure and production costs are reduced by reducing machining and recycling use of feedstock. Further research should be focused on processing conditions and their influence on the properties of final sintered parts, assuring satisfactory low-cost alternative for the production of piezoelectric ceramics with application specific properties.

4. Alumina for CIM

Aluminium oxide (Al_2O_3) is ceramics with high mechanical hardness, high electrical resistivity and thermal conductivity. It has good strength and stiffness, good wear and corrosion resistance, good thermal stability, low dielectric constant and loss tangent, low thermal expansion, low weight, etc. It is suitable for technical ceramic, electronic and medical products, etc. CIM alumina exhibits properties close to pressed and sintered samples (Hwang & Hsieh, 2005; Hausnerova et al., 2011; Krauss et al., 2005). The most common material that is being used for feedstock preparation is Al_2O_3 powder with 99.7% purity (Wei et al., 2000). Properties and scanning electron micrograph of 99.7% alumina powder are given in Table 10 and Fig. 12.

Multicomponent binder commonly used in feedstock preparation is 30wt% polypropilene, 65wt% paraffin wax and 5wt% stearic acid. After injection molding procedure, samples are being subjected to a debinding process (Table 11). After the debinding procedure, all samples should be inspected to ensure that all surfaces are free from visual defect. CIM alumina samples are then sintered in air at temperatures >1550°C.

Material	Al_2O_3 99.7%		
Typical composition			
Na_2O (0.05%)	SiO_2 (0.05%)	CaO (0.02%)	Fe_2O_3 (0.02%)
Typical properties of sintered parts			
Particle size, d_{50}	0.4-0.6 μm		
Theoretical density	3.85 g/cm³		
Density	96.7 %		
Purity	99.7 %		
Specific surface area	9.0 m²/g		

Table 10. Properties of Al_2O_3 powder.

Fig. 12. Scanning electron micrograph of alumina powder.

Immersion:	heptane, 3h at 60°C	
Thermal debinding:		
Ramping rate (°C/min)	Isothermal temperature (°C)	holding time (h)
2	200	0.3
2	250	2.0
5	450	0
10	1000	0.5
Cooling	-	-

Table 11. Typical debinding procedure for CIM alumina samples.

CIM alumina is the most widely used injection molded ceramic material. CIM alumina components (Fig. 13) have high surface finish quality even with extremely complex geometries. They have high hardness end mechanical strength, high wear and corrosion stability and good electrical insulation. CIM alumina components are also dimensionally stable and able to withstand high working temperatures. Since they combine good mechanical properties with low specific weight, CIM alumina components are being used in engineering (sensor covers, sensor tubes, micro electrodes for ultrasonic welding, etc.), textile industry (textile thread guides, wire guides, etc.), medical and dental applications (orthodontic brackets, dental implants, prosthetic replacements, etc.), watches (precision watch gears), metallurgy (ceramic casting cores), automotive components (valve components), electrical components (microwave dielectric components), office equipment (inkjet printheads), etc. Recent research activities proved that CIM alumina has a great potential because nowadays it is a commonly used material in both micro-CIM and 2C-CIM technology - advanced CIM technologies.

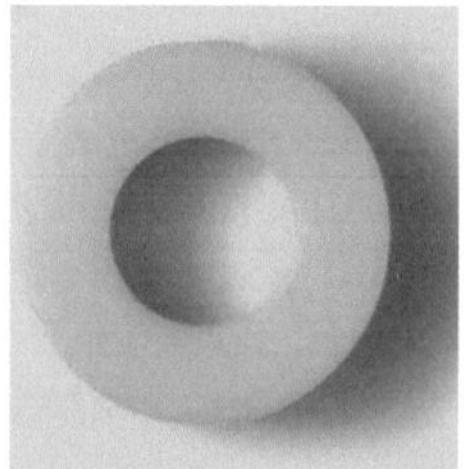

Fig. 13. CIM Al_2O_3 sample.

5. Advanced CIM technologies

Mass produced micro-parts are mainly being produced from ceramic materials which are readily available in submicron sizes because fine ceramic powders are easier to handle in comparison with metallic materials which are often pyrophoric in submicron sizes and for that reason difficult to handle. **Micro-CIM**, as an expanding technology for mass-production of micro-parts, emerged as a combination of plastic micro-injection molding technology and ceramic injection molding technology. It shares the same basic steps as the conventional CIM-technology, but it also exhibits special characteristics due to micro-size of the components (Liu et al., 2011; Piotter et al., 2003, 2010; Zauner, 2006). Micro-CIM parts can be formed using variety of ceramic materials such as ZrO_2, Al_2O_3, Si_3N_4, AlN and PZT and their main application fields are microsystem technology, microfluidics, biosensors, MEMS, medical technology, etc. (Fig. 14).

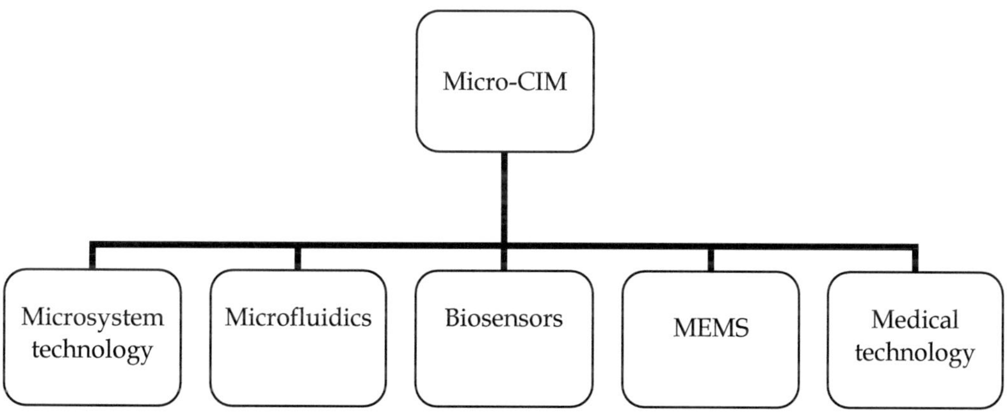

Fig. 14. Micro-CIM applications.

The increasing expansion of microsystem technology (MST) induced a great demand for the production of high-quality low-cost 3D micro-sized components such as micro-sensors, micro-reactors and micro-parts. The current microsystem production technologies (micro-cutting, laser ablation, LIGA, etc.) due to their high cost, low efficiency and limited materials are being replaced by micro-CIM technology that, as a miniaturized variant of CIM-technology, offers greater shape complexities, applicability to a wide range of materials and good mechanical properties. Micromechanical components made by micro-CIM are used to replace plastic parts and they especially benefit from ceramic material properties like corrosion resistance and high-temperature performance. Microfluidics and microreaction technology, biomedical industry and other growing markets give excellent opportunities for microparts. For some applications, such as reactions of highly corrosive media or high temperature gas phase reactions, micro-CIM components are of greatest interest due to their hardness and high chemical and thermal resistance. Also, there is a strong request for biocompatible materials such as ceramics and reliable technologies to produce complex shaped medical components.

The raw materials for micro-CIM technology are fine ceramic powders that allow production of micro-components with feature sizes down to 5μm. The powder has to be homogeneous and in order to obtain a fairly isotropic behavior the grain size of the sintered part should be at least one order of magnitude smaller than the minimum internal dimensions of the micro-part. From the aspect of surface quality, the best results can be achieved by using ceramic powders with mean particle diameter of 0.5μm or smaller. The viscosity of the melt should be sufficiently low to fill even the smallest structural details down to submicron range. For that reason, the molding tool should be heated near the melting point of the feedstock prior to injection into the tool. Because of the micro-part fragility highly precise tool movements are required. In order to control acceleration or slowing down of the injection molding process, ramps are being used. Micro injection molding machines use position regulated screws for that purpose. Also, micro components are considerably more difficult to handle from macroscopic components. They tend to stick to handling systems instead of dropping when electrostatic forces exceed gravitation force.

As an example of micro-CIM component, schematic presentation of zirconia (ZrO_2) micro gearwheel is shown in Fig. 15. It is a typical example of micro component for micromechanical applications. Micro gearwheels are demanding microstructures. Successful replication of structural details requires establishment of critical dimensions and determination of various physical properties such as densities, surface qualities, etc. However, geometry of the part is not a key factor when performances of the component are in question. The key factor is the surface quality of micro-CIM component. For production of ZrO_2 micro gearwheels high quality mold inserts are required. Cavities have to be scaled up by the certain percentage because of the shrinkage and have to be micro milled employing smallest mill cutters. Minimum edge radius within the mold and cutting depth affect the tooth shape. Besides restrictions related to manufacturing adequate mold inserts for micro gearwheel realization, there are also restrictions of molding as well as restrictions of sintering. Beside typical limits for ejection molding (aspect ratio), there are limitations related to design of the gate system. Demolding is also a challenge because the ejector pins have to be arranged very accurately due to the lack of space. When sintering process is in question, the shrinking, temperature and other process variables must be particularly taken into account. There are different temperatures in different places in the oven resulting in variable shrinking factors and therefore different sizes of micro gearwheels. Required tolerances for these components are in the same range as measurement accuracy (~1μm). This is of special interest when production of micro planetary gears is in question because it requires highly accurate micro assembly and micro measurements.

Fig. 15. Schematic of a ZrO_2 micro gearwheel (outer diameter 1200 μm).

Besides small dimensions, micro-components are often required to have several functions at the same time. For that reason two-component CIM technology (**2C-CIM**) can be successfully applied in production of advanced ceramic micro-components (Piotter et al., 2008; Yin et al., 2008). This technology uses different combinations of ceramic materials thus providing the major advantage of the 2C-CIM process: any assembly of micro-CIM parts after sintering is not necessary. The sintering rate control is crucial part of 2C-CIM process. Both components must sinter at similar rates and similar positions in the sintering temperature to avoid delaminating. The sintering behavior can be adjusted by lowering the powder content in one mix thus affecting its green density and risking high porosity of the component. Also, a non-sintering composite phase can be added. When 2C-CIM process is in question, chosen materials must have compatible sintering characteristics. The feedstock components can be injected sequentially or simultaneously depending on the design and size of the contact area of both ceramic materials. In order to secure high compound strength, it is essential that injection processes are performed in a quick succession. Feedstocks that are being used must have comparable shrinkage rates during co-debinding and co-sintering. The shrinking behavior depends on the type of ceramic powder used and its particle size distribution as well as on the solid content of the feedstock. If the onsets of shrinkage of both components are comparable and thermal expansion coefficients are almost the same, then the defect-free material compounds can be achieved and high cooling stresses can be avoided.

As an illustration of 2C-CIM process, a schematic presentation of two-component shaft-gear wheel combination with fixed shaft-to-collar connection is given in Fig. 16. As a suitable material for gear wheel, zirconia is selected because of its sufficiently high toughness. The shaft is composed of alumina because of its hardness. Each of two chosen materials has a characteristic behavior during processing. Zirconia has a thermal expansion coefficient of $10\text{-}11 \times 10^{-6}/K$ and therefore will show greater thermal expansion than alumina with a thermal expansion coefficient $7\text{-}8 \times 10^{-6}/K$. The shrinkage behavior of the two components

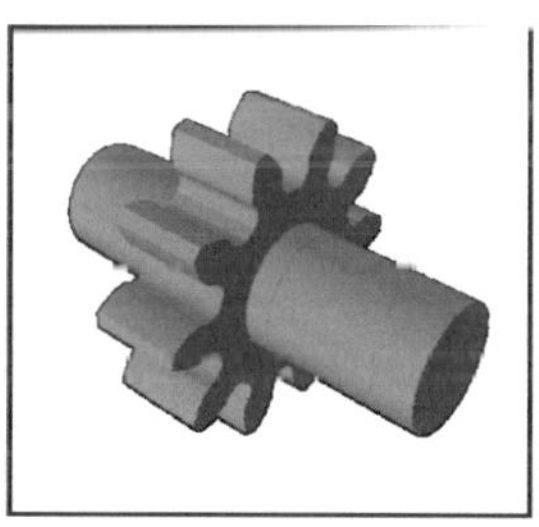

Fig. 16. Schematic of a 2C-CIM shaft-gear wheel combination (shaft: Al_2O_3; gear wheel: ZrO_2).

strongly depends on properties of selected zirconia and alumina powders. Since their sintering properties are affected by particle size, specific surface and additives, the shrinkage of both components should be adapted in such a way that dynamics of densification provide a shrink fit without cracking or destruction of the assembly. A different adaptation of parameters may lead to loosening of both components resulting in a formation of movable joint. Movable connections open new possibilities for 2C-CIM with a great reduction in mounting and assembly efforts and costs.

At the moment major growth areas in CIM are micro-CIM and 2C-CIM technologies (Fig. 17). Combination of the physical properties of the various materials is the key research activity when 2C-CIM technology is in question. Micro-CIM technology faces the challenge of developing suitable feedstock and high precision miniaturized mold cavities. Also, both technologies require computer modeling of molding, debinding and sintering in order to predict shape, size and reliability issues. Additionally, there are research activities ongoing to include higher functionality into micro-CIM components e.g. by the use of nano-sized powders.

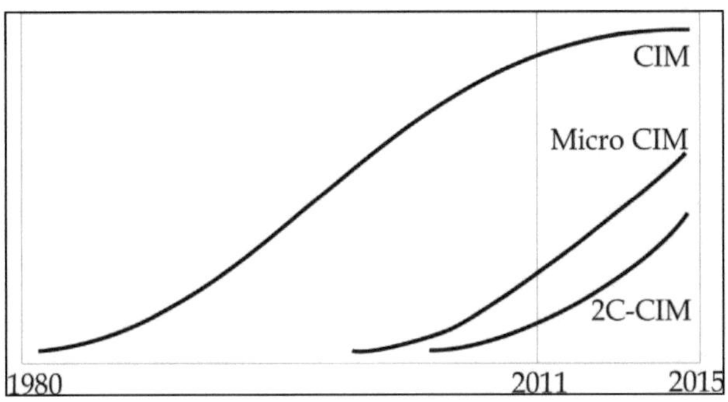

Fig. 17. Ceramic injection molding technology development.

6. Conclusion

In the recent years CIM technology has been accepted as one of the fundamental manufacturing techniques for mass production of ceramic components with complex geometries. Most commonly used CIM components are based on ferrite, piezoelectric and alumina ceramic powders. CIM components are widely used in automotive industry (mechanical parts), chemical applications (valves, membranes), medical applications (artificial bones), aerospace (mechanical parts, sensors and actuators), communications, oil and gas exploration (sensors, valves), etc. Advanced CIM technology, micro-CIM, offers microparts with structures in sub-millimetre range where materials like silicon and polymers rapidly reach their limits. Next step in micro-CIM technology is two-component micro-CIM that allows the production of multi-functional ceramic parts in just one processing step without any additional joining of the compound partners. Two-component injection molding allows one to combine materials with different properties, such as electrical conductivity/electrical insulation, high thermal conductivity/thermal insulation, or magnetic/non-magnetic properties. This opens new prospects for CIM technology and leads the way to new prospective market segments.

7. Acknowledgment

Authors are grateful for the partial support of the Ministry of Education and Science of Republic of Serbia (contracts III45007 and III44003).

8. References

Gu, Y.W.; Li, T.; Li, Q.F.; Pook, S.F. Pook, & Goh, C.W. (2008). Piezoelectric ceramics by powder processing. *SIMTech technical reports*, Volume 9, Number 4, (Oct-Dec 2008), pp. 189-194, Available from: http://www.simtech.astar.edu.sg/Research/TechnicalReports/STR_V9_N4_CD_Version/STR_V9_N4_02_FTG.pdf

Hausnerova, B; Marcanikova, L.; Filip, P. & Saha, P. (2011). Rheological Characterization of Powder Injection Moulding using Feedstock Based on Aluminium Oxide and Multicomponent Water-Soluble Polymer Binder. *Proceedings of Recent Advances in Fluid Mechanics and Heat & Mass Transfer*, pp. 245-250, ISBN: 978-1-61804-026-8, Florence, Italy, August 23-25, 2011

Hwang, K. S. & Hsieh, C. C. (2005). Injection-Molded Alumina Prepared with Mg-Containing Binders. *Journal of the American Ceramic Society*, 88 (9), (September 2005), pp. 2349–2353, DOI: 10.1111/j.1551-2916.2005.00370.x

Krauss, V.A.; Pires, E.N.; Klein, A.N. & Fredel, M.C. (2005). Rheological Properties of Alumina Injection Feedstocks. *Materials Research*, Vol.8, No.2, (April-June 2005), pp. 187-189, ISSN 1516-1439

Liu, L; Loh, N.H.; Tay, B.Y.; Tor, S.B.; Yin, H.Q. & Qu, X.H. (2011). Preparation and characterization of micro components fabricated by micro powder injection molding. *Materials Characterization*, 62 (6), (June 2011), pp. 615-620, DOI:10.1016/j.matchar.2011.04.009

Luo, J.S.; Yi, Z.Z.; Xiao, B.; Gao, Y.; Xie, Z.P.; Li, J.B. & Huang, Y. (2006). Injection molding of ultra-fine zirconia (Y-TZP) powders. *Journal of Ceramic Processing Research*, 7(1), (2006), pp. 14-19, ISSN: 1229-9162

Pigram, A. J. & Freer, R. (1994). The production of Mn-Zn ferrite ceramics by injection moulding. *Journal of Materials Science*, Vol. 29, No. 24, (January 1994), pp. 6420-6426, DOI: 10.1007/BF00353998

Piotter, V.; Gietzelt, T. & Merz, L. (2003). Micro powder-injection moulding of metals and ceramics. *Sadhana*, Vol. 28, Parts 1 & 2, (February/April 2003), pp. 299–306, ISSN: 02562499

Piotter, V.; Plewa, K.; Prokop, J.; Ruh, A.; Ritzhaupt-Kleissl, H.J. & Hausselt, J. (2008). Manufacturing of Versatile Ceramic or Metal Micro Components by Powder Injection Moulding. *Proc.of the 4th Internat.Conf. on Multi-Material Micro Manufacture*, pp. 69-72, ISBN 978-1-904445-76-0, Cardiff, GB, September 9-11, 2008, Dunbeath : Whittles Publ., 2008

Piotter, V.; Mueller, T.; Plewa, K.; Prokop, J.; Ritzhaupt-Kleissl, H.J. & Hausselt J. (2010). Manufacturing of complex-shaped ceramic components by micropowder injection molding. *The International Journal of Advanced Manufacturing Technology*, 46 (1-4), (January 2010), pp. 131-134, DOI: 10.1007/s00170 009 2095-7

Rak, Z.S. (1999). New trends in powder injection moulding. *Powder Metallurgy and Metal Ceramics*, Volume 38, Numbers 3-4, (March 1999), pp. 126-132, DOI: 10.1007/BF02676037

Rodríguez-Senín, E.; Várez, A.; Levenfeld, B.; Torralba, J. M. & París,M. A. (2005). Processing of Mn–Zn ferrites using mould casting with acrylic thermosetting binder. *Powder Metallurgy*, 48(3), (September 2005), pp. 249-253, DOI:10.1179/174329005X64117

Skolyszewska, B.; Tokarz, W.; Przybylski, K. & Kakol, Z. (2003). Preparation and magnetic properties of MgZn and MnZn ferrites. *Physica C*, 387(1-2), (May 2003), pp. 290-294, DOI: 10.1016/S0921-4534(03)00696-8

Stanimirović, I & Stanimirović, Z. (2010). Piezoelectric Ceramic by Powder Injection Molding. *Proceedings of 27th International Conference on Microelectronics MIEL 2010*, ISBN: 978-1-4244-7198-0, Niš, Republic of Serbia, May 16-19, 2010, IEEE

Stanimirović, Z & Stanimirović, I. (2010). Injection Molded Mn-Zn Ferrite Ceramics. *Proceedings of 27th International Conference on Microelectronics MIEL 2010*, ISBN: 978-1-4244-7198-0, Niš, Republic of Serbia, May 16-19, 2010, IEEE

Wang, S.; Li, J. F.; Wakabayashi, K.; Esashi, M. & Watanabe, R. (1999). Lost Silicon Mold Process for PZT Microstructures. *Adv. Mater.*, 11(10), (July 1999), pp. 873-876, DOI: 10.1002/(SICI)15214095(199907) 11:10<873::AID-ADMA873>3.0.CO;2-F

Wei, W.C.J.; Wu, R.Y. & Ho, S.J. (2000). Effects of pressure parameters on alumina made by powder injection moulding. *Journal of the European Ceramic Society*, Volume 20, Issue 9, (August 2000), pp. 1301-1310, DOI: 10.1016/S0955-2219(99)00295-2

Yin, HQ.; Jia, CC. & Qu, XH. (2008). Micro powder injection molding—large scale production technology for micro-sized components. *Science in China Series E: Technological Sci.*, 51 (2), (Feb. 2008), pp. 121-126, DOI: 10.1007/s11431-008-0023-y

Zauner, R. (2006). Micro powder injection moulding. *Microelectronic Engineering*, Volume 83, Issues 4-9, (April-September 2006), pp. 1442-1444, DOI:10.1016/j.mee.2006.01.170

Zlatkov, B. S.; Griesmayer, E.; Loibl, H.; Aleksić, O. S.; Danninger, H.; Gierl, C. & Lukić, L. S. (2008). Recent Advances in CIM Technology. *Science of Sintering*, 40(2), (2008), pp. 185-195, DOI:10.2298/SOS0802185Z

Zlatkov, B. S.; Mitrović, N.S.; Nikolić, M.V.; Marićić, A.M.; Danninger, H.; Aleksić, O. S. & Halwax, E. (2010). Properties of MnZn ferrites prepared by powder injection molding technology. *Materials Science and Engineering B*, 175(3), (December 2010), pp. 217-222, DOI:10.1016/ j.mseb.2010.07.031

7

Optimization and Simulation for Ceramic Injection Mould of ZrO_2 Fiber Ferrule

Bin Lin, Meiming Zhang, Chuhan Wu and Feng Liu
Tianjin University,
China

1. Introduction

Fiber ferrule is a crucial part for manufacturing fiber connectors. It is fairly difficult to produce fiber ferrule because that it requires high dimension accuracy. Currently, YTZ ceramic powder is the main material used to produce fiber ferrule and Ceramic Injection Molding (CIM) is a new fabricating method capable of producing ZrO_2 fiber ferrule (Fig. 1(a), Fig. 1(b)) with complex geometry and high accuracy. ZrO_2 fiber ferrule quality is significantly influenced by the process conditions of CIM. Therefore, the main focus of this paper is to optimize mould structure and processing parameters based on the simulation of CIM, which promotes solid load of ceramic powder and product quality. Optimal process conditions of Ceramic Injection Molding could be determined by analyzing the simulation results. It has been found that runner cross-section shape and runner system contribute to the efficiency and filling process significantly. Hence, optimal runner cross-section shape and runner system are proposed. Reducing the gravity influence on CIM is also suggested. Moreover, optimization of cooling system could be considered an effective way to improve the dimensional precision and surface quality of ZrO_2 fiber ferrule.

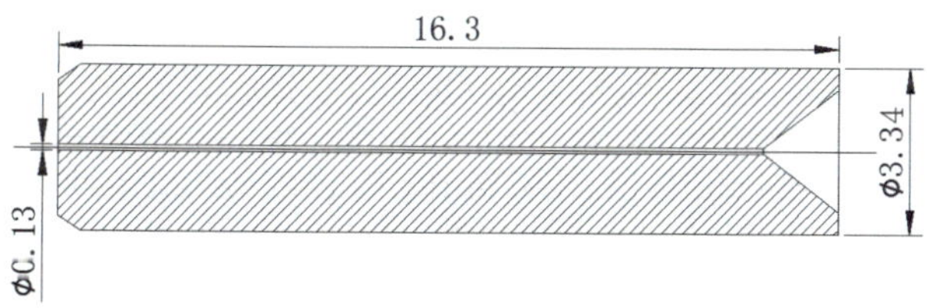

(a)

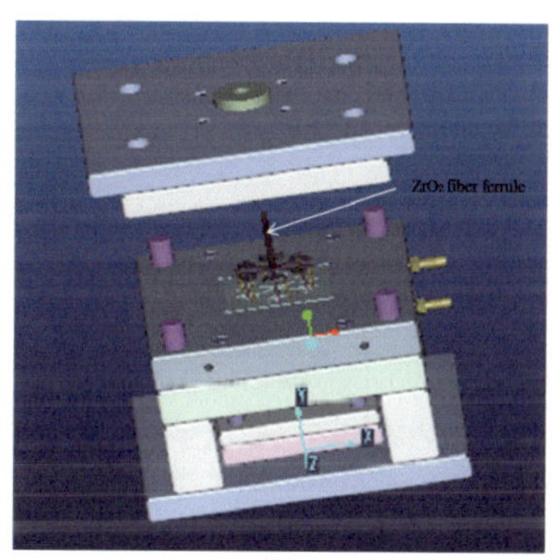

(b)

Fig. 1. ZrO_2 fiber ferrule; (a) Roughcast of the ZrO_2 fiber ferrule; (b) Geometry of the mould.

2. Optimization of runner cross-section shape

During filling stage, the melt is firstly injected into mould cavity and this stage is accomplished as the mould cavity is fully filled by melt. Therefore, it is of great importance to control melt flow pattern within mould cavity, which could promote the solid load of ceramic powder. Cooling time and many product defects like cavitations, weld line, short shot and product deformation are related to the melt flow pattern. The melt flow pattern is influenced by many factors like the dimension of runner cross-section shape and runner system arrangement. Hence, optimized runner cross-section shape and well-designed runner system would be beneficial to the Ceramic Injection Molding. In order to investigate the melt flow pattern, pressure changes, temperature, cavitations etc., we use Moldflow Plastic Insight (MPI) to simulate the melt flow pattern within the mould cavity. Also, potential defects would be predicted during this simulation. The optimal runner cross-section shape could be determined though analyzing simulation results of different runner cross-section shapes.

Runners should ensure that the melt ejected from injection machine can smoothly flow through runners and fully fill mould cavity. What is more, runners should adequately transfer pressure to all the positions of mould cavity to obtain high quality products during filling process.

Common runner cross-section shapes (Fig. 2) are circular, ladder, U-shape (combination of circular and ladder), semicircular and rectangular. It usually is recommended to use the first three runner cross-section shapes. Considering the ratio of volume to its surface area, circular cross-section shape is most suitable, with minimal pressure drop and heat loss. However, templates on both sides of circular runners need to be processed, which causes much higher cost. Furthermore, semicircles on both surfaces of those two templates of circular runners have to be aligned accurately.

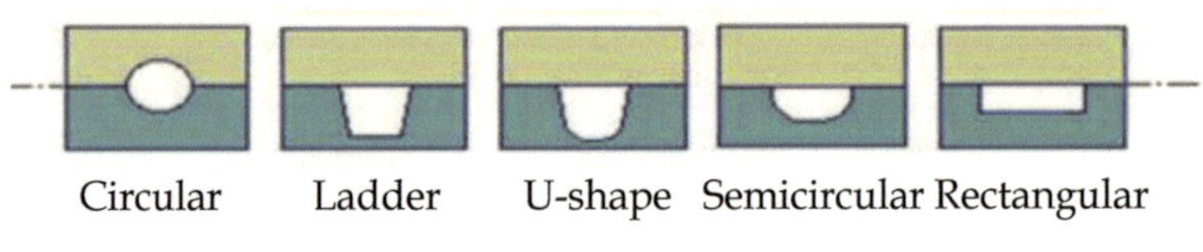

Fig. 2. Common runner cross-section shapes.

Ladder runner cross-section shape requires one processed template only, which still works well. Ladder runner is commonly used in three-plate mould. Circular runner cross-section shape is rarely adopted in three-plate mould for it may be difficult to demould and cause interference between runner and sliding part of templates. Different runner cross-section shapes could be compared by hydraulic diameter (Table 1) which is the index of flow resistance. The larger the hydraulic diameter is, the lower the flow resistance is. The definition of hydraulic diameter is described as equation 1.

$$D_h = \frac{4A}{P} \tag{1}$$

Where D_h is the hydraulic diameter, A is the section area and P is the perimeter.

Equivalent hydraulic diameters of various runner cross-section shapes are compared in Table 2.

Cross-section	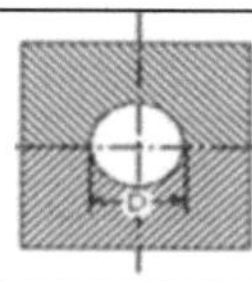	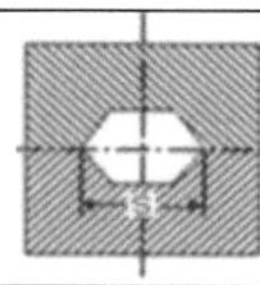	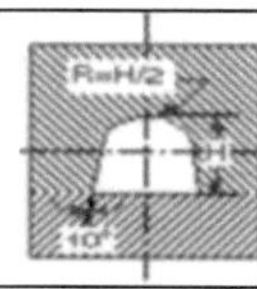	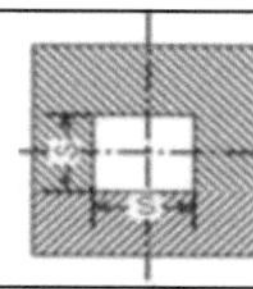
D_h	D	$0.9523D$	$0.9116D$	$0.8862D$
Cross-section	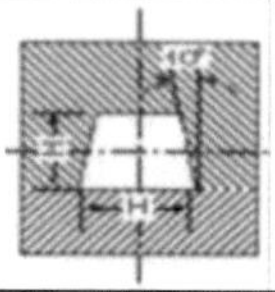	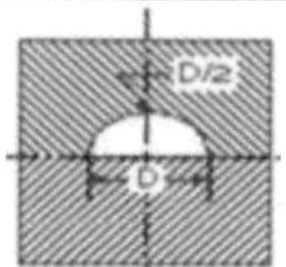	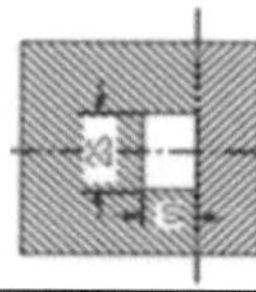	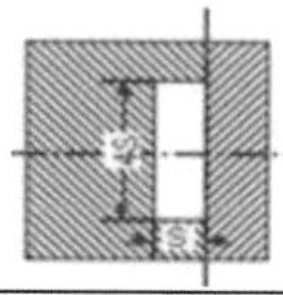
D_h	$0.8711D$	$0.8642D$	$0.8536D$	$0.7090D$

Table 1. Equivalent hydraulic diameters of various runner cross-section shapes.

	Cross-section Shapes	Equivalent Diameters
A	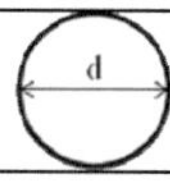	d
B	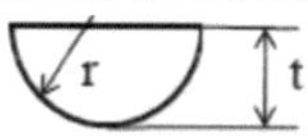	$2\sqrt{\frac{r^2\cos^{-1}x - r(r-t)\sin(\cos^{-1}x)}{\pi}}$ $x=\frac{r-t}{r}$
C	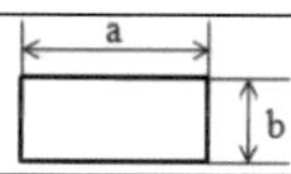	$2\sqrt{\frac{ab}{\pi}}$
D	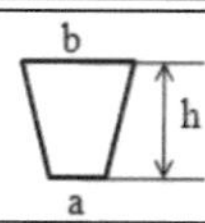	$\sqrt{\frac{2(a+b)h}{\pi}}$
E	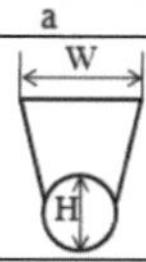	$\sqrt{\frac{2H}{\pi}}\sqrt{(W+H\arctan(\frac{H}{W}))}$

Table 2. Comparison between equivalent diameters of various runner cross-section shapes.

2.1 Analytical solutions of various runner cross-section shapes

The main factors influencing melt flow are injection pressure, melt temperature and viscosity, mould temperature, runner cross-section area and geometric cross-section shape. The geometric cross-section shape determines runner efficiency.

It requires the following conditions to improve runner efficiency (Pan et al., 1995). Runner resistance to the melt flow, which is normally caused by the friction between melt and inner runner surface, should be as low as possible to ensure that the melt can fully fill the mould cavity before solidifying.

The heat loss should be minimal as the melt flows through runner. When the melt with high temperature flows through runner with comparatively lower temperature, melt would transfer its heat to the mould, which increases mould temperature. On the other hand, the melt temperature decreases. Also, melt viscosity increases when melt temperature becomes lower, which makes melt fluidity worse. When the temperature decreases to a certain degree, temperature distribution within inner side of melt becomes significantly inhomogeneous that causes many defects in the products. In this case, we can improve the injection pressure moderately. However, it would affect clamp if injection pressure becomes too high.

For improving runner efficiency, many measurements could be adopted during designing and manufacturing stages. Firstly, the area of runner cross-section shape can be increased, which decreases the resistance to melt flow. However, it would waste energy and raw materials if the area of runner cross-section shape becomes oversize. Secondly, the contact area between runner and melt could be diminished by decreasing periphery length of runner cross-section shape. Thirdly, runner layout should be simple and its length should be minimal. Finally, the runner surface roughness can be decreased. Normally, R_a is between 0.8 and $1.8\mu m$. Also, mould temperature needs to be controlled within certain range.

Projected area of runner on the parting surface should be minimal. The injection area actually is decreased if we decrease projected area, which diminishes the opening force of mould. In this way, we can adequately use the clamp force of injection machine to clamp the mould.

The smaller the runner volume is the better. This would improve utilization ratio of raw materials and save energy. The runner volume with fixed length would become smaller as the runner cross-section shape area decreases. Hence, the runner cross-section shape area should not be oversize.

Runners with advantages mentioned above can be considered to have high efficiency. And runner efficiency is expressed by equation 2.

$$\eta = \frac{S}{L \bullet l} \tag{2}$$

where η is runner efficiency, S is runner cross-section shape area, L is peripheral length of runner cross-section shape, l is runner length and $L \bullet l$ is runner lateral area.

Equation 2 illustrates that runner efficiency is equal to the ratio of runner cross-section shape area to its lateral area. Also, the runner lateral area is equal to the peripheral length of runner cross-section shape multiplied by runner length. Therefore, increasing the runner cross-section shape area, decreasing the peripheral length of runner cross-section shape and reducing the runner length all can improve runner efficiency. In this paper, runner efficiency refers to the runner efficiency when runner length is unit length ($l = 1$). Its value is equal to the ratio of runner cross-section shape area to its peripheral length (equation 3).

$$\eta = \frac{S}{L} \tag{3}$$

Runner efficiency is influenced by its cross-section shape area and peripheral length. Therefore, runner efficiency is related to its geometric parameters of cross-section shape.

We compare various runners under the same conditions (Table 3). In the Table 3, η is runner efficiency and D is runner width. Cross-section shape areas of different runners are the same ($s = 1mm^2$) when calculating η and D. V is the runner volume when runner length is unit length and efficiency is 1 ($\eta = 1$). According to Table 3, the width of rectangular runner is minimal. Dimension of runner projection on the parting surface would be decreased if we reduce the width, which improves the mode-locking of the injection machine. This is the main advantage of rectangular runner. However, it is rather difficult to demould for rectangular runner without demoulding inclination. Therefore, we usually use gate with rectangular cross-section in practice.

Runners	Circular runner	Semicircular runner	Rectangular runner	Ladder runner	U-shape runner
Efficiency of the runner η	0.282	0.244	0.250	0.250	0.262
Width of the runner D	1.13	1.60	1.00	1.07	1.12
Volume of the runner with unit length V	12.57	16.83	16.00	16.04	14.61

Table 3. Comparison between the analytical solutions of various runners.

The efficiency of circular runner is highest and its runner volume to its unit length is minimal. Decreasing runner volume can promote the utilization of raw materials and save energy. However, if the circular runner is used in the cold-runner mould, circular runner needs to be divided into two semicircular runners on the parting surfaces of cover half and moving half respectively. These two semicircular runners have to be exactly the same. Therefore, it would be very difficult to manufacture mould with circular runners. Usually, the circular runners are used in the hot-runner mould rather than in the cool-runner mould.

Apparently, with the same cross-section shape area, efficiency of U-shape runner is highest compared with that of semicircular and ladder runners. The width of U-shape runner is smaller than that of circular runner. With the same efficiency and length (unit length), the volume of U-shape is minimal among the semicircular, ladder and U-shape runners and next to that of circular runner. Therefore, U-shape is most suitable in the cool-runner mould. Semicircular runner is not the perfect approach for its minimal efficiency and largest width and volume.

2.2 Modeling and mesh generation

Table 4 shows the parameters of various cross-section shapes with the same area ($S = 4\pi$).

Runners	Cross-section shapes	Parameters	Values (mm)
Circular	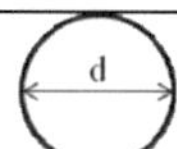	Diameter D ($2r$)	4
semicircular	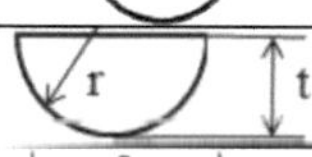	Diameter D ($2r$), height (t)	D=5.65; t=2.825
Rectangular	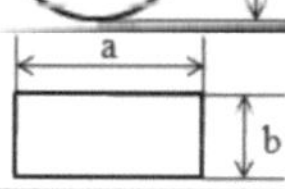	Width (a), height (b)	a=b=3.54
Ladder	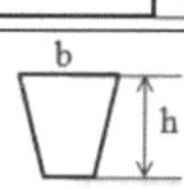	Upper line (b), lower line(a), height(h)	b=3.8; a=3.3; h=3.534
U-shape	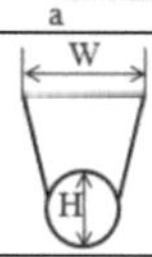	Width (W), height(H)	W=3.98; H=3.71

Table 4. Parameters of various cross-section shapes.

2.3 Feedstock rheology

Compositions of ZrO_2 feedstock are filler (solid paraffin), adhesive (vinyl acetate polymer) and lubricant (stearic acid). These compositions are shown as Table 5 (Wenjea et al., 1999) and the relationship between viscosity and shear rate is shown as Fig. 3.

	Zirconia (PSZ)	stearic acid (SA)	solid paraffin(PW)	Vinyl acetate polymer
Mass fraction (%)	86.7	0.9	7.0	5.4
Volume fraction (%)	50.0	3.3	28.0	18.7

Table 5. Compositions of ZrO_2 feedstock.

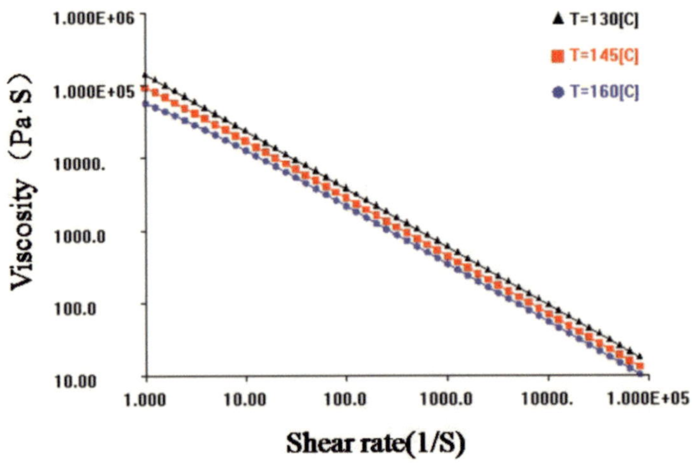

Fig. 3. Relationship between viscosity and shear rate.

2.4 Simulation outcomes

Pressure drop, filling time, temperature difference and clamp force determines the product quality and these are important parameters of injection machine. In this paper, we use the parameters mentioned above to discuss simulation outcomes of various cross-section shapes (Fig. 4, Fig. 5, Fig. 6, Fig. 7).

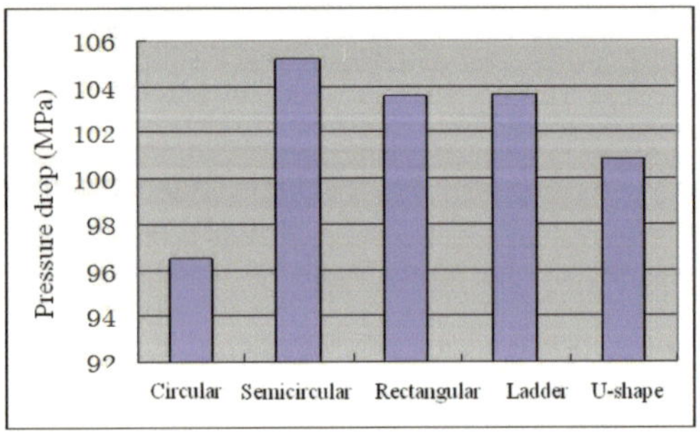

Fig. 4. Pressure drop simulation.

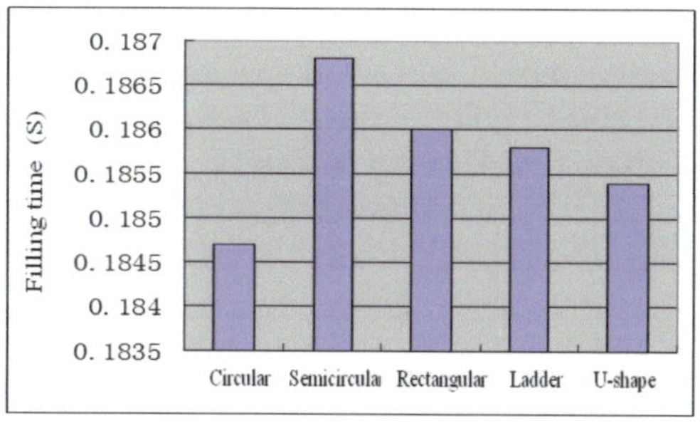

Fig. 5. Filling time simulation.

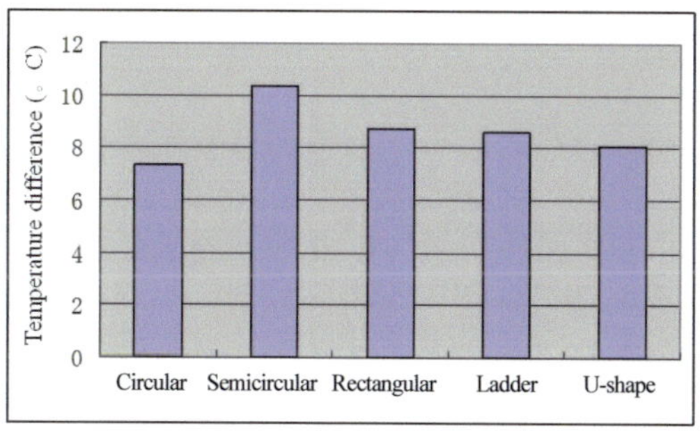

Fig. 6. Temperature difference simulation.

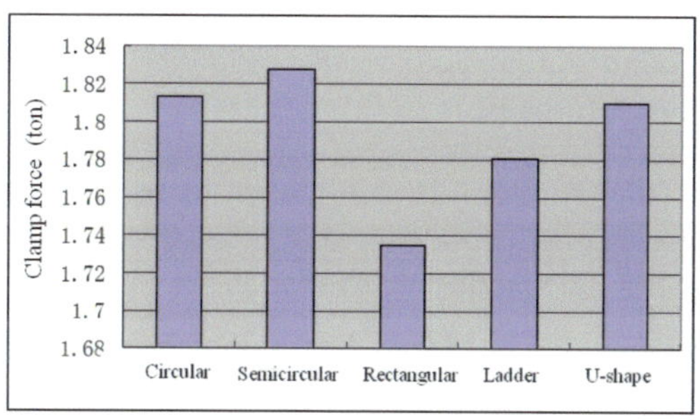

Fig. 7. Clamp force simulation.

During these simulations, injection temperature is 140° C and mould temperature is 60° C. Injection time, pressure and velocity are controlled by program. The simulation outcomes of various runner cross-section shapes are shown as Table 6.

Serial number	Cross-section shape	Pressure drop (MPa)	Filling time (s)	Temperature difference (°C)	Clamp force (ton)
A	Circular	96.492	0.1847	7.3	1.8128
B	Semicircular	105.177	0.1868	10.3	1.8277
C	Rectangular	103.587	0.1860	8.7	1.7344
D	Ladder	103.668	0.1858	8.6	1.7811
E	U-shape	100.792	0.1854	8.1	1.8101

Table 6. Analytical parameters of various runner cross-section shapes.

Injection pressure is an important, technical parameter of injection molding. It should not be too high or it may be hard to demould and cause raw edges on the product surface. What is more, the melt would not be able to fully fill mould cavity if the injection pressure is too high. Therefore, a suitable injection pressure is very necessary. Improving the injection pressure can improve the melt compression ratio and dimension accuracy (Liu et al., 2002). Runners with different cross-section shapes are compared under the same conditions (Table 7).

Runner	Circular	Semicircular	Rectangular	Ladder	U-shape
Cross-section shape	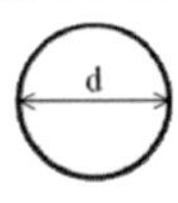	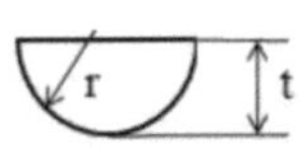	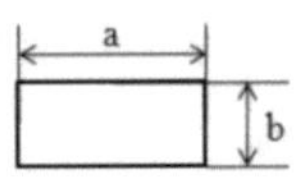	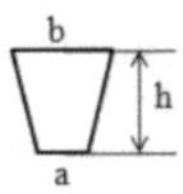	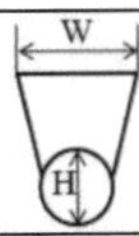
Pressure drop ranking	1	5	3	4	2
Filling time ranking	1	5	4	3	2
Temperature difference ranking	1	5	4	3	2
Clamp force ranking	4	5	1	2	3
Processability	Hard	Easy	Easy	Easy	Easy
Removing the cold material	Easy	Easy	Hard	Easy	Easy
Comprehensive evaluation	Suitable	Unsuitable	Rarely use in practice	Perfect	Most suitable for cold runner

Table 7. Comparison between runners.

According to Table 7, circular runner has minimal pressure drop, shortest filling time and smallest temperature difference, which means that it has the highest efficiency. On the other hand, rectangular runner has the smallest clamp force.

Compared with semicircular and ladder runners, U-shape runner has smallest pressure drop, shortest filling time, minimal temperature difference and highest efficiency. Therefore, runner with U-shape cross-section shape is the best choice.

3. Layout of runner system and optimization of technical parameters

During Ceramic Injection Molding stage, raw materials within staff canister firstly are heated to become melt which is driven quickly by the piston or screw into the closed cavity. Then, the melt within mould cavity compressed by mould and cools down to become product. The main concern of Ceramic Injection Molding is that the products should meet the quality requirement. Also, the solid load of ceramic powder, which is related to the product quality, should be as high as possible.

The conventional methodology optimizing process parameters requires experts use the trial and error method basing on their experience and professional knowledge. As the development of CAE technology, it becomes increasingly important to industry, especially to improve product quality and decrease cost etc. We combine CAE experiment and DOE methodology to get the best parameter combination based on the range analysis of orthogonal experiment. Also, the conclusion could be tested by the CAE comparison experiments.

3.1 Layout of runner system

The mould is composed of two templates with six cavities and common arrangement of multi-cavity system is radiated runner system. Melt can only be poured into the cavity through gates on both sides of mould for there is a fairly small hole in the axial center of products. Runner systems with rectangular and circular shunt are showed as Fig. 8 (Zhang, 2005, 2007).

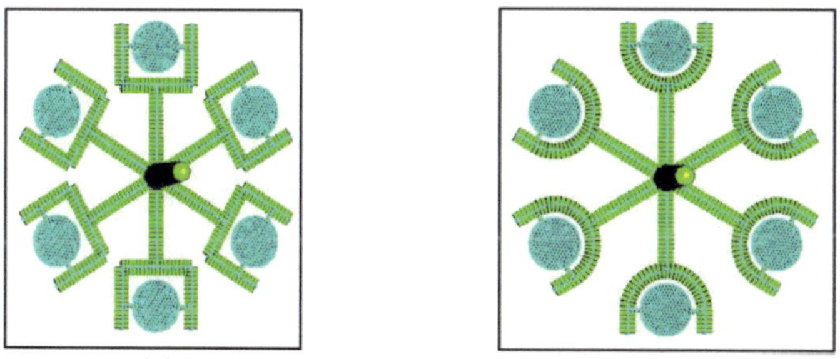

(a) (b)

Fig. 8. Two runner systems; (a) Runner system with rectangular shunt; (b) Runner system with circular shunt.

3.2 Orthogonal experiment methodology

Injection molding CAE technology uses finite element methodology, finite difference methodology and boundary element methodology to analyze the flow, dwell and cooling stage. It can calculate stress distribution within product and mould to predict product quality. Also, it can analyze the influences of process conditions, material parameters and mould structure on the products for the purpose of optimizing mould structure and process parameters.

Experiment design method (DOE) is mainly used to acquire the experimental data and analyze the experimental data and results scientifically. The main DOE application is the orthogonal experiment which designs the experiment based on data orthogonality (Yang et al., 2004). There are many distinct advantages of orthogonal experiment. Firstly, it can select a small number of experimental conditions, which are representative, from a large number of experimental conditions. Secondly, the best experimental conditions and manufacture process could be determined by analyzing experimental outcomes with those representative conditions. Finally, it would be much easier to process the data based on the orthogonal experiment.

Orthogonal table is the most important, basic tool and orthogonal experiments can easily calculate the effect of each condition on the results and display them by tables. Then, we can determine the best parameters after range analysis and comparison. All the calculations are done by tables and the whole processes are rather easy. Therefore, DOE is able to shorten the cycle of developing and designing new products, which is necessary to the manufacture and research. During injection molding, injection process parameters directly affect the product quality. Many researchers have designed experiments to research the relationship between them and got some useful conclusions (Skourlis et al., 1997; Jansen et al., 1998). However, conventional methodology requires a large number of experiments to research that relationship. Yet, research demonstrates that an economic method is to use the orthogonal experiment which in turn can instruct the injection molding process (Jin & Zhu., 2000).

3.3 Arrangement of orthogonal experiment

In this paper, we focus on these two runner systems mentioned above and use orthogonal experiment method to study the influence of two runner systems on products.

Mould temperature, injection temperature, screw velocity and gate dimension all are important parameters influencing product quality. Mould temperature A, injection temperature B, screw velocity C and gate dimension D all have three different values, namely A1, A2, A3, B1, B2, B3, C1,C2, C3, D1, D2, D3. We use L9 (3^4) orthogonal table to design experiments and use Moldflow software to investigate the influence of those parameters on product quality.

	factor	Level 1	Level 2	Level 3
A	Temperature of mould (°C)	35/40	45/50	60/65
B	Injection temperature(°C)	130/135	145/150	160/165
C	Screw velocity (%)	75	60	50
D	Width and length of the gates（mm）	1, 0.5	1.2, 0.6	1.5, 0.8

Table 8. Arrangement of factors and levels.

Finally, we can obtain the best parameter combination. Values of four parameters mentioned are shown as Table 8 and orthogonal table L9 is shown as Table 9. We consider the actual filling time, maximal injection pressure and temperature difference at the end of filling stage as the main parameters. By orthogonal experiment, we can know that when we just consider one condition with different values, this condition would have much more effect on the result if it has a larger range (Shen et al., 2001, 2002). In order to demonstrate the influence of each condition on the injection flow, we draw the relation graphs between them as Fig. 9, Fig. 10 and Fig. 11 (Zhang, 2005, 2007).

Experimental number	A	B	C	D
1	1	1	1	1
2	1	2	2	2
3	1	3	3	3
4	2	1	2	3
5	2	2	3	1
6	2	3	1	2
7	3	1	3	2
8	3	2	1	3
9	3	3	2	1

Table 9. Orthogonal table.

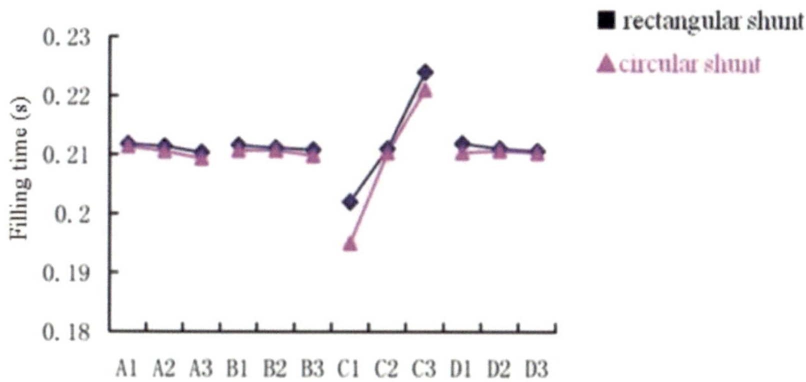

Fig. 9. Relationships between filling time and each factor.

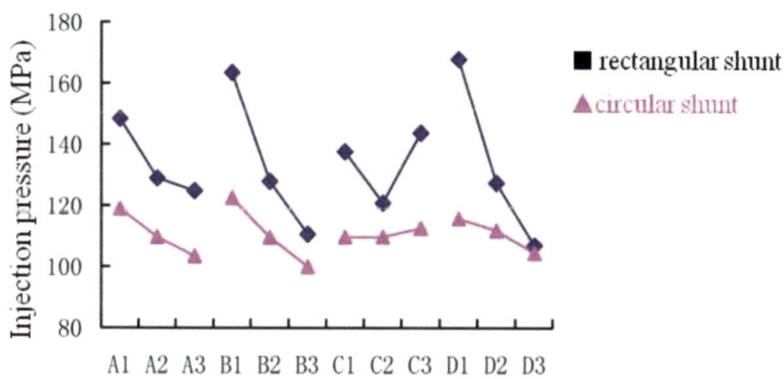

Fig. 10. Relationship between injection pressure and each factor.

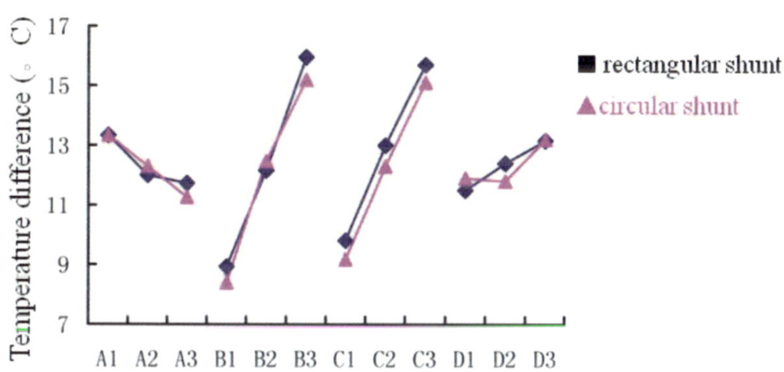

Fig. 11. Relationship between temperature difference and each factor.

3.4 Simulation results

Fig. 9 illustrates that screw velocity has largest influence on filling time. Therefore, increasing screw velocity can shorten filling time. Other factors influence filling time slightly.

Product quality could be improved by decreasing injection pressure. According to Fig. 10, we can see that injection temperature and gate dimension have largest effect on injection pressure. Therefore, improving injection temperature could decrease the injection pressure. Also, gates with too small dimension require comparatively large injection pressure.

Comparatively small temperature difference could be beneficial to the homogenous filling of powder and adhesive, which can prevent temperature gradient and density gradient caused by two-phase separation. Fig. 11 demonstrates that injection temperature and screw velocity have largest effect on temperature difference and next is the gate dimension and mould temperature. Hence, lower injection temperature and higher screw velocity contribute to decrease product surface temperature difference.

	Injection time (s)	Injection pressure(MPa)	Temperature difference (°C)
Rectangular shunt	0.195	108	9.1
Circular shunt	0.182	95.1	6.9

Table 10. Simulation outcomes of best parameter combination.

Considering all the factors, we can conclude that the best parameter combination of runner system with rectangular shunt is A2B2C2D3 and best parameter combination of runner system with circular shunt is A3B2C1D2. Simulation outcomes of two runner system are shown as Table 10. What is more, filling quality of runner system with circular shunt is much better than that of runner system with rectangular shunt and injection pressure the former runner system requires is 15MPa, less than that of the latter. Filling time and surface temperature difference of the former one are much smaller compared that of the latter one. Therefore, runner system with circular shunt is most suitable for ceramic injection molding.

4. Gravity influence on the ceramic injection mould

The melt fills five of six cavities well except the one on the top of mould where short shop happens. However, mould with six cavities is designed to have balance runner system, which means that the six cavities should all be filled well. Therefore, gravity should be taken into consideration for large runner length, zirconia density and viscosity.

In order to simplify the calculation and analysis, we select two cavities on the top and bottom parts of mould respectively as research objects. Fig. 12(a) demonstrates the simulation outcomes with conditions namely mould temperature (60° C), injection temperature (145° C), screw velocity (75%) and gate dimension (1.2mm and 0.6mm) when considering gravity influence. According to Fig. 12(a), we can see that cavities show difference in filling stage. Filling time of bottom die is much less than that of top die where the short shot happens. Also, bottom die quality is better than that of top die (Zhang, 2005, 2007).

4.1 Improvement

Two cavities on the top and bottom parts of mould show difference on the filling stage and short shot happens on the top cavity. Therefore, we increase runner diameter from 4mm to 4.17 mm. Fig. 12(b) illustrates that both the top and bottom dies with optimized balance runner system have same quality.

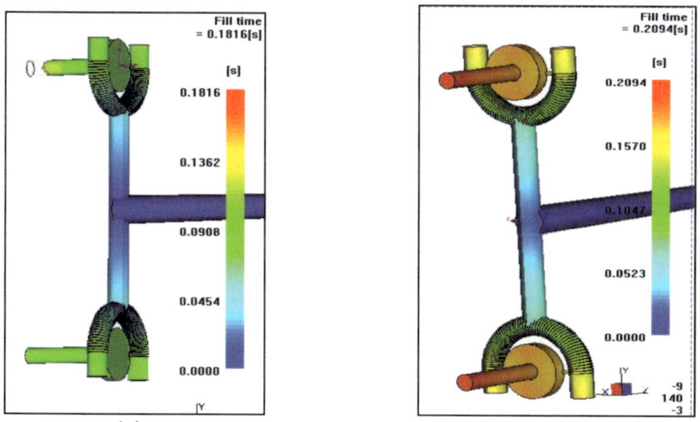

(a) (b)

Fig. 12. Simulation of gravity effect on ceramic injection molding; (a) runner diameter is 4mm; (b) runner diameter is 4.17mm.

5. Cooling simulations

Injection molding cooling refers to the stage after solidification to demould products from mould which occupies 3/4 of product cycle. Cavity temperature and uniformity directly influences product efficiency and quality. Injection molding temperature can be affected by various factors. Temperature control and regulation are mainly accomplished by cooling system. Cooling process parameters are composed of cooling pipeline dimension, connection and location etc. Physical parameters include cooling medium flow and gate temperature etc. The most important process parameter during cooling stage is cooling time and an efficient and balance cooling system could improve cooling efficiency and decrease residual stress. The purpose of cooling analysis is to determine cooling system though simulating the cooling process which predicts the surface temperature of mould and cooling time etc.

5.1 Summary of cooling simulation

The main stages of injection molding cycle are filling, dwell and cooling stages. The heat transfer process of injection molding shows that inner part of melt with high temperature transfers heat to the mould and the heat is taken by cooling medium. Therefore, balance cooling could prevent hot streak on product surface and decrease warpage and residual stress within product.

Injection molding cooling is mainly controlled and regulated by cooling system. The main purpose of cooling system is to cool the product fast and evenly. Cooling system parameters are composed of geometric and process parameters like cooling hole location, dimension, cooling medium flow and gate temperature. Cooling stage simulation could predict the cavity and core temperature, temperature difference distribution and cooling time with given parameters (Chen et al., 2002).

5.2 Establishing the mathematical model

5.2.1 Basic assumption and controlling equation

Physical process of cooling stage is fairly complex and we need simplify physical process before constructing controlling equation. Firstly, we assume that the mould work state is stable without considering periodic temperature changes of die well. Secondly, we assume that the heat flow only propagates along the normal direction of inner cavity surface. Thirdly, we assume that the product surfaces and die well have the same temperature and the product contacts cavity surface completely.

Based on the assumption mentioned above, we consider the injection molding cooling to be steady heat conduction without heat source. And controlling equation is equation 4(Li et al., 2001).

$$\frac{\partial^2 T}{\partial x^2}+\frac{\partial^2 T}{\partial y^2}+\frac{\partial^2 T}{\partial z^2}=0 \quad (x,y,z)\in V \tag{4}$$

Where V is the region enclosed by outer surface of mould, inner surface of cavity and surface of cooling gates.

5.2.2 Boundary conditions

Boundary condition on the cavity surface is equation 5 (Li et al., 2001).

$$-K_w\frac{\partial T}{\partial u}=\bar{q} \tag{5}$$

where u is out normal direction of cavity surface, K_w is thermal conductivity of mould and $\bar{q}$ is average heat flux which is defined by equation 6.

$$\bar{q}=\frac{1}{t_c+t_p}\left\{\int_0^{t_c} q_1(t)dt+\int_0^{t_c+t_p} q_2(t)dt\right\} \tag{6}$$

where t_c and t_p are cooling and demoulding time respectively, $q_1(t)$ and $q_2(t)$ are the instantaneous heat flux during cooling and demoulding stage. The cooling time t_c and $q_2(t)$ could be obtained by solving the one dimension transient heat conduction equation (equation 7).

$$\rho C_p\frac{\partial T}{\partial t}=\frac{\partial}{\partial s}(K_p\frac{\partial T}{\partial s}) \tag{7}$$

where t is the time, T is the melt temperature, ρ, K_p and C_p are the density, heat conductivity and equivalent specific heat respectively and s is local coordinate along the product thickness direction. When analyzing one dimension transient heat conduction of injection mould, we consider the injection temperature or melt temperature distribution at the end of filling stage to be the initial condition. Also, we select the cavity surface temperature as the boundary condition.

The boundary condition of gate surface is defined as equation 8 (Li et al., 2001).

$$-K_m \frac{\partial T}{\partial u} = h(T - T_c) \tag{8}$$

where u is outer normal direction of gate surface, T_c is cooling medium temperature, h is heat transfer coefficient between mould and coolant (equation 9).

$$h = 0.23 \frac{kc}{D} R_e^{0.8} P_r^{0.4} \tag{9}$$

Where $R_e = 4Q / \pi D\nu$ is Reynolds number, $P_r = \nu / a$ is Prandtl number, Q is the coolant volume, D is the cooling hole diameter, ν, a and kc are the kinematic viscosity, thermal diffusivity and heat conductivity respectively.

The outer surface heat exchange of mould normally does not have much effect on the temperature distribution of cavity surface, which means that it is unnecessary to calculate outer surface temperature distribution of mould. Therefore, we can consider the outer surface of mould to be an infinite, adiabatic sphere.

5.3 Cooling analysis and moldflow software application

Regulating and keeping the mould temperature could decrease product deformation and improve mechanical properties and dimension accuracy. Therefore, it is necessary to design the cooling system perfectly for injection molding. Researchers have done a lot of research related to the cooling system and got many simplified and empirical formula. MPI/Cool can analyze the effect of cooling system on the mould and optimize arrangement of cooling system.

5.3.1 Summary of MPI/cool software

Many factors affecting the injection molding cooling are product shape, cooling medium type, temperature, velocity, geometric parameters and arrangement of cooling pipe, mould material, melt temperature, ejected temperature, mould temperature and thermal cycling interaction between production and mould etc. These factors interact and relate with each other, which means that the best methodology is to combine these parameters. Yet, it only can be achieved by CAE analysis rather than by conventional simplified and empirical formula.

MPI/Cool software simulates this three dimension temperature field by boundary element method. Analytical solution could be used to calculate temperature field along the product thickness direction. What is more, MPI/Cool can obtain the interactive solution between mould temperature field and temperature field along the product thickness direction. Also, MPI/Cool can calculate the interface temperature between product and mould by the simultaneous energy equation of mould temperature field. Furthermore, we consider the influence of cavity and core asymmetry along the thickness direction on the product temperature distribution.

MPI/Cool can simulate the cooling pipe (separator pipe, jet pipe and connecting hose), insert, various mould materials, cool runner and hot runner, parting surface and product temperature. This can provide information for optimizing the cooling system.

MPI/Cool can not only analyze the neutral plane and fusion mould but also analyze 3D mould. Also, the dynamic analysis of injection process could be obtained by combining MPI/Cool and MPI/Flow.

5.4 Cooling simulation

Product mould could be constructed by Pro/E and UG etc. which can be read into MPI by STL file format. Then, cooling system and gating system are built in MPI. Three different cooling systems are shown Table 11 (Liu et al., 2010).

		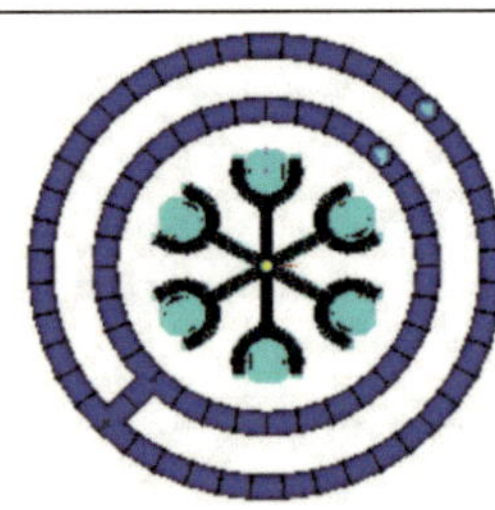
Cooling system one	Cooling system two	Cooling system three

Table 11. Arrangement of cooling systems.

5.4.1 Input the process parameters

We set the melt temperature, cavity temperature and cooling pipe diameter to be 150° C, 40° C and 8mm respectively. The coolant is water (25° C) and Reynolds number is 10000. Finally, we use t software to calculate cooling time.

5.4.2 Simulation results of cooling time

We obtain coolant temperature, coolant velocity, cooling pipe temperature and Reynolds number of coolant (Fig. 13, Fig. 14, Fig. 15) after analyzing cooling process.

According to Fig. 13, Fig. 14, Fig. 15, we find that pipe and coolant temperature distribution in the third cooling system is much more homogeneous than that in the former two. In order to compare the cooling efficiency of three different cooling systems, we can calculate the cooling time of three cooling system by software. The calculated cooling time is shown as Table. 12. According to Table. 12, the first cooling system has longest cooling system, next is the second cooling system and the third cooling system has shortest cooling time.

Therefore, the third cooling system has best heat balance and cooling efficiency.

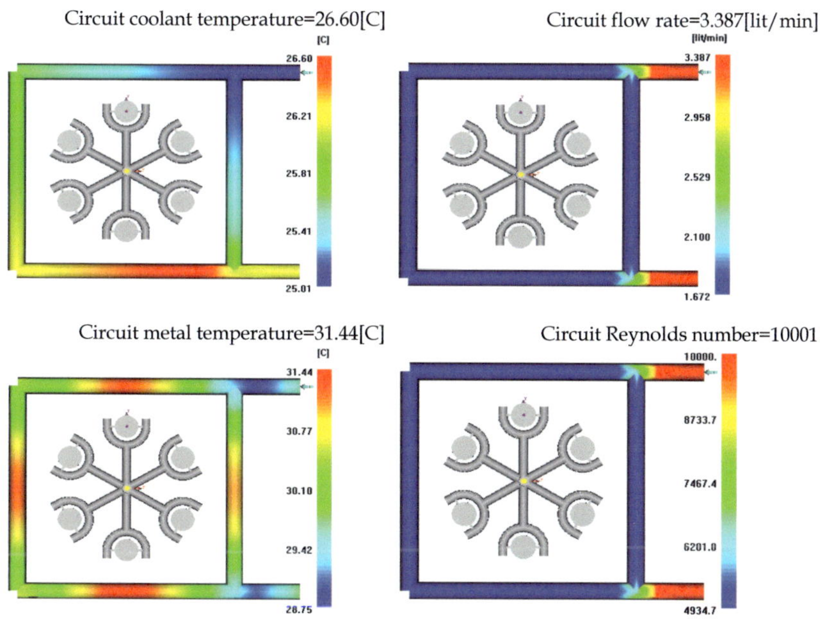

Fig. 13. Simulation outcomes of first cooling system.

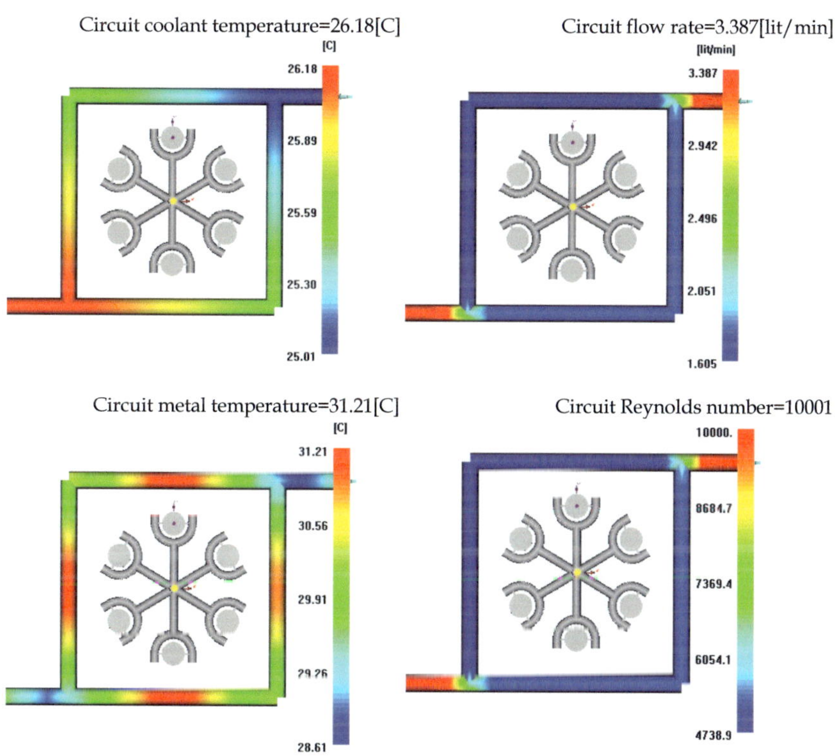

Fig. 14. Simulation outcomes of second cooling system.

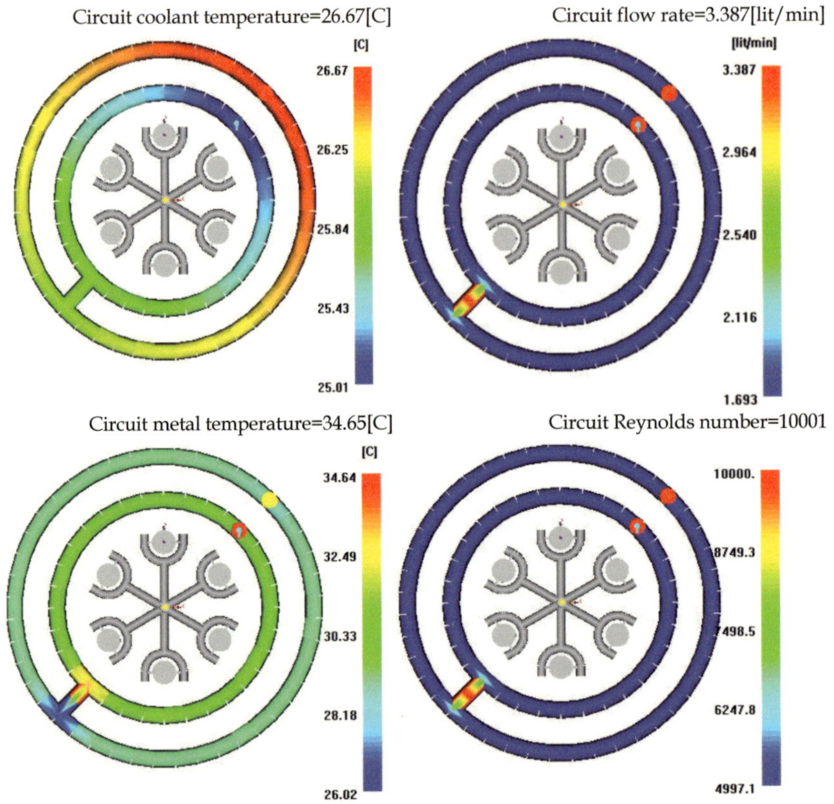

Fig. 15. Simulation outcomes of third cooling system.

	Cooling system one	Cooling system second	Cooling system three
Cycle time (S)	38.9850	37.9800	32.5205

Table 12. Cycle time of different cooling system.

5.5 Comparison between simulation results of these three cooling system

There are two kinds of cooling analysis, namely manual and automatic cooling analysis. We need to set the cooling time when using manual cooling analysis. Cooling time is calculated by software in the automatic cooling analysis. We use the automatic approach to analyze the cooling efficiency of three different cooling systems in the former chapter. In order to compare their cooling efficiency with given cooling time, we use the manual cooling analysis and set the cycle period to be 35s, 30s, 25s respectively.

During cooling process, six cavities are cooled unevenly for the different arrangement of cooling pipes. We number the six cavities according to clockwise direction (Fig. 16 (a)) and select the top, middle and bottom parts of every cavity (Fig. 16(b)) to analyze the temperature distribution of each cavity with different cooling condition.

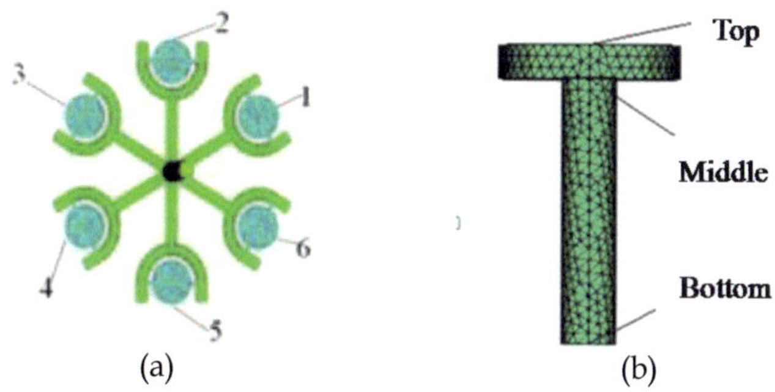

Fig. 16. Cavity labels; (a) Number of cavities; (b) Three locations in each cavity.

5.5.1 Simulation outcomes with 35s cycle period

Table 13 demonstrates the general temperature difference of product in three cooling systems respectively when cycle period is 35s. According to Table 13, product temperature difference in the first cooling system is similar to that in the second cooling system. Also, temperature difference in the third cooling system is comparatively higher for it cools much faster, which causes local parts cool significantly. However, product temperature differences in these cooling systems all are acceptable.

	Maximum temperature	Minimal temperature	Temperature difference
Cooling system one	46.13	37.34	8.79
Cooling system two	45.93	37.15	8.78
Cooling system three	45.19	36.35	8.84

Table 13. General temperature difference of product when cycle period is 35s (°C).

Table 14 illustrates the temperature simulation of three different parts in the cavity axis in different cooling systems when the cycle period is 35s. According to Table 14, the given part temperature in certain cavity in the first cooling system is the highest. Next is the second cooling system. The given part temperature in certain cavity in the third cooling system is the lowest.

Number of cavities		1	2	3	4	5	6	Maximal difference
Cooling system one	Top	41.90	41.94	41.97	41.98	41.96	41.92	0.09
	Middle	46.10	46.19	46.21	46.26	46.25	46.15	0.16
	Bottom	37.74	37.80	37.9	37.91	37.89	37.85	0.17
Cooling system two	Top	41.68	41.68	41.75	41.70	41.72	41.76	0.08
	Middle	45.53	45.55	45.66	45.60	45.59	45.58	0.13
	Bottom	37.53	37.54	37.65	37.68	37.60	37.68	0.15
Cooling System three	Top	40.99	41.01	41.05	41.03	41.03	41.04	0.06
	Middle	44.69	44.71	44.79	44.75	44.80	44.74	0.11
	Bottom	36.67	36.70	36.75	36.73	36.75	36.76	0.09

Table 14. Temperature simulation of three different parts in the cavity axis in different cooling systems when cycle period is 35s (°C).

The cooling effect of given position varies in different cavities. In the first cooling system, Cavity one and two near the coolant inlet have better cooling effect compared with cavity three and four. In the second cooling system, cavity one and six have better cooling effect compared with cavity four and five. In the third cooling system, Cavity one has the lowest temperature. However, the general temperature of six cavities is approximately the same.

In these cooling systems, the extreme value of temperature difference of each cavity is shown as Fig. 17. It demonstrates that the extreme value of temperature difference of top, middle and bottom parts in the third cooling system is much lower than that in the former two, which means that the third cooling system has the best cooling efficiency.

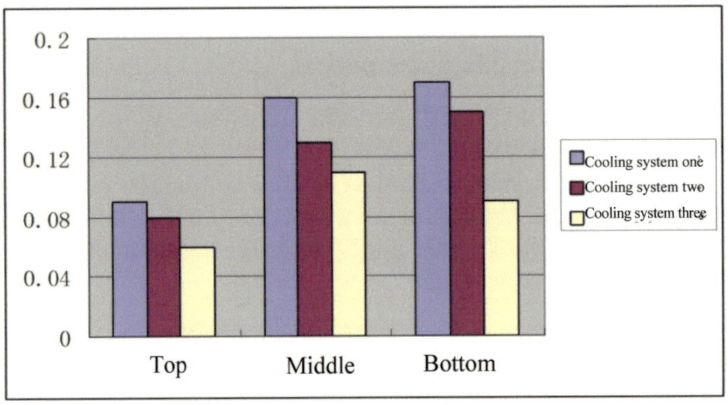

Fig. 17. Extreme values of temperature difference of various positions in each cavity when cycle period is 35s (°C).

5.5.2 Simulation result with 30s cycle period

Table 15 demonstrates the general temperature difference of product in these three cooling systems when the cycle period is 30s.

According to Table 15, product temperature difference in the first cooling system is similar to that in the second cooling system. On the other hand, product temperature in the third cooling system is comparatively higher. Therefore, simulation results are similar to that in three cooling systems when the cycle period is 35s.

	Maximum temperature	Minimal temperature	Temperature difference
Cooling system one	48.84	38.98	9.86
Cooling system two	48.62	38.76	9.86
Cooling system three	47.80	37.87	9.93

Table 15. General temperature difference of products when cycle period is 30s (°C).

Table 16 illustrates temperature simulation outcomes of three different positions in six cavities in three cooling system when the cycle period is 30s. In the first cooling system, cooling effect of given position in product varies in different cavities. Cavity one and six near the water inlet have better cooling efficiency. Next is cavity two and five. Positions in cavity three and four have comparatively high temperature. Cavity three and four do not have ideal cooling efficiency. In the second cooling system, product temperature in the cavity one and six is much lower than that in the cavity three and five. In the third cooling system, cavity one and six have much lower temperature. However, temperature distribution in six cavities is even.

Number of cavities		1	2	3	4	5	6	Maximal difference
Cooling system one	Top	44.10	44.11	44.23	44.25	44.19	44.15	0.15
	Middle	48.15	48.19	48.37	48.36	48.28	18.20	0.22
	Bottom	39.49	39.50	39.52	39.69	39.62	39.56	0.20
Cooling system two	Top	43.95	43.98	44.09	44.03	44.05	44.08	0.14
	Middle	48.99	45.00	49.15	49.09	49.09	49.18	0.19
	Bottom	39.33	39.33	39.50	39.4	39.45	39.50	0.17
Cooling system three	Top	43.12	43.15	43.20	43.19	43.18	43.23	0.11
	Middle	47.32	47.34	47.49	47.41	47.43	47.45	0.17
	Bottom	38.38	38.41	38.51	38.46	38.49	38.55	0.16

Table 16. Temperature simulation outcomes of three different positions of six cavities in three cooling system when cycle period is 30s (°C).

Extreme values of temperature difference of various positions in each cavity when cycle period is 30s are shown as Fig. 18. The extreme values of temperature difference of top, middle and bottom part in each cavity in the third cooling system are smaller than that in the former two.

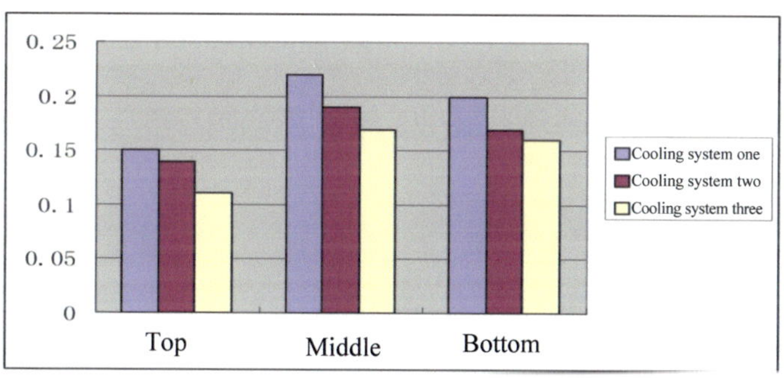

Fig. 18. Extreme values of temperature difference of various positions in each cavity when cycle period is 30s (°C).

5.5.3 Simulation outcomes when cycle period is 25s

Table 17 demonstrates the temperature difference of product with three cooling system when cycle period 25s. According to Table 17, temperature difference of product in the first cooling system is similar to that in the second cooling system. The general temperature difference of product in the third cooling system is the highest. Simulation results are similar to the outcomes obtained above.

	Maximal temperature	Minimal temperature	Temperature difference
Cooling system one	54.64	42.57	12.07
Cooling system two	54.37	42.30	12.07
Cooling system three	53.38	41.21	12.17

Table 17. General temperature difference of product when cycle period is 25s (°C).

Table 18 shows the temperature simulation of three different positions in different cooling system when the cycle period is 25s. Apparently, cooling efficiency of third cooling system is the best and temperature distribution of each cavity is even in the third cooling system.

Cavity Number		1	2	3	4	5	6	Maximal difference
Cooling system one	Top	49.10	49.08	49.27	49.31	49.19	48.17	0.23
	Middle	53.98	54.07	54.21	54.29	54.22	54.12	0.31
	Bottom	37.74	37.75	37.99	38.02	37.91	37.81	0.28
Cooling system Two	Top	48.83	48.85	48.99	48.92	49.02	48.91	0.19
	Middle	53.69	53.78	53.97	53.86	53.91	53.75	0.28
	Bottom	42.82	42.80	43.04	42.96	42.91	43.01	0.24
Cooling system three	Top	47.76	47.79	47.89	47.83	47.91	47.86	0.15
	Middle	52.78	52.83	52.99	52.89	52.90	52.93	0.21
	Bottom	41.88	41.90	42.07	41.98	42.01	42.03	0.19

Table 18. Temperature simulation of three different positions with different cooling system when cycle period is 25s (° C).

Extreme values of temperature difference of various positions in each cavity are shown as Fig. 19 when cycle period is 25s. The extreme values of temperature difference of top, middle and bottom parts in each cavity in the third cooling system are smaller than that in the former two.

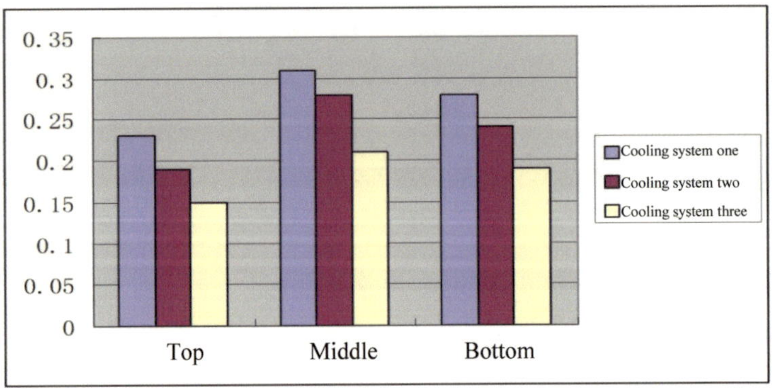

Fig. 19. Extreme values of temperature difference of various positions in each cavity when cycle period is 25s (° C).

6. Conclusion

Firstly, we used Moldflow to calculate pressure drop, filling time, temperature difference and clamp force of five different cross-section shapes. Outcomes demonstrate that U-shape runner has smallest pressure drop, shortest filling time, minimal temperature difference and highest efficiency. Therefore, U-shape runner is most suitable for cool-runner mould rather than circular or other kinds of runners.

Secondly, we investigated runner systems with rectangular and circular shunt respectively by orthogonal table. Also, we researched influence of mould temperature, injection temperature, screw velocity and gate dimension on products. Results show that runner system with circular shunt is most suitable for Ceramic Injection Molding. Furthermore, we considered the gravity influence on Ceramic Injection Molding and found that short shot tends to happen on the top cavity when runner diameter is 4mm. All six cavities are filled well after increasing runner diameter to 4.17 mm.

Finally, we simulated cooling efficiency of three cooling systems and results show that the third cooling system has shortest cooling cycle and best cooling efficiency, which can cool product as fast as possible. Cooling efficiency in six cavities is not the same for the cooling system arrangement and for the inlet and outlet location. Temperature extreme values of top, middle and bottom positions in each cavity in the third cooling system are smaller than that in former two cooling system.

7. References

Chen, J. B.; Shen, C. Y.; Wang, Z. F. (2002). Numerical simulation of the cooling process of Injection mould. *Polymer Materials Science & Engineering*, Vol.18, No. 4, pp.21-25, ISSN 1000-7555

Jansen, K. M. B.; Van, D. J.; Husselman, M. H. (1998). Effect of processing conditions on shrinkage in injection molding. *Polymer Engineering and Science*, Vol.38, No. 5, pp.838~846, ISSN 0032-3888

Jin, X. M.; Zhu, X. F. (2000). Investigating the influence of processing parameters of Injection Mould on the quality of the product by the orthogonal method. *Journal of south china university of technology*, Vol.28, No. 9, pp. 77-81, ISSN 1000-565X

Li, Q.; D, B. B.; Yang, M. S. et al. (2001). The Application of cooling simulation technology in the Injection mould and designing the processing parameters. *Engineering Plastics Application*, Vol.29, No. 4, pp.34-37, ISSN 1001-3539

Liu, F.; Lin, B.; Zhang, M. M.; Li, L. J. (2010). Redesign and optimization for ceramic injection mould of ZrO_2 fiber ferrule. Key Engineering Materials, Vol.434-435, pp.840-843, ISSN: 1662-9795

Liu, F. Q.; Yan, L. T.; Hu, H. Q. et al. (2002). Optimization of the refrigerator accouterment runner. *Engineering Plastics Application*, Vol.30, No. 4, pp. 39-42, ISSN 1001-3539

Pan, H. J.; Cui Q.; Pan, Y. J. (1995). Analysis of cross-section shape area of the runner of the Injection Mould. *China Plastics*, Vol.30, No. 4, pp. 79-87, ISSN 1001-9278

Shen, Y. K.; Yeh, P. H.; Wu, J. S. (2001). Numerical simulation for thin wall injection molding of fiber-reinforced thermoplastics. *International Communications in Heat and Mass Transfer*, Vol.28, No. 8, pp.1035~1042, ISSN 0735-1933

Shen, Y. K.; Liu, J. J.; Chang, C. T.; Chiu, C. Y. (2002). Comparison of the results for semisolid and plastic injection molding process. *International Communications in Heat and Mass Transfer*, Vol.29, No. 1, pp.97~105, ISSN 0735-1933

Skourlis, T. P.; Mohapatra, B.; Chassapi, C. et al. (1997). Evaluation of the processing parameters on the properties of advanced styrenic resins-A design of experiments approach. *Advances in Polymer Technology*, Vol.16, No. 2, pp.117-128, ISSN: 0730-6679

Wenjea J. T.; Liu, D.M.; Hsu, C.K. (1999). Influence of stearic acid on suspension structure and green microstructure of injection-molded zirconia ceramics. Ceramics International, Vol.25, No. 2, pp.191-195, ISSN 0272-8842

Yang, W.; Hu, S. G.; Jin, G. (2004). Optimal allocation of process parameters of Injection Mould based on the CAE, DOE and fuzzy weight. *Mechanical & Electrical Engineering Magazine*, Vol.21, No. 5, pp.37-40, ISSN 1001-4551

Zhang, M. M. (2005). Master Thesis, SME, Tianjin University, China

Zhang, M. M.; Lin, B. (2007). Simulation of ceramic injection molding for zirconia optical ferrule. Key Engineering Materials, Vol.336-338, pp.997-1000, ISSN: 1662-9795

Part 4

Microcellular Injection Molding

8

Microcellular Foam Injection Molding Process

Hu Guanghong and Wang Yue
National Engineering Research Center of Die & Mold CAD,
Shanghai Jiao Tong University, Shanghai,
China

1. Introduction

In recent years, the polymer resin price is rising due to the petroleum shortage. How to save plastics on the premise to ensure the plastics part quality is one of the research hotspots. Microcellular foam injection molding process is developed in this background. Microcellular foam technology was invented by MIT in the early 1980's [1]. The traditional foaming processes, which produce bubbles larger than 0.25mm, are not feasible due to excessive loss of strength. Thus, the idea was born to create microcellular foam to both save plastics and have reasonable strength.

Generally, microcellular foam process takes advantage of supercritical fluid (SCF) as physical blowing agent. CO_2 and N_2 are usually used as agent. The microcellular foam parts have uniform cell diameters of 1 to100 microns and cell density of 10^9 to 10^{15} cells per cubic centimeters. Figure 1-1 shows the scanning electron micrographs of microcellular polystyrene sample [2].

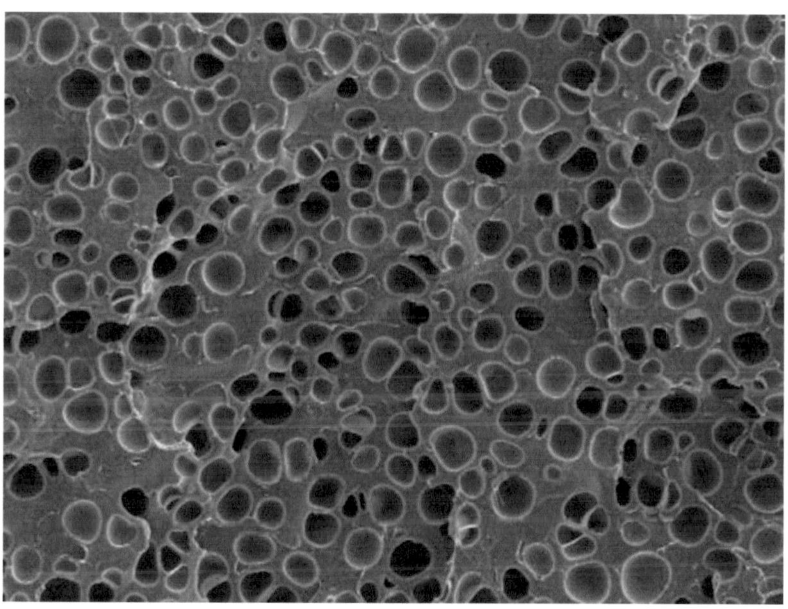

Fig. 1-1. Electron micrographs of microcellular polystyrene scanning sample [2].

Now microcellular foam technology is extended into many other plastics forming process such as extrusion, injection, blowing process. And microcellular foam technology is widely used in the homework appliance, aerospace and auto industry *etc.* In this book, microcellular foam injection molding process is mainly discussed.

1.1 Microcellular foam injection process principle

During microcellular foam injection molding process, SCF is injected into the polymer melt. And the single phase of polymer- SCF mixed solution is obtained under certain temperature and pressure. When the mixer is injected into the mold, the pressure of the single-phase solution is dropped from microcellular process pressure (MPP) to atmospheric pressure. The nucleation phenomena occur due to the gas separated out of the mixer. Then these nuclei finally grow up to stable bubbles.

Figure 1-2 shows the microcellular foam injection molding process. And generally the microcellular foam injection molding process is described as following four steps.

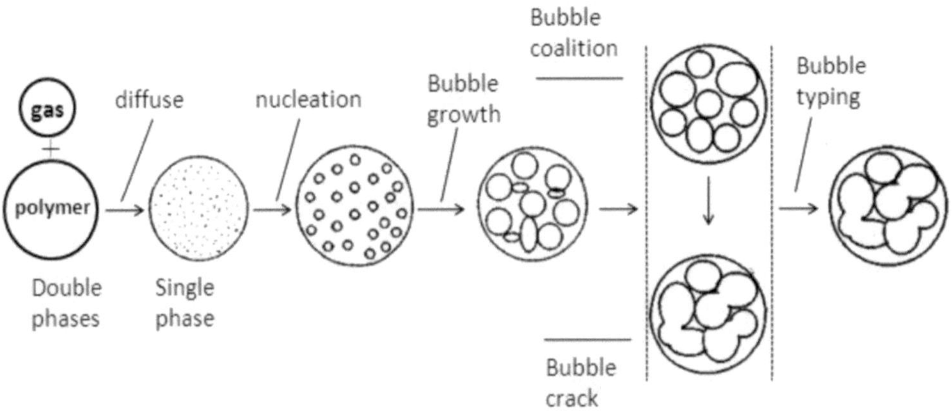

Fig. 1-2. Illustration of microcellular polymer foaming process [3].

Polymer-SCF single phase generation

During microcellular foam injection molding process, the supercritical nitrogen (N_2) or carbon dioxide (CO_2) is injected into plastics injection machine barrel and dissolved into polymer melt. Then a single phase polymer-SCF solution is generated under the definite temperature and pressure. In this stage, the concentration of SCF is determined by saturation, microcellular process pressure (MPP) and the mixer temperature. These parameters also significantly affect the final bubbles size.

Homogeneous nucleation

Theoretically, only when the polymer-SCF mixer is in the thermodynamics equilibrium and millions of nuclei are generated at the same time, homogeneous nucleation will be possible. When the polymer-SCF single phase mixer is injected into mold cavity, the mixer pressure is changed from MPP to the atmospheric pressure. Thus a rapid pressures unloading occurs. Then, SCF separates from the single phase mixer, and a large number of nuclei are generated. With the nucleus growing, free energy of the mixer is also increasing. Only when the nucleus size is bigger than the critical one, the nucleus will be stable. And the bubble growing can be possible. Thus, the mixer temperature, MPP and SCF concentration affect the nuclei process and the final nucleus density.

Bubbles growth

When millions of nuclei are generated and the nucleus is stable, the bubbles growth start. SCF concentration of mixer is higher than the SCF concentration inside bubbles. Due to the concentration difference SCF in the mixer enters the bubbles. And the gas bubbles grow up. Until the SCF concentration inside bubbles equals to the outside one or the melt is frozen, the gas bubbles will keep growing up. Thus, the final bubble morphology is determined by the SCF concentration and injection process parameters.

Product typing

Along with mold cooling, the melt temperature is decreased and the melt freezes up. The bubbles stop growing up. And the shape of part is fixed.

From above microcellular foam injection molding process, the properties of microcellular foam injection molding parts are determined by nucleation process and final bubble morphology besides tradition injection process condition such as part shape, the kind of polymer, mold structure, process parameters. Thus microcellular foam injection molding process has distinct characters comparing to the traditional plastics injection.

1.2 Microcellular foam injection molding process characters

Due to SCF injected into the polymer melt, it is great affect the polymer melt viscosity, injection molding process cycle, part weight, mechanical properties and surface quality *etc.*

1.2.1 Melt viscosity

Due to the SCF dissolved in the polymer melt, the glass transition temperature of polymer melt becomes lower. So the polymer viscosity is decreased and the melt fluidity becomes better. Thus, the required injection pressure is lower than tradition injection and the requirement of injection machine properties is less. Figure 1-3 shows the effect of SCF on the PA, PBT melt viscosity [4]. The results indicate that the viscosity is decreased after the SCF is added.

It should be pointed out that the effect of SCF on the polymer viscosity is determined by the polymer kind and filler. Because SCF can't be dissolved into the filler, it will not affect the filler viscosity. Thus comparing to the pure polymer, the effect of SCF on the viscosity of polymer with filler is less.

1.2.2 Injection cycle time

Microcellular foam injection molding technology can reduce the cycle time. The reasons mainly include: (1). because the gas in the bubbles can provide the packing pressure, the packing and holding phase can be eliminated. (2). When the millions of nucleus are generated and bubbles grow up, they are all endothermic reaction. So the cooling time is saved. (3). Due to the bubbles in the part, the part weight is reduced. The cooling time is also saved. (4). The lower viscosity means higher filling speed. The filling time becomes short. Generally 20%~50% cycle time can be saved by microcellular foam injection molding

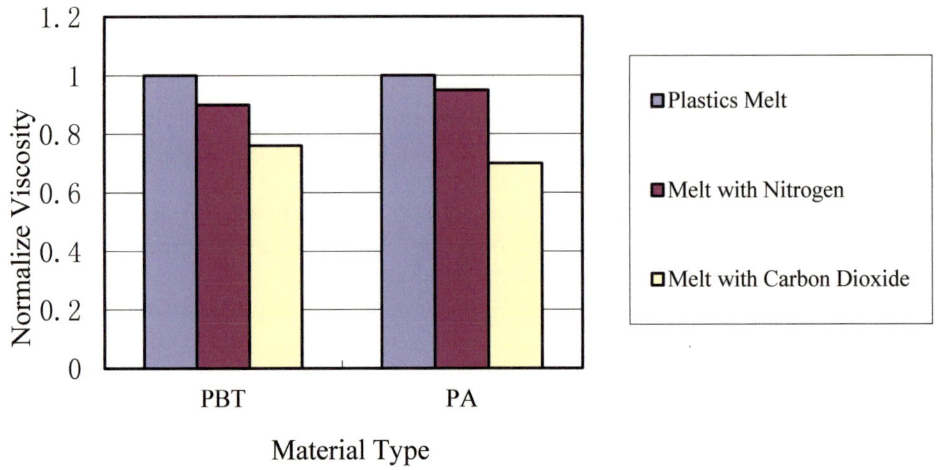

Fig. 1-3. Effect of SCF on melt viscosity [4].

process. Figure 1-4 shows the comparison between microcellular foam injection molding process cycle and traditional injection one.

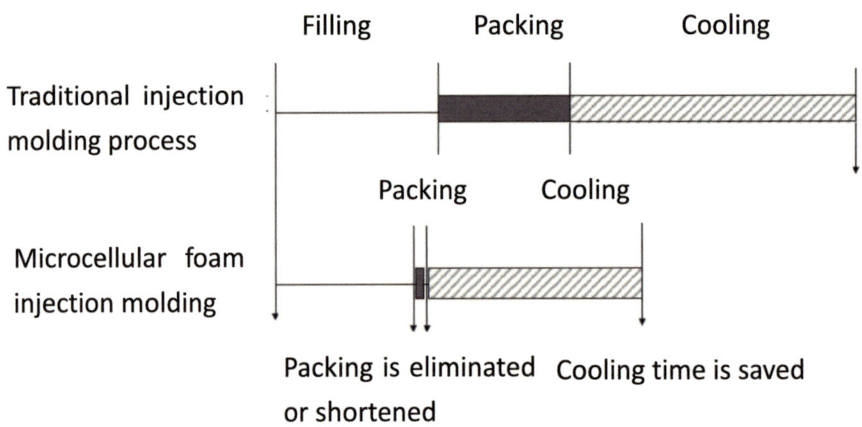

Fig. 1-4. Comparison between microcellular foam injection molding process cycle and traditional injection process.

1.2.3 Part weight

Due to the bubbles in the part, the polymer obviously can be saved. Generally the part weight can be reduced as 0.5mm thickness weight by microcellular foam injection molding process. At the same time, all kinds of polymer, even including the high temperature polyphenylsulfone, can be formed by this technology. The effect of microcellular foam injection molding process on the weight reduction is shown in the Table 1-1.

Polymer	Part thickness(mm)	Weight reduction (%)
Polyphenylsulfone	5	50
PS	1.5	30
Acetal	1.5	15
PET	5	30
TPE	1.5	20
PP (30%Talc)	2.1	25
HDPE	5	60
PC/ABS	2.1	23
PA	1.2	9
PA(40% Filler)	2	15
PC	7.2	45

Table 1-1. Effect of microcellular foam process on weight reduction [5].

1.2.4 Part mechanical properites

Also, the parts mechanical properties are changed due to the bubbles. The former researches indicate that the part bend strength of microcellular foam polymer is almost same as the solid polymer. Thus microcellular foam technology can be used to produce the inner structure part. However it is quite different situation for the part tensile strength. The tensile property data shows that the tensile strength of microcellular foams decreases in proportion to the foam density. It means that a 50% relative density foam can be expected to have 50% of the strength of the solid polymer. To the part impact strength, it is more sensitive to variation from polymer to polymer. And the results cannot be generalized. However the Gardner impact strength of PVC foam experiment results show that the impact strength decreases linearly with foam density. It should be pointed out that the impact strength of

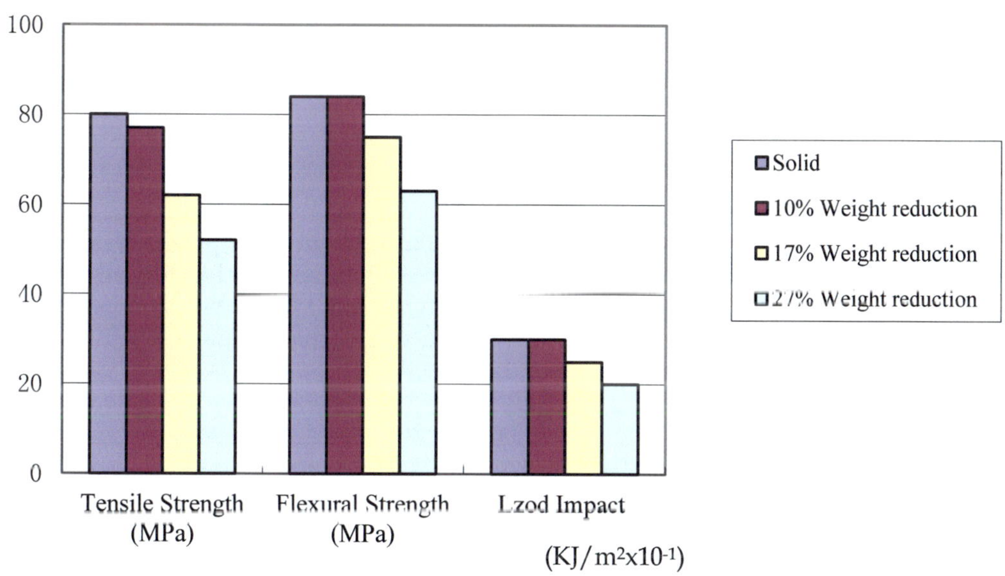

Fig. 1-5. PBT mechanical properties on the different weight reduction ratio.

polymer added filler is decreased less than one without any filler. The main reason is that the filler properties and content percent great affect the part impact strength. And SCF has no effect on the fillers[6-13]. Figure 1-5 shows the bend strength, tensile strength and impact strength of PBT (30% GF) on the solid polymer and different weight reduction ratio. The results present the almost same rules as above.

1.2.5 Surface quality

As said above, microcellular foam injection molding process presents nice formability and lots of advantages. But still due to SCF, the part surface quality is worse than tradition process. Typical surface defects are swirl marks, silver streak, surface blistering, post-blow and large surface roughness. These defects limit the application scope of microcellular foam injection process seriously. Figure 1-6 shows main surface defects of microcellular foam injection molding parts.

(a) (b) (c) (d)

Fig. 1-6. Surface defects of microcellular foam injection molding parts (a) swirl mark[14]; (b) silver streak[14]; (c) surface blistering[15]; (d) post-blow[15].

Swirl marks

Grooves on the part surface are caused by the trapped gas on the mold surface when the polymer-SCF mixer begins to solidify. And the area of grooves surface shows positive correlation. The shape of these grooves is slender along the flow direction, and the aspect ratio of grooves indicates the size of shear strength which is caused by the polymer-SCF mixer filling behavior in the mold cavity. Swirl marks are these grooves whose shapes are curled (see Figure 1-5a).

Yoon propose that the glass transition temperature (for the amorphous polymer) or the melt temperature (for the crystalline polymer) is one of the important effect factors on the swirl mark forming [16]. Zhang YT points out that swirl marks always appear near the gate [17]. While the polymer-SCF mixer is injected into mold cavity, many parameters in different mold cavity area are varied. Generally near the gate, the temperature is higher, viscosity of the polymer-SCF is smaller, and melt strength is lower. So the gas near the gate is easy to diffuse to the mixer surface, and the bubbles near the surface break up easily.

Silver streak

Silver streak is a defect that shows silver gloss in the sunlight (see Figure 1-5b). Silver streak of microcellular foam injection parts shows two different appearances. One is called silver thread because its boundary looks like a thread. This defect is caused by the broken bubbles at the surface of melt. The other is called silver strip because it looks like a strip which parallels the flow direction. The difference between them is that there are no broken bubbles at the surface to the latter.

Michaeli and Cramer point out that the silver streaks are flow marks of the polymer-SCF mixer on the mold cavity surface. It's the shear deformation of the bubbles that are close to the surface. Because of different bubble sizes, the depth of silver threads is different and then the parts surface roughness is different. Compared with silver trips, silver threads will cause larger surface roughness [18].

Surface blistering

When many tiny bubbles converge at the part thin wall place, it makes a thin polymer layer separate from the main part body. This phenomenon is called surface blistering. (see Figure 1-5c). Surface blistering most likely appears in the parts that are made by crystalline polymer without filler such as POM. Surface blistering can be eliminated by adjusting the microcellular foam injection process parameters and improving the mold design.

Post-blow

Post-blow is similar to the internal blistering and always appears at the place of hot spots (see Figure 1-5d). The post-blow defect is caused by following two factors. One is that the cooling is not enough at the hotspots; the other is that too much gas enters the some certain bubbles due to the high SCF concentration and form some large size bubbles. When the pressure inside the bubbles is higher than the outside one, the post-blow will happen. So the method to eliminate this defect is to enhancing cooling at the hot spots and adjusting SCF concentration.

Surface roughness

In addition to the above serious defects, surface roughness is another problem that limits the application scope of microcellular foam injection molding process. During bubbles growing up, some small bubbles break up near the surface, and the gas is trapped on the mold surface when the polymer-SCF mixer begins to solidify. So the surface roughness of microcellular foam injection parts is higher than that of traditional injection parts.

2. Microcellular foam injection molding theories

According to above chapters, all the advantages and disadvantages are all caused by the SCF injected into the polymer melt. Before introduction microcellular foam injection molding theories, supercritical fluid is firstly discussed.

2.1 Supercritical fluid

Supercritical fluid is any substance at certain temperature and pressure above its critical point, where distinct liquid and gas phases do not exist. It can effuse through solids like gas, and dissolve materials like liquid. In addition, close to the critical point, small changes in pressure or temperature result in large changes in density, and allowing many properties of supercritical fluid to be "fine-tuned". Supercritical fluids are suitable as a substitute for organic solvents in a range of industrial and laboratory processes. Carbon dioxide and nitrogen are the most commonly used supercritical fluids for microcellular foam injection molding. Figure 2-1 shows the Carbon dioxide pressure-temperature phase diagram.

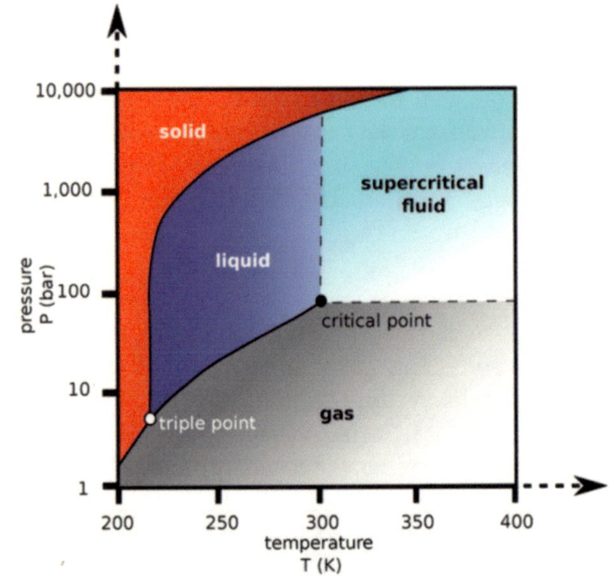

Fig. 2-1. Carbon dioxide pressure-temperature phase diagram [30].

In Figure 2-1, the boiling separates the gas and liquid region and ends in the critical point, where the liquid and gas phases disappear to become a single supercritical phase.

In general terms, supercritical fluids have properties between those of gas and liquid. In the Table 2-1, the critical properties are shown for some components, which are commonly used as supercritical fluids.

Solvent	Molecular weight g/mol	Critical temperature K	Critical pressure MPa (atm)	Critical density g/cm3
CO_2	44.01	304.1	7.38 (72.8)	0.469
N_2	28	126.2	3.4 (33.6)	--
H_2O	18.015	647.096	22.064 (217.755)	0.322
CH_4	16.04	190.4	4.60 (45.5)	0.162
C_2H_6	30.07	305.3	4.87 (48.1)	0.203
C_3H_8	44.09	369.8	4.25 (41.9)	0.217
C_2H_4	28.05	282.4	5.04 (49.7)	0.215
C_3H_6	42.08	364.9	4.60 (45.4)	0.232
CH_3OH	32.04	512.6	8.09 (79.8)	0.272
C_2H_5OH	46.07	513.9	6.14 (60.6)	0.276
C_3H_6O	58.08	508.1	4.70 (46.4)	0.278

Table 2-1.Critical properties of various solvents [30].

2.1.1 Nitrogen vs carbon dioxide

Both nitrogen and carbon dioxide are widely used in microcellular foam processing. However, the choice of blowing agent affects the final parts bubble morphology. Therefore, the choice should be made depending on what microcellular foam bubble morphology is desired rather than on ease of use or blowing agent costs.

Table 2-2 shows that carbon dioxide generally has much greater solubility in molten polymers than nitrogen. It indicates that more carbon dioxide can be added to the polymer melt in microcellular foam processing than nitrogen. The result of higher blowing agent concentration in the polymer melt means more density reduction. Table 2-2 shows CO_2 and N_2 maximum solubility in different polymer melt at 200℃ temperature and 27.6MPa pressure[1].

Polymer	Carbon dioxide (%)	Nitrogen (%)
PE	14	3
PP	11	4
PS	11	2
PMMA	13	1

Table 2-2. Estimated Maximum Gas Solubility at 200℃/27.6MPa[1].

However, because of the similar diffusion rates of nitrogen and carbon dioxide in polymers melt, as shown in the Table 2-3, nitrogen lends to generate smaller cells at the same concentration in polymer melt than carbon dioxide. And the driving force of nitrogen to devolve from the polymer-SCF single phase solution is greater than carbon dioxide. Thus more nucleation sites can be formed in the polymer-nitrogen mixer. Because the diffusion rates are similar, all nucleation sites grow at the same rate whatever nitrogen or carbon dioxide is the blown agent. Thus nitrogen has smaller cell sizes.

Polymer	Carbon Dioxide(cm^2/s)	Nitrogen(cm^2/s)
PS	1.3×10^{-5}	1.5×10^{-5}
PE	2.6×10^{-6}	8.8×10^{-7}
HDPE	2.4×10^{-5}	2.5×10^{-5}
LDPE	1.1×10^{-4}	1.5×10^{-4}
PTFE	7.0×10^{-6}	8.3×10^{-6}
PVC	3.8×10^{-5}	4.3×10^{-5}

Table 2-3. Estimated Diffusion Coefficient at 200°C [1].

2.2 Nucleation theory

2.2.1 Theories of nucleation processing

The nucleation theory was established by Gibbs in early 20th century. Colton [31] proposed the classic nucleation theory, which should be classified into three types: the homogeneous nucleation, heterogeneous nucleation and cavity nucleation.

The main concern of classical nucleation theory is a thermodynamic description of initial stage of nucleation from embryo to nucleus with a little larger size than the critical one.

Homogeneous nucleation occurs in single phase solution system that has no impurity. During the pressure unloading process, every gas molecules will be a nucleation point. So theoretically the largest nucleation density and the smallest bubble size in the final parts will be obtained by homogeneous nucleation. However, due to the purity system, it need more energy to overcome the "energy barrier" to create stable and effective nucleus. Thus there should be more super saturation in the polymer-SCF system.

Heterogeneous nucleation considers that there will be some impurity dispersed in the polymer-SCF mixer. Because there will be more interfacial energy at the impurity solid surface, the nucleation driving force at the impurity solid surface is bigger than the other places. It means that less free energy should be overcome for the nucleus generation. Compared with homogeneous nucleation, heterogeneous nucleation is easier to generate nuclei.

Cavity nucleation is that many nuclei are generated at the cavity places. The gas will be absorbed in the cavity by the nucleating agent or any other micro impurities. Polymer melt can't enter the split wedges at the roughness surface. However the gas will be trapped in these split wedges. During the nucleation process, the gas is tended to enter these cavities to form the nuclei. At the same time, these cavities can save the nucleation energy. And then the stable nucleus can be generated easily.

In this chapter, based on the classical homogeneous nucleation, the microcellular foam nucleation theory is introduced.

2.2.2 Homogeneous nucleation

Classical homogeneous nucleation [19]

The main concern of classical homogeneous nucleation theory has been a thermodynamic description of initial stage of nucleation from embryo to nucleus. When the thermodynamic equilibrium is broken and the change of free energy of mixer is more than the "energy barrier", the phase transition occurs and the nuclei are generated. When the nuclei are bigger than the critical one, the nuclei become stable and continue to grow up to bubbles. The rate of homogeneous nucleation can be described by the following Equation 2-1.

$$N_{homo} = C_0 f_0 exp\left(\frac{-\Delta G}{KT}\right) \tag{2-1}$$

where N_{homo} is the number of nuclei generated per cm^3 per second. C_0 is the concentration of the gas (number of molecules per cm^3). f_0 is the frequency factor of the gas molecules. K is the Boltzmann's constant. And T is absolute temperature. The term ΔG is the "energy barrier" for homogeneous nucleation. ΔG can be calculated by Equation 2-2:

$$\Delta G = \frac{16\pi\gamma^3}{3\Delta P^2} \tag{2-2}$$

where ΔP is magnitude of the quench pressure and γ is the surface energy of the bubble interface.

The frequency factor of gas molecules in the Equation 2-1, f_0 , can be expressed as:

$$f_0 = Z\beta \tag{2-3}$$

where, Z, the Zeldovich factor, accounts for the fact that a large number of nuclei never grow, but rather dissolve. The rate at which the molecules are added to the critical nucleus, β, can be calculated as surface area of the critical nucleus times the rate of impingement of gas molecules per unit area. The calculation method can be expressed as Equation 2-4.

$$\beta = (4\pi r_c^3) R_{impingement} \tag{2-4}$$

Substituting Equation 2-4 into Equation 2-3:

$$f_0 = Z(4\pi r_c^3)R_{impingement} \tag{2-5}$$

Equation (2-5) shows that the frequency factor of the gas molecules joining a nucleus to make it stable varies with the surface area of the nucleus. Generally, $ZR_{impingement}$ can be regarded as a fitted parameter.

Knowing the surface energy of the system as a function of pressure and temperature, the critical size of the nuclei can be calculated at any conditions by Equation 2-6.

$$r_c = \frac{2\gamma}{\Delta P} \tag{2-6}$$

Where r_c is the radius of the critical nucleus.

Equations 2-1, 2-2, 2-5, 2-6 form a complete set of the nucleation model for polymer-SCF solution.

In order to calculate the total number of nuclei generated in the system at given saturation conditions. The rate of nucleation needs to be integrated over the time period of nucleation. Generally the gas pressure falls as a function of time. Thus the starting saturation pressure (P_{sat}) and the pressure at which the polymer vitrifies (P_g) define the time scale over which the rate of nucleation should be integrated. Therefore, the total number of nuclei, N_{total}, can be calculated by Equation 2-7.

$$N_{total} = \int_0^t N_{homo} dt = \int_{P_{sat}}^{P_g} N_{homo} \frac{dP}{\left(dP/_{dt} \right)} \tag{2-7}$$

2.2.3 Effect of nucleation process conditions on bubble morphology

Based on the above nucleation model, the main nucleation process parameters include saturation pressure, mixer temperature and SCF concentration. In this chapter, the effect of the three parameters and the interaction among them on the part cell morphology will be discussed.

2.2.3.1 Simulation experimental model and Taguchi method

The simulation experimental model is a thin box. The size is 15.5mm×14mm×13mm. the thickness varies from 0.35mm to 1.8mm. Figure 2-2 shows the cavity distribution, gate system and cooling channels. The characteristic point position is also selected near the gate.

The PS/CO_2 foam system is bulit and PS brade is Vestgran 620. The each level of three process parameters are shown in the Table 2-4. Besides the studied three parameters, the initial values of other process parameters are set in the Table 2-5.

Factors	Level1	Level2	Level3
(A)Saturation pressure/ MPa	11	16	21
(B) Melt temperature/ °C	220	240	260
(C) Gas concentration/ %	0.3	0.55	0.8

Table 2-4. Level of process parameters.

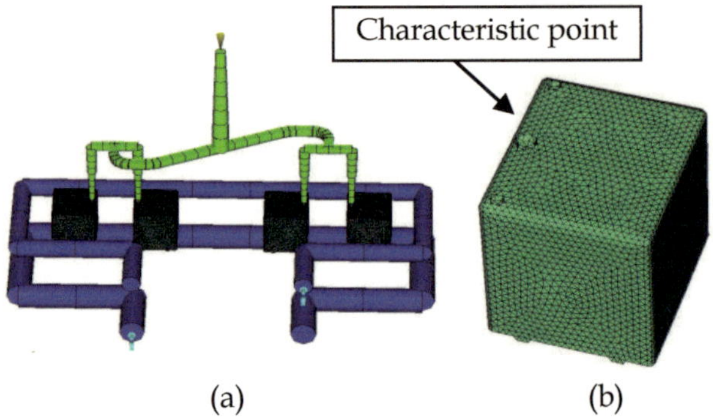

Fig. 2-2. Experimental model and characteristic point position (a): CAE analysis model; (b): Characteristic point position.

Process parameters	Value
Mold temperature/ °C	50
Injection time/ s	0.6
Cooling time/ s	35
Open mold time/ s	5

Table 2-5. Othe rocess parameters list.

2.2.3.2 Taguchi method

Taguchi method is used as an experiment arrangement and parameters optimization method. Based on the setup of parameters and levels, the $L_9(3^4)$ orthogonal array is selected to arrange the experiments. Table 2-6 shows the orthogonal array. The variable analysis is used to calculate the effect order of each process parameters on the cell size and obtain the process parameters optimization combination. At the same time, the experimental results are directly analyzed, that is to calculate the average value of cell size under the three levels of the each process parameter. Here, the cell size is considered that the smaller is better. Therefore it is a minimum value issue. The calculation formula is shown as Equation 2-8 [20]:

$$m = \frac{1}{n}\sum_{i=1}^{n} y_i \tag{2-8}$$

where m is the average value of process parameter under a certain level, n is the number of the level, y_i is the result value of the process parameter under the level. Then the difference R_{diff} of each process parameter can be calculated by the maximum average value subtracting the minimum average one. Based on the R_{diff} value, the effect of process parameter on the cell size can be achieved.

2.2.3.3 Results and discussion

Experiment result and signal-to-noise analysis

The simulation experiments are arranged according to $L_{27}(3^{13})$ orthogonal table. At the same time, each experiment's cell size at characteristic point is obtained. The results are shown in the Table 2-6.

	A(MPa)	B(°C)	x1x2		C(%)	x1x3		x2x3		Cell Size(um)
1	11	220	1	1	0.3	1	1	1	1	29.6
2	11	220	1	1	0.55	2	2	2	2	94.6
3	11	220	1	1	0.8	3	3	3	3	32.4
4	11	240	2	2	0.3	1	1	2	3	109.0
5	11	240	2	2	0.55	2	2	3	1	77.8
6	11	240	2	2	0.8	3	3	1	2	37.2
7	11	260	3	3	0.3	1	1	3	2	41.4
8	11	260	3	3	0.55	2	2	1	3	49.4
9	11	260	3	3	0.8	3	3	2	1	35.0
10	16	220	2	3	0.3	2	3	1	1	18.8
11	16	220	2	3	0.55	3	1	2	2	14.2
12	16	220	2	3	0.8	1	2	3	3	11.4
13	16	240	3	1	0.3	2	3	2	3	13.0
14	16	240	3	1	0.55	3	1	3	1	12.8
15	16	240	3	1	0.8	1	2	1	2	10.2
16	16	260	1	2	0.3	2	3	3	2	15.0
17	16	260	1	2	0.55	3	1	1	3	10.6
18	16	260	1	2	0.8	1	2	2	1	12.8
19	21	220	3	2	0.3	3	2	1	1	9.0
20	21	220	3	2	0.55	1	3	2	2	7.2
21	21	220	3	2	0.8	2	1	3	3	7.8
22	21	240	1	3	0.3	3	2	2	3	9.2
23	21	240	1	3	0.55	1	3	3	1	9.6
24	21	240	1	3	0.8	2	1	1	2	6.2
25	21	260	2	1	0.3	3	2	3	2	10.2
26	21	260	2	1	0.55	1	3	1	3	6
27	21	260	2	1	0.8	2	1	2	1	7.2

Table 2-6. $L_{27}(3^{13})$ Orthogonal table and experimental results.

According to Table 2-6, the S/N is calculated and the effect trend of each factors on the S/N also are gotten. Figure 2-3 shows the details. According to Figure 2-3, the significance order from big to small of the effect of each process parameters on cell size is saturation pressure (A), SCF concentration (C) and mixer temperature (B).

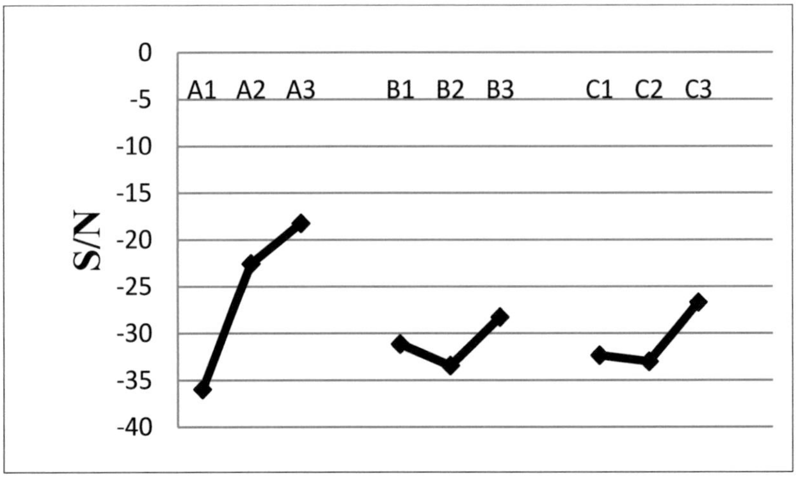

Fig. 2-3. Effect of each factors on S/N ratio.

ANOVA analysis

In order to further analyze the effect of each factors and the interaction among these factors on the cell morphology, ANOVA analysis is calculated according to above S/N results and experiment values. The calculation results are shown in the Table 2-7.

	Degree of freedom	Sum of square of deviations	mean square error	F value	Significance	Significance
A	2	12620.1	6310.05	24.1638	67.93%	***
B	2	536.500	268.250	1.02724	2.89%	
C	2	912.518	456.259	1.74721	4.91%	Δ
A×B	4	1158.04	289.510	1.10865	3.11%	
A×C	4	6880.48	1720.12	6.58706	18.51%	**
B×C	4	979.170	244.792	0.93741	2.64%	
Error	8	2089.08	261.135			
Sum	26	23086.8				

Table 2-7. ANOVA analysis results.

According to Table 2-7, the conclusion of effect of saturation pressure (*A*), SCF concentration (*B*) and mixer temperature (*B*) on the cell morphology is same as the S/N results. However the interaction among the three factors is taken into account in the ANOVA analysis. Also according to Table 27, the significance order is: saturation pressure (*A*) possess 67.93%, the interaction between saturation pressure (*A*) and SCF concentration (*C*) posses 18.51%, SCF concentration (*C*) is 4.91%, Mixer temperature (*B*) is 2.89%. Compared with the S/N results, the interaction between saturation pressure (*A*) and SCF concentration (*C*) is also a very important factor to affect the cell morphology. According to F value, the effect of other factors on the cell morphology is less. So these factors belong to the error range.

Therefore, the optimization parameters combination is mainly determined by the factor *A* and *A*×*C*. Because the smaller cell size is better, the value of *A* and *B* should be the *A*3 and *B*3 in the optimization combination. Due to three levels of *C*, the *A*3×*C* combination has three arrays. And every combination has three experimental results. The average value of each three experimental results is shown in the Table 2-8.

	C1	C2	C3
*A*3	12.53	9.47	34.86

Table 2-8. *A*3×*C* combination table.

According to Table 2-8, the smallest cell size is in the *A*3*C*2 array. Thus the optimization parameters combination is *A*3*B*3*C*2. And the experiment result is validated in the Figure 2-4.

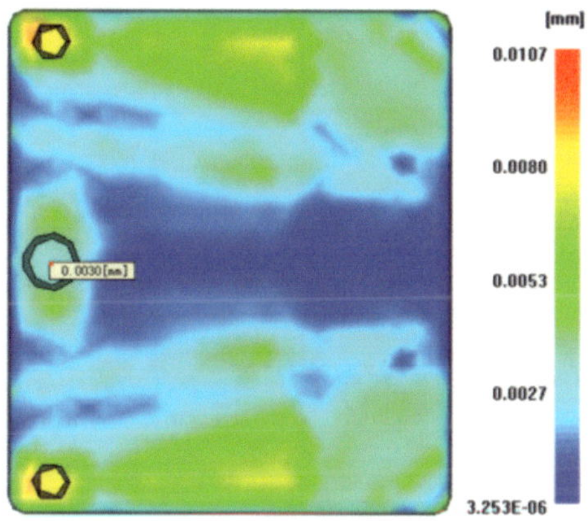

Fig. 2-4. Cell size distribution based on the optimized process parameters combination.

From the Figure 2-4, the cell radio at the characteristic point is 3 um. And the cell size on the part is between 5 um and 10 um. It means that the cell size in the part is acceptable and the distribution is reasonable. Thus the optimization parameters combination is suitable.

2.3 Bubble growth process

When the nucleation is completed, bubbles begin to grow up. Because the pressure of the mixer is higher than the pressure inside bubbles, SCF in the mixer diffuses into the bubbles and the bubbles grow up. Until the pressure inside the bubbles equals to the outside one or the melt is frozen, the bubbles will keep growing up.

2.3.1 Classic bubble growth model

Initially, the growth and collapse of gas bubbles in both viscous Newtonian and viscoelastic non-Newtonian fluids has been investigated to research on the effect of mass transfer, and the hydrodynamic interaction between the bubble and the liquid was neglected. Barlow et al. [21] are the first to study the phenomenon of diffusion-induced bubble growth in a viscous Newtonian fluid with both mass and momentum transfer. To predict the diffusion of the dissolved gas in the viscous liquid, they used a thin shell approximation. It is assumed that the gas concentration outside the shell always remained equal to the initial concentration. The simplified diffusion equation and an analytical solution were obtained to describe the initial stage of the growth at low Reynolds numbers.

Classic bubble growth model was constructed to illustrate bubble growth in foam processing after bubble nucleation. Considered a bubble concentrically surrounded by a shell of polymer melt with a constant mass, the gas dissolved in the melt shell uniformly distributes in a saturation state at the initial time and only diffuses between the melt shell and the bubble during bubble growth. Figure 2-5 shows the configuration of the bubble and the melt shell surrounding the bubble. The spherical coordinate is selected with the center of the bubble as the origin. In Figure 2-5, R is the bubble radius, S is the outer radius of the melt shell, and c is the concentration of the dissolved gas in the mixer.

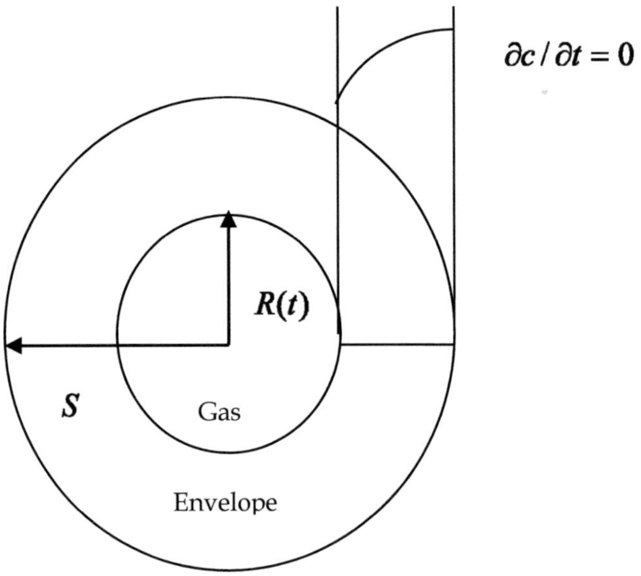

Fig. 2-5. Schematic of the unit cell model.

Before analyzing the bubble growth, the following assumptions are made.

1. Bubble and melt shell have the same and fixed sphere center during the bubble growth.
2. The gravity and inertia effects are ignored because of the highly viscous polymer melt.
3. The polymer melt is incompressible. The volume of dissolved gas in the melt is ignored.
4. Because the timescale of the bubble expansion is much shorter than the cooling time, the growth process is considered to be isothermal.
5. The dissolved gas in the polymer melt is in the uniformly supersaturated state before bubble growth.
6. The dissolved gas does not go in and out at the outer boundary of the analyzed region.

At the same time, it is also assumed that the cell shape is spherical, the initial radius is R_0,the internal gas pressure P_{g0} equals to the melt plasticization pressure and the gas in the cell is ideal gas. Thus the change rate of the radius of the cell, R, is controlled by Equation 2-9 [22]:

$$\frac{\mathrm{d}R}{\mathrm{d}t} = \frac{1}{4\eta}\left[\left(P_{\mathrm{g}} - P\right)R - 2\sigma\right] \tag{2-9}$$

where η is the melt viscosity, P_g is the gas pressure in the micro-cell, σ is the surface tension at the interface of the melt and the gas, P is the pressure of the melt at the outer boundary of the cell.

2.3.1.1 Gas diffusion

Based on the dynamics principle of cell model, the value of P_g decreases when R becomes larger. At the same time, the gas only diffuses into the cell. The gas diffusion is determined by the gas dissolution grads. The diffusion will be going on until the driven power disappears or the melt is frozen. Thus, the relationship between R and P_g can be governed by Fick´s law of diffusion:

$$\frac{\partial c}{\partial t}+v_r\frac{\partial c}{\partial r}=D\left[\frac{1}{r^2}\frac{\partial}{\partial r}\left(r^2\frac{\partial c}{\partial r}\right)\right] \quad R(t)\le r\le S \tag{2-10}$$

where c is the concentration of the dissolved gas in the mixer, v_r is the gas diffusion velocity in the radius direction of spherical coordinates, D is the diffusion coefficient in the single solution, r is the radial coordinate, t is the time, S is the radius of the cell.

The left of Equation 2-10 shows the change rate of gas concentration, while the right shows the gas mass diffusion. Where v_r can be estimated by Equation 2-11 [22]:

$$v_r=\frac{R^2}{r^2}\frac{\mathbf{d}R}{\mathbf{d}t} \tag{2-11}$$

when $t=0$, $c(r,t=0)=c_0$.

It assumes that the cell size in the same area is consistent. Thus,

when $r=S$: $\frac{\partial c}{\partial r}=0$.

And the gas concentration c between S and $R(t)$ can be calculated by Henry law:

$$c(R,t)=k_{\mathrm{h}}P_{\mathrm{g}}(t)\ , \tag{2-12}$$

where k_h is the Henry law coefficient. It is determined by the plastics and gas type and governed by Equation 2-13:

$$\ln k_{\mathrm{h}}=-2.1+0.0074T_{\mathrm{cr}}\ , \tag{2-13}$$

where T_{cr} is the critical temperature.

As said above, the gas in the cell is assumed as ideal gas. Thus $P_g(t)$ can be calculated by Equation 2-14 [23]:

$$P_{\mathrm{g}}(t)=\left[\frac{1000R_{\mathrm{g}}T}{M_{\mathrm{w}}}\right]\rho_{\mathrm{g}}(t)=A\rho_{\mathrm{g}}(t)T \tag{2-14}$$

where R_g is gas constant, $R_g=8.3145\ \mathrm{J/(mol\cdot K)}$, T is the temperature (K), M_w is gas molecular weight, ρ_g is the gas density in the cell. Here $A=1000R_g/M_W$, thus A is constant for a certain gas.

The gas diffusion coefficient D in the Equation 2-10 can be calculated by Equation 2-15:

$$D = df_1 \exp(\frac{df_2}{T}) \tag{2-15}$$

where T is the temperature, df_1 and df_2 are given coefficients.

2.3.1.2 Material properties

The melt viscosity η can be proposed by Cross-WLF model. Thus the melt viscosity η in the Equation 2-9 can be expressed by Equation 2-16 [23]:

$$\eta(T, \dot{\gamma}, p) = \eta_0(T, p) f(\phi) \left[1 + \left(\frac{\eta_0 \dot{\gamma}}{\tau^*} \right)^{(1-n)} \right]^{-1} \tag{2-16}$$

where η is the viscosity of the polymer-gas single solution, T is the temperature, $\dot{\gamma}$ is the shear rate, p is the pressure, n is the power coefficient, τ^* the critical shear stress, $\eta_0(T, p)$ is the viscosity under zero shear rate. Because SCF is added into the melt, the effect of SCF on the plastics viscosity can be expressed by $f(\phi)$. The following equation can be used to describe $f(\phi)$ [23]:

$$f(\phi) = (1-\phi)^{\alpha} \tag{2-17}$$

where α is an empirical constant. here $\alpha = 2$. ϕ is the volume fraction of the gas. It can be calculated by Equation 2-18:

$$\phi = \frac{4\pi R^3 / 3}{1/(\rho N_{\mathrm{m}}) + 4\pi R^3 / 3} \tag{2-18}$$

where N_{m} is the cell number in unit volume.

The surface tension σ at the interface between melt and gas in the Equation 2-9 can be calculated by Equation 2-19 [22]:

$$\sigma(T) = \sigma(298) \left(\frac{\bar{\rho}}{\bar{\rho}(298)} \right)^4 \tag{2-19}$$

where $\sigma(298), \sigma(T)$ are the surface tensions at the room temperature and at process temperature respectively, and $\bar{\rho}(298), \bar{\rho}$ are the densities of the single solution at the room temperature and at process temperature respectively.

2.3.2 Effect of process conditions on bubble morphology

2.3.2.1 Simulation experimental model

Based on the above mathematic model, the pre-filled volume, initial cell diameter, cell density and SCF concentration are necessary as boundary conditions besides the process parameters required by traditional injection process simulation. Finite element method and finite difference method are combined to solve the equations. To ensure the accuracy of simulation results, plastics properties used in the simulation must be recalculated based on the material model described in the "Material properties" section.

2.3.2.2 Simulation experimental model

A flat part, with the size of 320mm×280mm×2mm, is selected for simulation. Figure 2-6 shows the part geometries, gate and cooling systems. As known from the former research, the difference of cell size near the gate between the true value and simulation result is smaller. So the characteristic point near the gate is selected to study the effect of process parameters on the cell size. The position is also shown in the Figure 2-6.

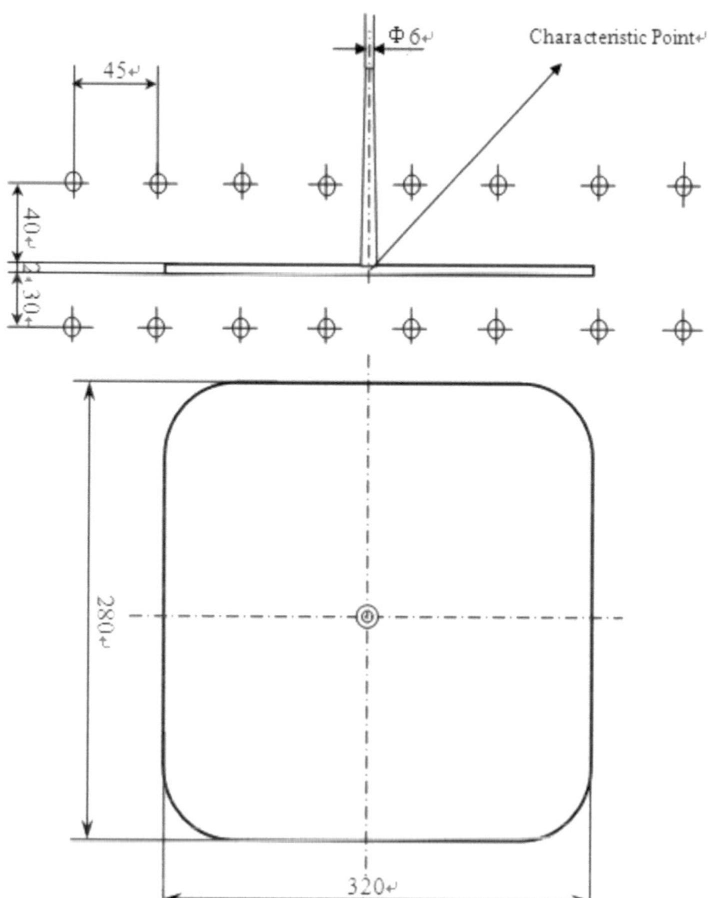

Fig. 2-6. Experiment model.

Polypropylene material is used in the simulation. Its main properties are shown in Table 2-9. Nitrogen is used as blow agent. Its main properties are as follows. $M_W = 28$, $\sigma(298) = 5\times10^{-5}$ **N/mm**, $kh_1 = 1.346\times10^{-9}$, $kh_2 = -1.709\times10^{5}$, $df_1 = 3.819\times10^{-7}$, $df_2 = -2803.5$ [24].

Main properties	Value
Eject temperature /°C	122
Max. melt temperature /°C	250
Special heat /J/kg•°C	3531
Thermal conductivity /W/m•°C	0.17
Melt density /g/cm³	0.814

Table 2-9. Polypropylene properties.

The effect of mold and melt temperatures, injection time and pre-filled volume on the cell size is studied. Based on the cooling and gate systems, recommended material parameters and the initial simulation results, the selected levels for each process parameter are shown in Table 2-10. Besides the studied four parameters, the initial values of other parameters are set as the following. nucleation density $\rho = 2 \times 10^{11}\,\mathbf{/m^3}$, initial gas concentration $c_0 = 0.25\%$, initial cell radius $R_0 = 1.0 \times 10^{-6}$ m [25].

2.3.2.3 Results and discussion

The $L_9(3^4)$ orthogonal array is used to arrange the simulation experiments. The cell sizes values at the characteristic point are calculated. Table 2-10 shows the experiment arrangement order and the simulation results. Based on the Equation 2-8, the average values of each process parameter at each level are calculated. The R_{diff} values are also achieved after the max. and min. average values are gotten. Table 2-11 shows the details.

No.	Column				Mold temp. (°C)	Melt temp. (°C)	Inj. time (s)	Pre-filled vol. (%)	Cell size (um)
	1	2	3	4					
1	1	1	1	1	10	180	1	85	28
2	1	2	2	2	10	210	1.5	90	43
3	1	3	3	3	10	240	2	95	40
4	2	1	2	3	20	180	1.5	95	25
5	2	2	3	1	20	210	2	85	47
6	2	3	1	2	20	240	1	90	42
7	3	1	3	2	30	180	2	90	34
8	3	2	1	3	30	210	1	95	37
9	3	3	2	1	30	240	1.5	85	47

Table 2-10. $L_9(3^4)$ orthogonal array, experiment arrangement and results.

	Mold temp. (°C)	Melt temp. (°C)	Inj. time (s)	Pre-filled vol. (%)
m_1	37.0	29.0	35.7	40.7
m_2	38.0	42.3	38.3	39.7
m_3	39.3	43.0	40.3	34.0
R_{diff}	2.3	14.0	4.6	6.7

Table 2-11. Direct analysis of process parameters.

According to Table 2-11, the effect of process parameters on the cell size is showed in the Figure 2-7.

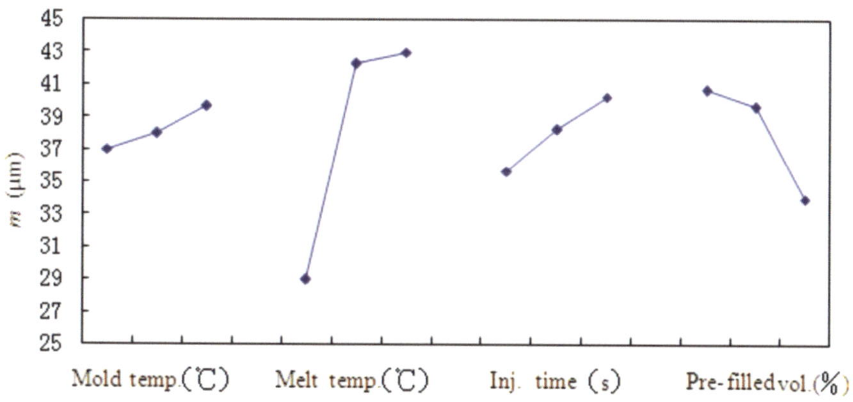

Fig. 2-7. Effect of each process parameter on cell size.

According to the R_{diff} values in Table 2-11, the effect order from big to small of each process parameter on the cell size is melt temperature, pre-filled volume, injection time and mold temperature and the optimization parameters combination is mold temperature(10°C), melt temperature(180°C), injection time(1s) and pre-filled volume(95%). Based on the above combination, the cell size distribution is shown in Figure 2-8. The cell radius at the characteristic point is 7 μm and the cell size in whole part is between 5μm and 40μm. The smaller cell size can avoid some part defects such as dimples etc. Obviously this cell size distribution can be accepted.

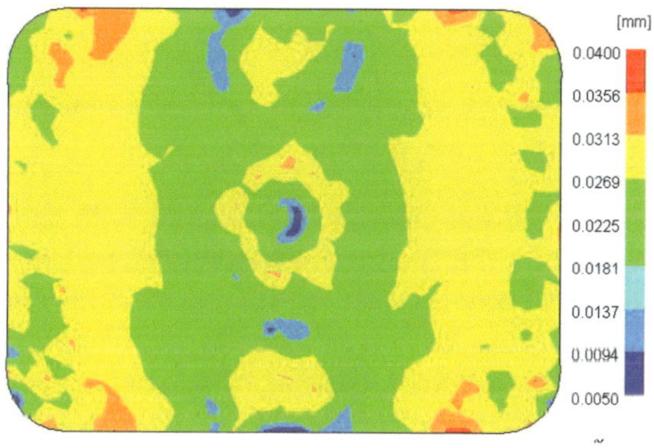

Fig. 2-8. Cell size distribution.

According to Figure 2-7, appropriately reducing the melt temperature and increasing the pre-filled volume can optimize the cell size. However the effect of injection time and mold temperature on cell size was less significant. In order to further research the effect trend of each process parameters on the cell size. More simulation experiments are done. Because the mutually effect among the selected process parameters is not taken into account, the further research is done by adjusting one of the four parameters and fixing the other parameters.

When one of the four parameters is studied, other parameters are set according to the optimization result. Table 2-10 shows the adjusted values of each parameter. So the effect trend of melt temperature, pre-filled volume, injection time and mold temperature on the cell size are achieved. Figure 2-9 shows the effect trend.

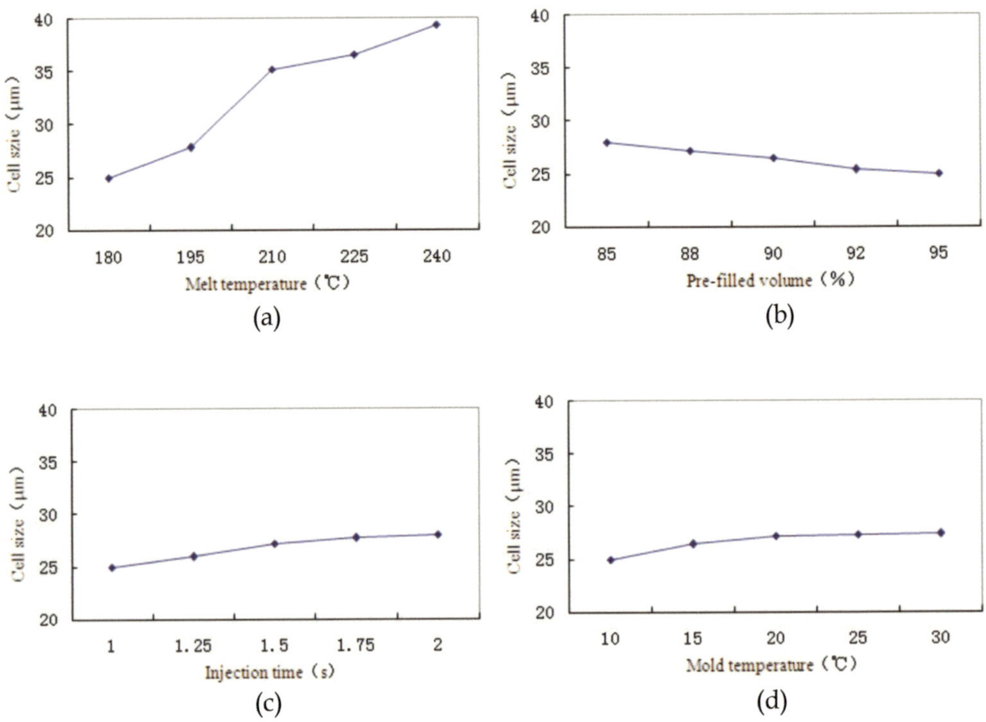

Fig. 2-9. Effect trend of each process parameters on cell size (a) melt temperature; (b) pre-filled volume; (c) injection time; (d) mold temperature.

According to Figure 2-9 (A), cell size changes largely along with temperature drop. Because of the lower melt temperature, the cooling time is shorter and the cell growth time also becomes shorter. The cell size becomes smaller. At the same time, due to the shorter cooling time, the cell growth can be controlled easily. Thus the smaller and evener cell size can be produced. With the pre-filling volume increasing, the foaming space becomes smaller. At the same time, the number of nucleation points per volume is certain. So the cell size becomes smaller. However, on the other hand, the more part weight can not be reduced with the pre-filled volume increasing. Figure 2-9 (B) shows the cell size change trend along with the pre-filled volume change. When the injection time is increased, the cell growth time also becomes longer. Thus the cell size becomes bigger. However Figure 2-9(C) shows the effect of injection time on the cell size is inferior to melt temperature and pre-filled volume. At last, according to the mathematic model of cell growth, the effect of mold temperature on the cell size is little. Figure 2-9 (D) also shows that the cell size changes little with mold temperature decreasing.

3. Microcellular foam injection molding products surface defects and solutions

As said in above chapter, microcellular foam injection molding parts have many advantages such as saving material and energy, reducing cycle time, and parts excellent dimensional stability. Despite these advantages, the low parts' surface quality limits its application scope seriously. Typical defects are swirl marks, silver streak, surface blistering, post-blow, large surface roughness. The details are introduced on above chapter.

3.1 Technologies to improve surface quality

Until now, many technologies for improving surface quality have been studied. The typical technologies include Gas Counter Pressure (GCP), Rapid Heating Cycle Molding (RHCM) and Film Insulation which is derived from RHCM.

Gas Counter Pressure (GCP)

When polymer-SCF mixer is injected into the cavity, counter pressure can prevent bubbles growth due to the high cavity pressure. When the injection is completed, the high cavity pressure is released, and then the bubbles begin to grow up. However, the surface melt has solidified at that time. So the parts surface quality can be as satisfied as traditional injection parts'.

GCP method can control the bubbles growth and remove the swirl marks. But it is not suitable for mass production due to the complex mold structure and high cost.

Rapid Heating Cycle Molding (RHCM)

Compared with conventional injection molding process, RHCM process is that the mold is rapidly heated before filling stage. The heated mold temperature is higher than the polymer thermal deformation temperature. Then the filling and packing process are going. Afterward, the mold is rapidly cooled. Finally, the products are ejected from the mold. So RHCM process circle is finished [18].

RHCM technology is widely used to improve the surface quality of injection molding parts. For example, to improve optical transparence and decrease birefringence of polystyrene, radiation heating on injection mold is proposed to directly control the temperature during the filling stage. A polycarbonate lens with a variation thickness from 1.5 mm to 7 mm can be successfully produced by electric heaters combined with chilly water cooling method.

Previous discussions about microcellular foam injection parts surface defects show that the melt temperature on the cavity surface affects the parts surface quality obviously. RHCM can meet the temperature requirement. On Oct., 2010, Trexel Inc., the supplier of the MuCell microcellular foaming technology, announced to promote MuCell for injection molding parts with Class-A/high-gloss surface finish at a global licensing agreement with Ono Sangyo Co. Ltd.. Chen SC and Li HM has successfully demonstrated the usefulness of a variable mold temperature in improving parts surface quality during microcellular foam injection molding process [14]. Figure 3-1 shows their experimental results.

Figure 3-2 shows that the effect of mold temperature on the surface roughness is very insignificant when the mold surface temperature is below 100°C. The surface roughness

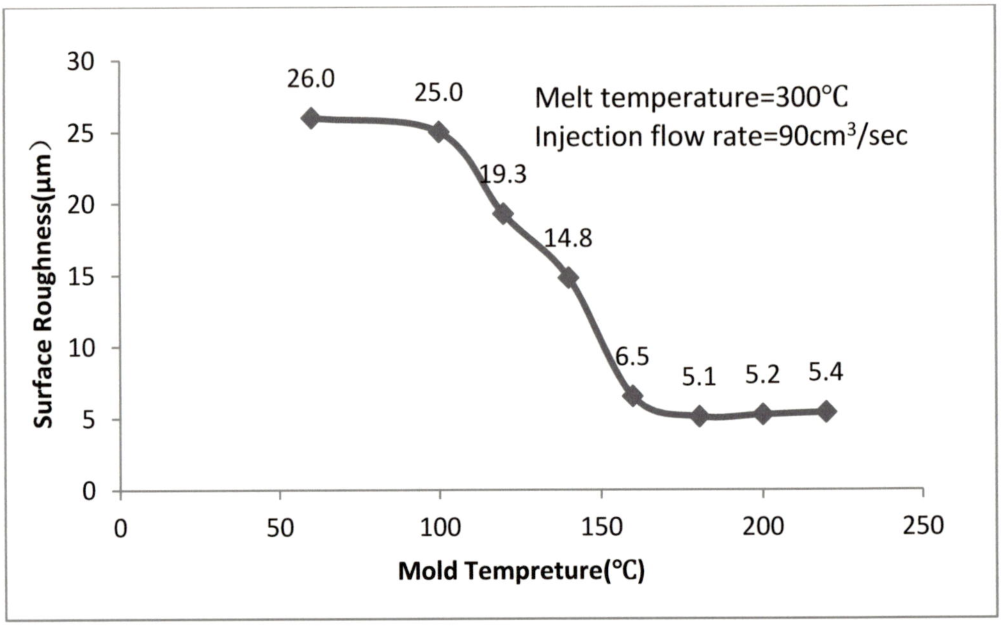

Fig. 3-1. Effect of mold temperature on the surface roughness of microcellular foam injection molded parts [26].

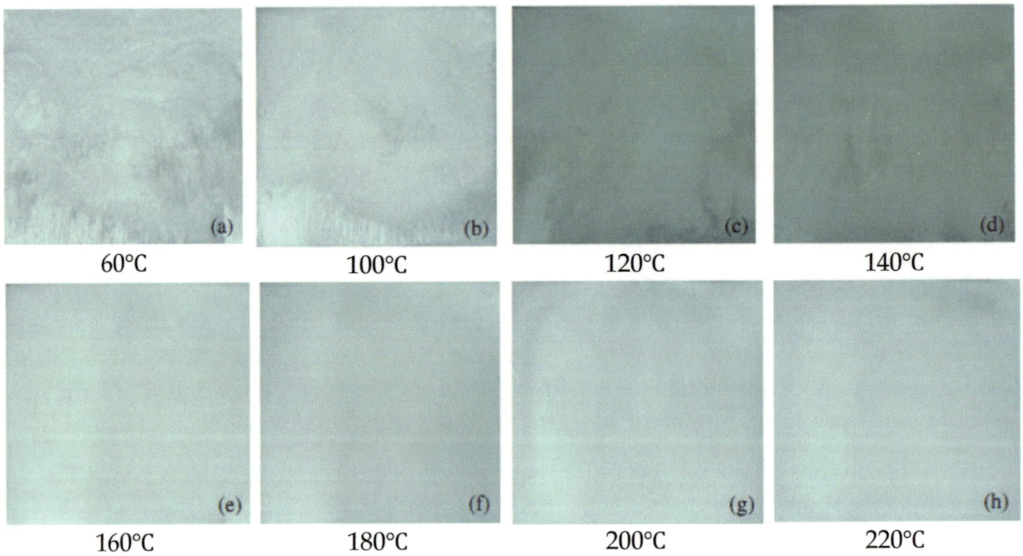

Fig. 3-2. Surface visual quality molded under different mold temperatures [26].

decreases from 25μm to 6.5μm when the mold surface temperature increases from 100°C to 160°C. When the mold temperature reached a critical value of approximately 180°C, the surface roughness begins to level off at 5μm.

Figure 3-2 reveals that visible surface flow marks were eliminated with the mold temperatures higher than 160°C. The reason is that when the temperature of the polymer-

SCF mixer is higher than its glass transition temperature or the melting point (140°C for the PC resin), gas bubbles flow marks do not form on the surface of the microcellular foam injection parts. So to improve the microcellular foam injection surface quality, RHMC process is one of useful methods.

Film insulation

RHCM technology can evidently improve the surface quality, but the heating equipment is necessary and complicated and the mold should be surface finish, good corrosion resistance and excellent hot strength. These will lead to more cost. Based on the theory of RHCM, the insulated films is stick on the surface of mold core to control the melt temperature on the cavity surface. This method is called Film Insulation. At present, the reported materials that can be used as insulation film include PEEK, PTFE, PET/PC, and so on[27-28].

Polymer film (82%PET+18%PC) is used as insulated film to improve surface quality. Table 3-1 is the experiment results [14].

Film thickness [mm]	Surface roughness [μm]	Improved efficiency [%]
0	26	0
0.125	5.6	78
0.188	1.8	93

Table 3-1. Surface toughness and improved efficiency under different thicknesses of films used for molding [14].

Table 3-1 shows that the surface roughness decreases obviously from 5.6μm to 1.8μm when the film layer thickness increases from 0.125mm to 0.188 mm. Compared with parts molded at mold temperature of 60 ℃ without film layer, the surface quality can be greatly improved without a significant cycle time increase.

PTFE insulated film is also used [29]. And the experiment results in terms of surface roughness, surface profile of conventional and microcellular injection molded parts with and without the insulated film are discussed. Table 3-2 shows the thermal analysis of the corresponding microcellular foam injection molding experiments.

Thickness of PTFE [μm]	Interfacial Temperature [°C]	Heat Fluxes [kW/m^2]
75	59	113
125	76	102
175	90	93.5
225	104	85.8

Table 3-2. Predicted interfacial temperatures and heat fluxes with different thickness of PTFE [29].

The experiment results show that the swirl marks are eliminated under the condition of the film thickness bigger than 175μm. Because of the excellent properties about low thermal conductivity (k=0.25W/(m·K)), low coefficient of friction(<0.1) and high melting point (327°C), PTFE is very suitable to be used as insulated film.

Film Insulation method makes the interfacial temperature with a thin layer of insulated film higher than that of the conventional injection mold. These results show that Film Insulation is as acceptable as RHCM.

4. Summary

In this chapter, microcellular foam injection molding process is introduced. Based on the analysis of the characters of microcellular foam injection molding process, the nucleation theory and bubble growth model are described. Then the effect of process parameters on the cell morphology is detailed studied. At last, the part surface defects of microcellular foam injection molding process are introduced. At the same time, the methods to overcome such defects are referred.

5. References

[1] Kelvin T. Okamoto. Microcellular Processing [M]. Cincinnati: Hanser Gardner publications Inc., 2003

[2] University of Washington, Microcellular Plastics Lab. [online] Available: http:// faculty. washington.edu/ vkumar/microcel/

[3] Zhai Wentao,Yu Jian,He Jiasong. Research progresses in preparation of microcellular polymers by supercritical fluid technique [J]. Chinese polymer bulletin. 2009, 3: 1-10

[4] Hyde LJ, Kishbaugh LA. The MuCell® Injection Molding Process: A Strategic Cost Savings Technology for Electronic Connectors[C]. Trexel Inc, IICIT Annual symposium, USA, 2003

[5] Bill N, Mark B. The Supercritical Fluid (SCF) Delivery System[J/OL]. Trexel. Inc

[6] Williams JM, Wrobleski DA. Microstructures and properties of some microcellular foams [J]. Journal of materials science, 1989, 24 (11):4062-4067

[7] Matuana LM, Park CB, Balatinecz JJ. Structures and mechanical properties of microcellular foamed polyvinyl chloride [J]. Cellular polymers, 1998, 17(1): 1-16

[8] Kumar V, Juntunen RP, Barlow C. Impact strength of high relative density solid state carbon dioxide blown crystallizable poly(ethylene terephthalate) microcellular foams [J]. Cellular polymers. 2000, 19(1): 25-37

[9] Ozkul MR, Mark JE, Aubert JH. Elastic and plastic mechanical responses of microcellular foams [J]. Journal of applied polymer science. 1993, 48(5): 767-774

[10] Ozkan E. Thermal and mechanical properties of cellular polystyrene and polystyrene and polyurethane insulation materials aged on a flat roof in hot-dry climate [J]. Journal of testing and evaluation, 1994, 22(2): 149-160

[11] Nimmer RP, Stokes VK, Ysseldyke DA. Mechanical properties of rigid thermoplastic foams-Part II: Stiffness and strength data for modified polyphenyleneoxide forms [J]. Polymer engineering and science. 1988, 28(22): 1501-1508

[12] Progelhof RC, Kumar S, Throne JL. High speed puncture impact studies of three low pressure styrene thermoplastic structural foam plaques [J]. Advances in polymer technology. 1983, 3(l): 15-22

[13] Yin Z, Heath RJ, Hourston DJ. Morphology, mechanical properties, and thermal stability of polyurethane-epoxide resin interpenetrating polymer network rigid foams [J]. Journal of applied polymer science. 2000, 75(3): 406-416
[14] Chen Shiachung, Li Haimei, Hwang Shyhshin, *et al.* Passive mold temperature control by a hybrid filming-microcellular injection molding processing [J]. International Communications in Heat and Mass Transfer.2008, 35: 822–827
[15] Hu Guanghong: Research on key technologies of microcellular foam injection molding process [D]. Shanghai: Shanghai Jiao Tong University, 2009
[16] Yoon JD, Hong SK, Kim JH, et al. A mold surface treatment for improving surface finish of injection molded microcellular parts [J]. Cellular polymer, 2004, 23(1): 39–47
[17] Zhang Yatao, Li Haimei, Hwang Stanley, et al.Surface Defects and Morphology of Microcellular Injection Molded PC Parts [J].Polymer materials science & engineering. 2010, 4: 127-130
[18] Wang Yuea, Hu Guanghong.Research progress of improving surface quality of microcellular foam injection parts [J]. Applied Mechanics and Materials. 2011, 66-68: 2010-2016
[19] Goel SK, Beckman EJ. Generation of microcellular polymeric foams using supercritical carbon dioxide. I: Effect of pressure and temperature on nucleation [J]. Polymer engineering and science, 1994, 34(14): 1137-1147
[20] Fang Kaitai, Orthogonal and Symmetrical Experiment Design [M]. Beijing: Scientific Publishing House, 2001: 1-17 (in Chinese).
[21] Barlow EJ, Langlois WE. Diffusion of Gas from a Liquid into an Expanding Bubble [J]. Journal of Research and Development. 1962, 6(3): 329-337
[22] Han S J, Kennedy P, Zheng R, *et al*, Numerical analysis of microcellular injection molding [J], Journal of Cellular Plastics, 2003,39(11):475-485.
[23] Gramann P, Turng L S, Chandra A *et al*, Modeling cell nucleation during microcellular injection molding [C]. SPE's Annual Technical Conference, Wisconsin, 2003:569-575.
[24] Moldflow Pty. Ltd. Moldflow Plastics Insight Training Manual [M]. Moldflow Pty. Ltd , USA, 2003.
[25] Osorio A, Turng L S, Mathematical modeling and numerical simulation of cell growth in injection molding of microcellular plastics, [J] Polymer Engineering Science, 2004 ,44(12): 2274-2286.
[26] Chun Shiachuang, Lin Yuwan, Chien Reander, *et al.* Variable Mold Temperature to improve Surface Quality of microcellular injection molded parts using induction heating technology [J].Advances in polymer technology.2008, 27(4): 224–232
[27] Takada M, Tanigaki M, Ohshima M. Effects of CO2 on crystallization kinetics of polypropylene [J]. Polymer Science and Engineering. 2001, 41(11): 1938-1946
[28] Turng LS, Kharbas H. Development of a hybrid solid-microcellular co-injection molding process [J] International polymer processing. 2004, 19(1): 77-86
[29] Chen Huili, Chien Reander, Chen Shiachung. Chen.Using thermally insulated polymer film for mold temperature control to improve surface quality of microcellular injection molded parts [J].International Communications in Heat and Mass Transfer. 2008, 35: 991-994

[30] Wikipedia, the free encyclopedia. Supercritical fluid. [online] Available: http://en.wikipedia. org/wiki/Supercritical_fluid
[31] Colton J.S,The nucleation of microcellular thermoplastic foam[D]. Massachusetts: MIT, 1985.

Part 5

Other Topics

9

Insert Molding Process Employing Vapour Chamber

Jung-Chang Wang[1], Tien-Li Chang[2] and Ya-Wei Lee[3]
[1]*National Taiwan Ocean University,*
[2]*National Taiwan Normal University,*
[3]*National Defense University,*
Taiwan, R.O.C.

1. Introduction

Insert molding process is a simplified injection molding method that eliminates secondary processing and assembly. In this process, the metal inserts are firstly formed, and placed in the mold during injection molding (the metal inserts can be designed into a grooved pattern, allowing them to be connected closely with the plastics), and then the mold is closed for injection molding. Although insert molding process can greatly improve the assembly and manufacturing procedure, the joining of two materials is the main problem yet to be solved. Because the molten plastic drives the air out of the mold cavity from the vent resulting in welding lines of a plastic part during the filling step in an injection molding process, then there will be a V-notch formed between the plastic and the mold wall if the air cannot exhaust before the melted plastic fronts meet. Thus, this chapter concerns about a V-notch found on the exterior surfaces of welding lines, which will form between the plastic and the mold wall. Not only are they appearance defects, but they also decrease the mechanical strength of the parts. Once the plastics are filled, the temperature of plastics bypassing one side of the inserts may decline more quickly than that of plastics contacting the temperature side of the mold wall, so a weld line may be formed when meeting again after bypassing the inserts. The strength at the position of weld line is generally lower than that at the region of plastics; moreover, the metal inserts are generally located at the stress region when the product is utilized. Hence, the rupture of plastics often occurs from the weld line at the rear of the metal inserts, leading to damage of products (Wang & Tsai 2011). The key approach is to rapidly and uniformly increase the temperature of inserts before the plastics enter into the mold cavity.

Some studies indicated that, during the injection molding process, the defect of weld line could be resolved by adjusting the mold temperature. The locations of the welding lines are usually decided by the part shapes and the gate locations. Welding lines can be eliminated by the following three methods. The first method is increasing the molten plastic temperature. The viscosity of the molten plastic fluid decreases with the increasing temperature, which improves the flow pattern of the plastic, and reduces the depth of the V-notch of the welding lines. However, degradation of the material strength sometimes occurred if the melting temperature is too high. The second method is increasing the number of the vents. Increasing the number of the vents (eg. ejecting pin or inserts) at the vicinity of the welding line will dispel air from mold cavity more easily. Nevertheless, these extra vents will leave marks on the product surfaces, especially for transparent plastics. The third method is raising the mold temperature. Raising the mold temperature will increase the viscosity of the material, which in turn can reduce the depth of the V-notch. On the other hand, raised mold temperature also increases cycle time of the injection molding process, thus increases the production costs. If increased cycle time is not a problem then raise mold temperature is the simplest way for elimination of the welding lines among the three methods mentioned above.

There are several methods for raising the mold temperature. For example, replacing the water in the mold temperature controller for oil can increase the mold temperature for about 100˚C. But the time required for the mold to rise to the designed temperature is too long to be practical. The mold heating is achieved by using high pressure vapour as a media in this method. When the filling stage in the injection molding process was finished, high pressure vapour was introduced for cooling. The technology does achieve a very fast heating and cooling cycle, however, its application was limited by the mold size. Wang et al. (2011a) demonstrated a vapour chamber of a two-phase flow heat transfer device consists of sealed container, wick structure and working fluid can occupy the technology easily. Its working theorem is as same as the heat pipe and generates high pressure vapour as the process of Rapid Heat Cycle Moulding (R.H.C.M.). Vapour chamber (V.C.) is verified through experiment and theory that the temperature difference between the center and edge of vapour chamber-based thermal module is within 1˚C and this temperature difference for the copper Integral Heat Spreader (IHS) with the same size at the same experimental parameters are more than 6˚C (Wang, 2011a). Therefore, vapour chamber represents better temperature homogeneity. In recent years, technical development related with the application of two-phase flow heat transfer assembly to thermal modules has become mature and heat pipe-based two-phase flow heat transfer module is one of the best choices (Wang & Wang, 2011). The thermal performance of a heat sink with embedded heat pipes has been developed a Windows program for rapidly calculating through Visual Basic commercial software (Wang et al., 2010b). Moreover, one set of risers of the L-shape heat pipes were functioning as the evaporating section while the other set acted as condensing section. Six L-type heat pipes are arranged vertically in such a way that the bottom acts as the evaporating section and the risers act as the condensing section (Wang, 2011c). The temperature difference is under 1° C while the percentage of the non-condensation gas is less than 8×10^{-5}%, and the single heat pipe has the maximum heat capacity (Wang, 2011b). To establish a practical quick methodology that can effectively and efficiently determine the thermal performances of heat pipes so as to substitute the use of the conventional steady-state test. This is much more efficient than the steady-state test and would be greatly beneficial to the notebook PC industry or other heat dissipation technologies that use heat pipes.

Fig. 1 showed a heat pipe / thermal pin employed in the injection molding process over several decades. Liquid thermoplastics require the cooling inside the injected cavity. The mold must be equipped to provide the cooling surroundings. Thermal pins had become standard components to cool areas that are difficult to reach with water lines. A commercially available thermal pin with heat conductivity several magnitudes higher than core, which accomplishes better cooling effect and more uniform temperature distribution. However, thermal pin is just only one-dimensional heat conducting path. Vapour chamber

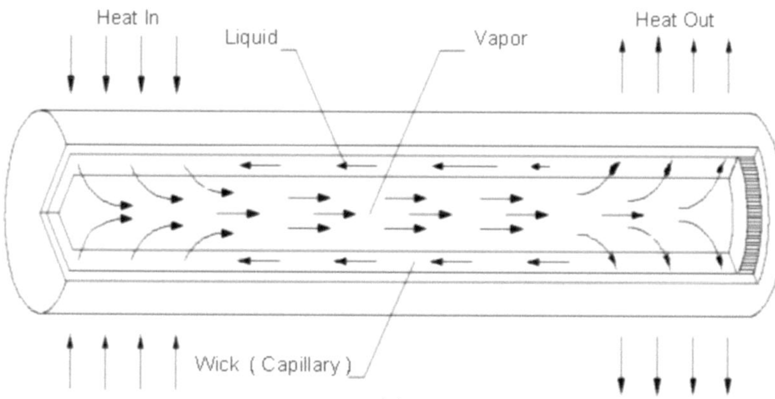

(a)

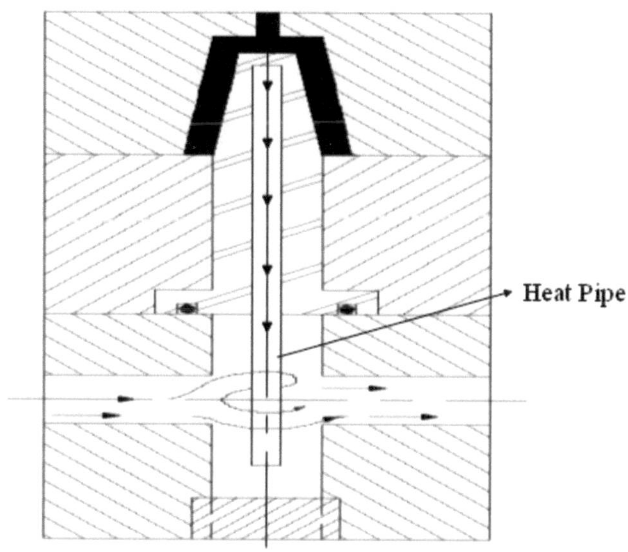

(b)

Fig. 1. Heat pipe / thermal pin for mold cooling.

may regard as many thermal pins to heat spreading area. Wang et al. (2011a) uses the local heating mechanism, along with the excellent thermal performance of vapour chamber, to analyze and enhance the strength of products formed after insert molding process. Finally, it has been found that a vapour chamber based rapid heating and cooling system for injection mold is proposed for reducing the welding lines of the transparent plastic products and enhancing their strengths through these results.

Vapour has advantages of fast, large amount and safety. The effectiveness and better thermal performance of vapour chamber has been already confirmed according up-to-date researches and mass production application in server system and VGA thermal module. Moreover, vapour chamber-based thermal module has existed in the thermal-module industry for a year or so especially in server application (Wang & Chen 2009; Wang 2010; Wang et al. 2011b). A novel formula for effective thermal conductivity of vapour chamber has been developed by use of modified dimension analysis in combination with thermal-performance experimental method (Wang & Wang 2011). It is deduced from the novel formula that the maximum effective thermal conductivity is above 800 W/m°C, and comparing it with the experimental value, the calculating error is no more than ±5%. It respectively discussed these values of one, two and three-dimensional effective thermal conductivity and compared them with that of metallic heat spreader. For metallic materials as the heat spreaders, their thermal conductivities have constant values when the operating temperature is slight change. The thermal conductivities of pure cooper and aluminum as heat spreaders are 401 W/m°C and 237 W/m°C at operating temperature of 27 °C, respectively. When the operating temperature is 127 °C, they are 393 W/m°C and 240 W/m°C, respectively. The maximum heat flux of the vapour chamber is over 800,000 W/m^2, and its effective thermal conductivity will increase with input power increasing. Thermal performance of V.C. is closely relate to its dimensions and heat-source flux, in the case of small area vapour chamber and small heat-source flux, the thermal performance will be less than that of pure copper material.

A vapour chamber is a two-phase heat transfer components with a function of spreading and transferring uniformly heat capacity so that it is ideal for use in non-uniform heating conditions especially in LEDs (Wang 2011d). The solid-state light emitting diode (SSLED) has attracted attention on outdoor and indoor lighting lamp in recent years. LEDs will be a great benefit to the saving-energy and environmental protection in the lighting lamps region. Wang et al. (2010a) introduce a thermal-performance experiment with the illumination-analysis method to discuss the green illumination techniques requesting on LEDs as solid-state luminescence source application in relative light lamps. The temperatures of LED dies are lower the lifetime of lighting lamps to be longer until many decades. The thermal performance of the LED vapour chamber-based plate (LED-VCPCB) is many times than that of LED copper- and aluminium-based plates (Wang & Huang 2010; Wang 2011d). Wang et al. (2010a) utilizes experimental analysis with window program VCTM V1.0 to investigate the thermal performance of the vapour chamber and apply to 30 Watt high-power LEDs. Results show that the maximum effective thermal conductivity is 965 W/m°C at 187.5 W/cm^2. Anyway, the chapter is divided into two parts; first part examines a vapour chamber to rapid heating and cooling system for injection molding process. Second part employs the vapour chamber system with numerical method and performance experiment to investigate the key factors affecting the performance of the Self-Contained Underwater Breathing Apparatus (SCUBA) system utilizing the insert molding process (Tsai et al. 2011).

2. Vapour chamber for rapid heating and cooling system (VC_RHCS)

Vapour Chamber (V.C.) is a two-phase heat transfer device inside vapour-liquid working, which has better thermal performance than metallic material in a large footprint heat sink. The overall operating principle of V.C. is defined as follows: at the very beginning, the interior of the vapour chamber is in the vacuum, after the wall face of the cavity absorbing the heat from its source, the working fluid in the interior will be rapidly transformed into vapour under the evaporating or boiling mechanism and fill up the whole interior of the cavity, and the resultant vapour will be condensed into liquid by the cooling action resulted from the heat convection effect between the fins and fan on the outer wall of the cavity, and reflow to the place of the heat source along the capillary structure. Thus, vapour chamber spreads large amount heat flow rapidly and is feasible to utilize in the insert molding process, which the metal insert is firstly placed into the mold, and then formed into an embedded plastic product by injection molding. Therefore, the products can be formed in a single molding process, helping to shorten the processing time and reduce the possible human error arising from several procedures. These common defects are often found on the exterior surfaces of welding lines, which are not only appearance defects; they also decrease the mechanical strength of the parts. When the metal inserts are placed in the mold for injection molding, a weld line may be formed from the plastics after bypassing the inserts. Traditionally, the inserts are placed in the mold at room temperature, but the temperature of inserts is lower than that of the mold (the temperature of inserts cannot rise sharply within a very short period) when the filling is conducted after mold sealing.

2.1 Verification of VC_RHCS

Fig. 2 reveals the apparatus of VC_RHCS employing vapour chamber to rapid heating and cooling system for injection molding process. A local heating mechanism of vapour chamber proposed that the mold is a moveable slider, at inner side of which a heating rod is installed; a vapour chamber is mounted at the joint of the frontal edge of the slider and the mold wall. When the inserts are placed in the mold and the mold is closed, the heated slider will move forward, and contact the vapour chamber, so that the inserts are heated up rapidly and quickly with the vapour chamber. A heating and cooling system with vapour chamber was developed in this chapter. A vapour chamber was installed between the mold cavity and the heating block. The heating and cooling system used in the experiment was 50 x 50 x 80mm^3. It can be incorporated with the mold at any location regardless of the mold size. The heating cycle is activated by a lever mechanism which pushes the vapour chamber to contact with the mold at the filling stage. The lever pusher is a hydraulic cylinder. Two electrical heating tubes are provided a mold steel block named P20 material and thermocouples are embedded to measure the temperature of the heat insert device. The heat source for vapour chamber is a low density cartridge heater which contact with the vapour chamber only when the lever is activated. When the filling completed, the heat source separates form the vapour chamber.

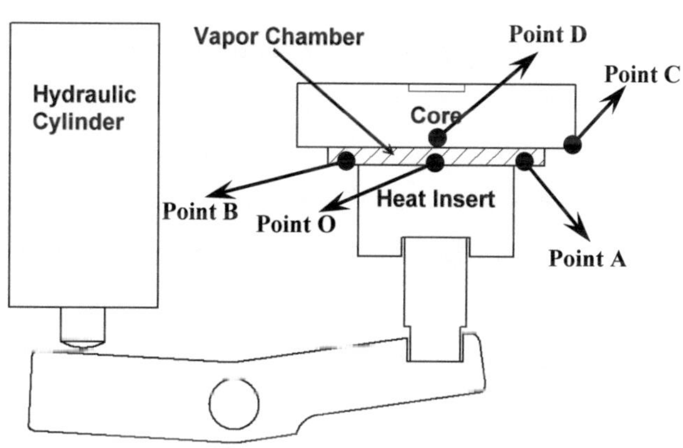

Fig. 2. Mechanics of heating and cooling cycle system with vapour chamber (VC_RHCS).

There are five thermal couples placed on the surface of cavity to measure the temperature of vapour chamber as shown in Fig. 2. Points A and B are measured the temperatures of the opposite to the position of the heat insert device on the surface of cavity. Points C and D are measured the temperature of the opposite to the position of vapour chamber on the surface of cavity. Point O is measured the central temperature of cavity and vapour chamber. And the point C is the farthest distance away from point O. Mold temperature was raised to above the glass transition temperature of the plastic before the filling stage. Cooling of the mold was then started at the beginning of the packing stage. The entire heating and cooling device was incorporated with the mold. The capacity and size of the heating and cooling system can be changed to accommodate the mold. The plastic was ABS (Chi-Mei PA-758) in the experiment, its glass transition temperature is 109°C. The material for mold base is JIS S50C. The material for mold core is ASSAB 718. There are two injection molds estimated the products using vapour chamber and the testing material is ABS. The VC_RHCS can raise the tensile strength and reduce the defect of the welding lines of a plastic product because of rapid-uniform heating and cooling cycle with vapour chamber.

These temperature curves of zero to sixty seconds with / without vapour chamber are shown in Fig.s 3. In the Fig. 3(a), the temperature of point O is 73.9°C and the farthest corner away from point O is 34.7°C of point C at 60 seconds. Furthermore, temperature differences of the other points A, B, D are large values about 10°C each other at 60 seconds. In the Fig. 3(b), the temperature of point O and point C is 83.7°C and 67.8°C at 60 seconds, respectively. In addition, the temperatures of the other points A, B, D are almost 80°C at 60 seconds. The temperature curve of point O of Fig. 3(a) is faster raise than that of Fig. 3(b) at zero to thirty seconds resulting from the heater direct contact and slows down after thirty seconds. In other words, the temperature of point O of Fig. 3(a) is higher than that of Fig. 3(b) before thirty seconds. The mean temperature of these five points with vapour chamber and without vapour chamber is 78.3°C and 56.1°C, respectively. Rapid uniform temperature effect employing vapour chamber heating system is better even if the heater is direct touching point O on the surface of cavity for no vapour chamber system.

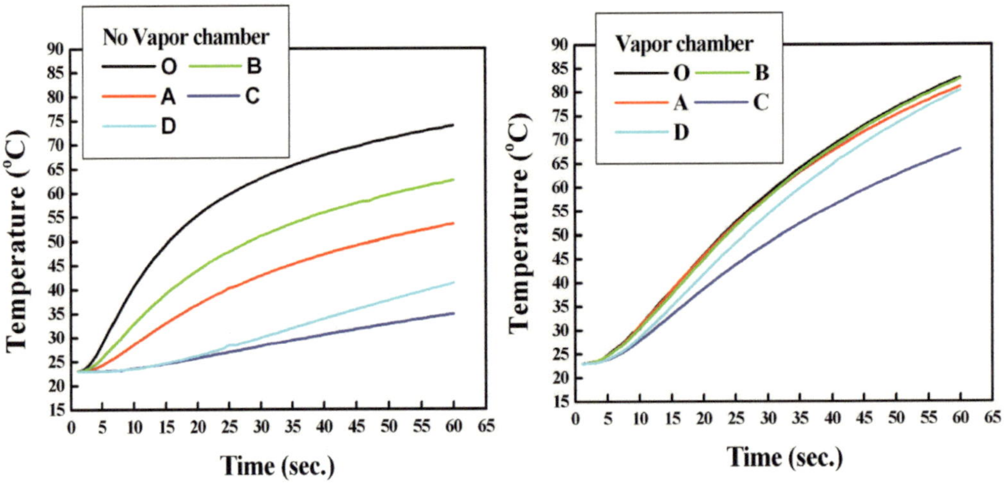

(a) (b)

Fig. 3. Relationships of the temperatures with the heating time with/without vapour chamber.

2.2 Employing VC_RHCS in tensile testing parts

The products of the injection molding experiments were tensile testing parts. They are evaluated the effectiveness of the proposed heating and cooling system with vapour chamber and without vapour chamber. There are two different mold designs namely one gate mold and two opposite gates mold were selected for the experiments. Fig. 3 shows the relationships of the temperature with the heating time with/without vapour chamber. If there are no the effect of vapour chamber, these five thermal couples show non-uniform on the surface of cavity. However, the dimensions of these tensile testing parts are 215.9 x 12.7 x 25.4mm^3 with 3.175mm thickness.

Table 1 shows the tensile strength of the products in different situation of one gate and two opposite gates with and without vapour chamber. The tensile strength supposes maximum for one gate without vapour chamber heating system because of on welding line from the Fig. 4. Fig. 4 shows the test specimen and SEM images of welding line. One gate without vapour chamber system experiments resulted tensile test pars without any welding lines. Two gates without vapour chamber heating system appears apparent welding line and its tensile strength of testing part is lower 11.1% than the part of one gate from table 1 and Fig. 4.

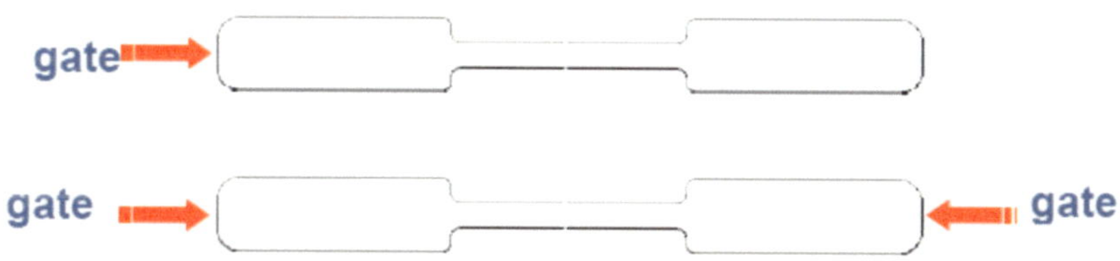

(a) Different design for tensile test specimen

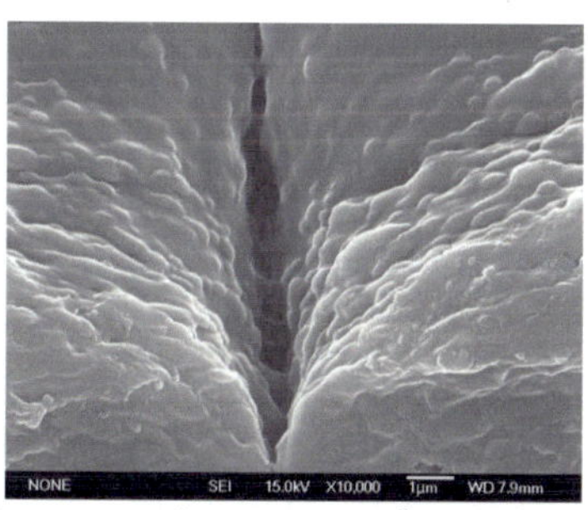

(b) Two opposite gates at the temperature of Point O is 75°C without vapour chamber system

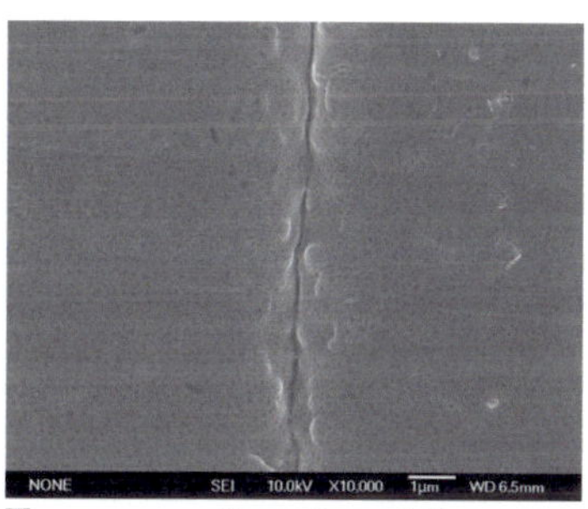

(c) Two opposite gates at the temperature of Point O is 75°C with vapour chamber system

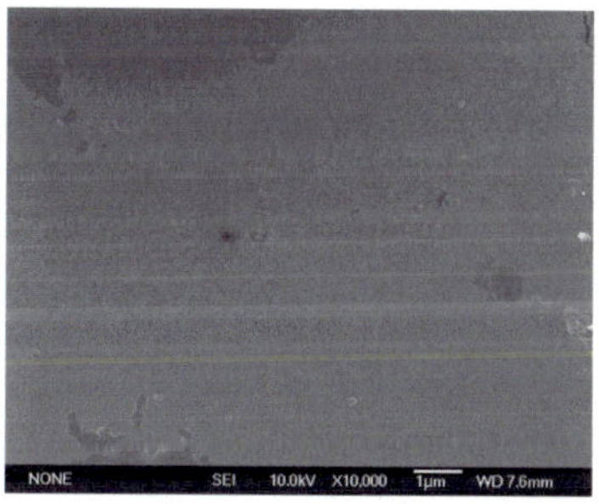

(d) Two opposite gates at the temperature of Point O is 110°C with vapour chamber system

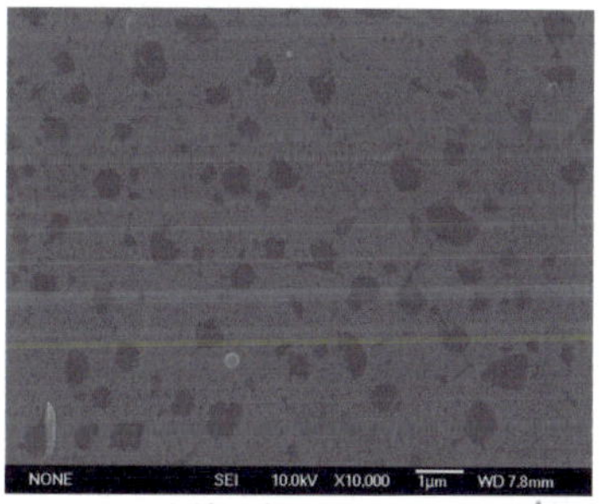

(e) One gate at the temperature of Point O is 75°C without vapour chamber system

Fig. 4. SEM pictures of one gate and two opposite gates.

Condition	Max. stress (Kgf / cm^2)	Strength (%)
Two gate without vapour chamber Cavity temp. = 75°C	165	88.9
Two gate with vapour chamber Cavity temp. = 75°C	178	95.7
Two gates with vapour chamber Cavity temp. = 110°C	184	98.9
One gate without vapour chamber Cavity temp. = 75°C	186	100.0

Table 1. The results of the tensile test.

Two gates with vapour chamber heating system appears light welding line and its tensile strength of testing part is higher 6.8% than the part of two gates without vapour chamber system. If increasing the preheating temperature from 75°C to 110°C with two gates vapour chamber system, the tensile strength can again add 3.2%.

2.3 Employing VC_RHCS in eight holes test part

Another test part is the multi-holes products tested to evaluate the effectiveness of the system with vapour chamber in different situation of cavity and core temperatures. These multi-holes products have eight holes with four 10mm and 5mm diameter holes, respectively. The dimensions of eight holes products are 110 x 53 x 3.175mm^3. Three temperature combinations were tested in the experiments. Case 1 is the conditions of cavity temperature 60°C and core temperature 60°C. Case 2 is the conditions of cavity temperature 60°C and core temperature 130°C. Case 3 is the conditions of cavity temperature 80°C and core temperature 130°C.

Figs. 5 show the specification of the eight holes test part and SEM images of V-notch. The means is that there are many welding lines on the surface of the transparent parts. The depth of the V-notch is deeper the welding line is more obvious for transparent. From the Figs. 5, the depths of the V-notch found on each case are 12μm, 2μm, and 0.5μm respectively. The product of the case 1 shows a V-notch 24 times deeper than the product of case 3. The effects of cavity and core temperatures are also important for welding line. Finally, utilizing VC_RHCS shows that the temperature differences of cavity employing vapour chamber are smaller than that without vapour chamber and increasing preheating temperature can add the tensile strength for two opposite gates resulting from extending enough fluid flow. And the new VC_RHCS can reduce the depth of v-notch as much as 24 times.

3. SCUBA system utilizing VC_RHCS

The underwater environment imposes many physical and mental stressors on those working in the modern SCUBA (Self-Contained Underwater Breathing Apparatus) diving equipment. A scuba system consists of a high pressure compressed air tank and a pressure regulator. The development of SCUBA diving is based on the invention of the regulator. The high pressure air carried by diver must be reduced to the pressure in the ambient environment by the regulator before the diver can breathe it. The life of the diver thus depends on the performance and stability of the regulator. The SCUBA is based on the

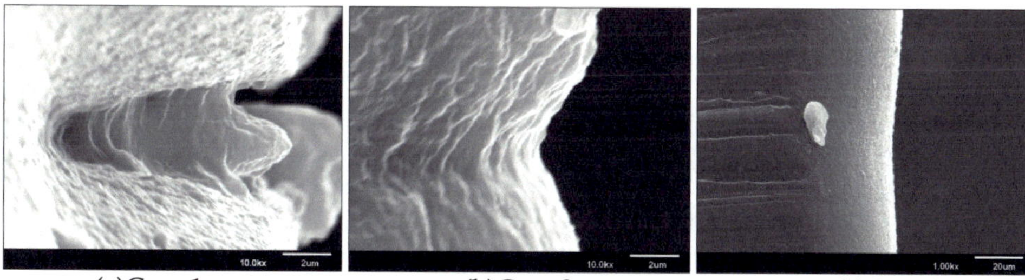

(a)Case 1 (b)Case 2 (c)Case 3

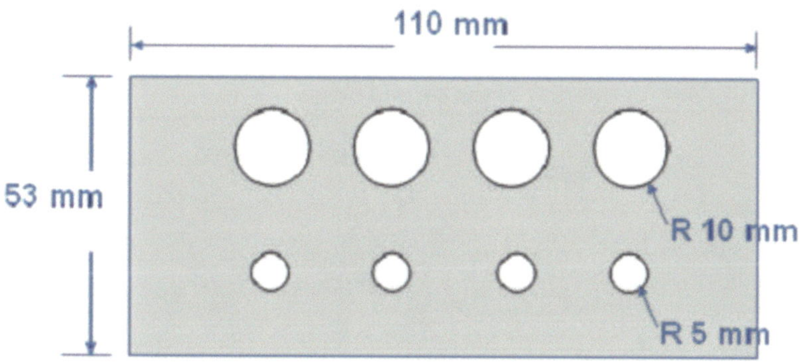

(d) Specification of the eight holes test part

Fig. 5. SEM picture of the eight holes plate.

invention of breathing regulators. Basically, a regulator can be divided into the first and second stages. The first stage regulates air output pressure at a stable value of 14atm, reducing from the compressed air cylinder pressure of 200atm to 20atm, and then supplies that air to the second stage. The second stage is connected to the divers' mask and supplies air at ambient pressure based on the diver's respiration.

Underwater movement requires a higher expenditure of energy and an increased rate of respiration than the same movement on land. This is caused by the higher density of water, and the need to overcome water resistance or drag. The second stage supplies air at a pressure equivalent to water pressure, which in turn depends on the depth, in order to lessen the diver's lung burden. A structural representation of the second stage is shown in Fig.6, while Fig. 7 gives a schematic of the second stage. The second stage of the regulator thus affects the smoothness of diver breathing, a key factor in the divers' ability to function underwater. During this experiment, the relationship between characteristic factors and performance of breathing regulators will be explored through numerical methods, to enable designers to create more efficient breathing regulators.

Breathing resistance is directly related to the diver's ability to receive sufficient air to safely perform in the underwater environment and must be considered a primary factor in the design of breathing apparatuses. An ideal breathing regulator allows divers to breathe without consuming additional energy when breathing under water. In other words, in the ideal regulator, respiratory work rate and respiratory impedance are zero. Many patents (Belloni 2001; Brown & Brown 2000 & 2002; Christianson 1987; Ferguson 1997; Garraffa 1997; Garraffe 1996; Hansen & Lingenfelter 1987; Houston 1981; Toth1985) involve optimum designs for breathing regulator. These patents discussed above provide researchers with a

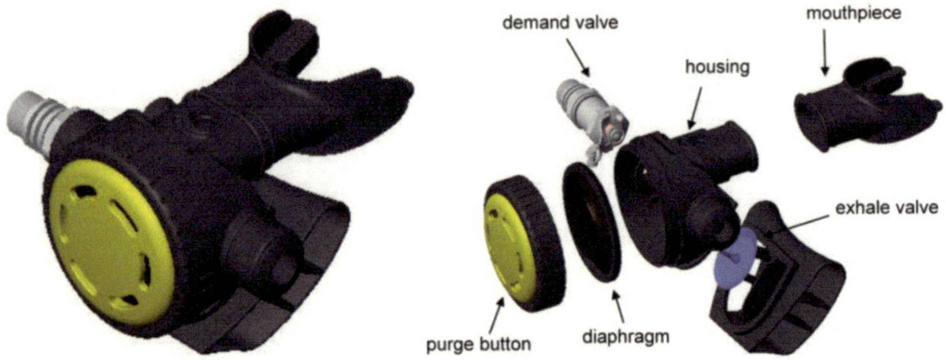

Fig. 6. The structural representation of the second stage.

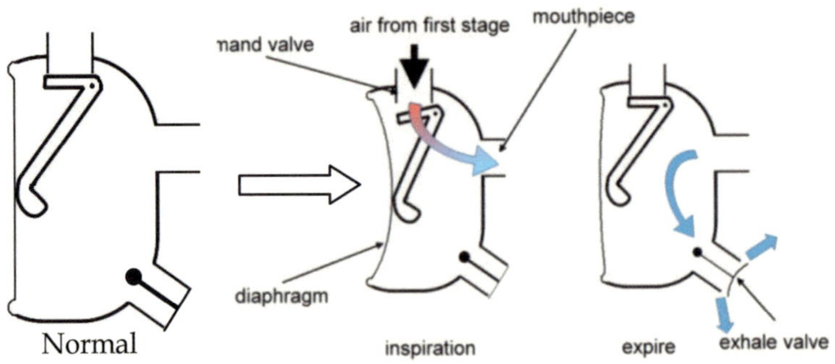

Fig. 7. The schematic diagram of the second stage.

reference for developing new breathing regulators. However, in previous studies, the parameters affecting the breathing regulator's performance have not been identified. This research identifies the design parameters affecting regulator performance through experimental and numerical methods.

The analysis is divided into two steps in this section. In the first step, the mechanical characteristics of samples are varied and the relevant performance parameters of the breathing regulators are tested using demand regulator testers made by ANSTI Co., as shown in Fig. 8. In the second step, we modify the characteristic internal shapes of the regulators and explore how these changes influence regulator functions using commercial numerical simulation software, FLOW-3D® from Flow Science Inc. Company. There are several important considerations to design of breathing regulators for diving apparatus. A regulator must meet the requirements for easy breathing and a stable air supply at normal conditions. Easy breathing is the most key index of a regulator. However, to a great extent, the above factors are shaped by the subjective feeling of the user. In order to accurately describe breathing smoothness, designers have defined two parameters, namely respiratory work rate and respiratory impedance.

The unit of respiratory work rate is Joule/Liter and its value represents the average energy consumed by breathing in and out one liter of air. Respiratory impedance can be regarded as the pressure supplied by a diver when breathing underwater. For example, in order to smoothly discharge gas through a regulator when diving, a diver must apply a pressure using the thoracic cavity and mouth to force the valves to open. To breathe in, additional

Fig. 8. The regulator testers made by ANSTI Co.

negative pressure must be produced to open the valves. These pressures are cumulatively known as respiratory impedance. Respiratory impedance values are widely used to evaluate various breathing mask products, such as filter-type dust masks and gas masks. A picture of the regulators under this study is shown in Fig. 9. A deflecting plate is located at the front end of the air inlets, whose angle can be adjusted by the user. There are two spray holes, one in the upper part and the other in the lower part, directing air flow towards the diaphragm, inside the housing. The factors under consideration in this research are the deflecting plate and the spray holes.

Fig. 9. A picture of the test regulator.

For the numerical analysis, the entire analytical model is established by utilizing file conversion skill between CAD/CFD. The overall dimension of simulation analytical model is about 30 x 30 x 25 cm^3. Schematic process of numerical simulation analysis adopted in the article can be divided into pre-processing, numerical solving and post-processing. With regard to pre-processing, first of all, a geometrical model is established for 3D CFD module. Generally, in order to reduce computation grid elements and time taken for simulation and solving, some minor characteristics without influence or with a little influence will be ignored when establishing 3D geometrical model. And input the boundary conditions and thermo-physical properties, which the ambient temperature is set to 27 °C, the input pressure is set to 9 atm and initial pressure is set to be the surrounding conditions, turbulent model is the k-ε two-equations, the grid pattern is structural one and the entire simulation analysis type is transient time state. For the entire module, about 1,500 thousands grid elements are used, time step is about 10^{-9}, iterations is about 50 per time-step and it will take about 14 days to simulate every scenario.

The following paragraphs will describe the functions of the three parts.

3.1 Flow channel

A flow channel functions to direct gas from the valves into the diver's mouth. Flow channel design is thus closely related to respiratory impedance, respiratory work rate, and gas flow velocity. Fig. 10 shows three types of flow channel design for the simulation models. Model B1 represents the original design; Model B2 is a lengthened flow channel which completely covers the entire suction port, while Model B3 is short flow channel design.

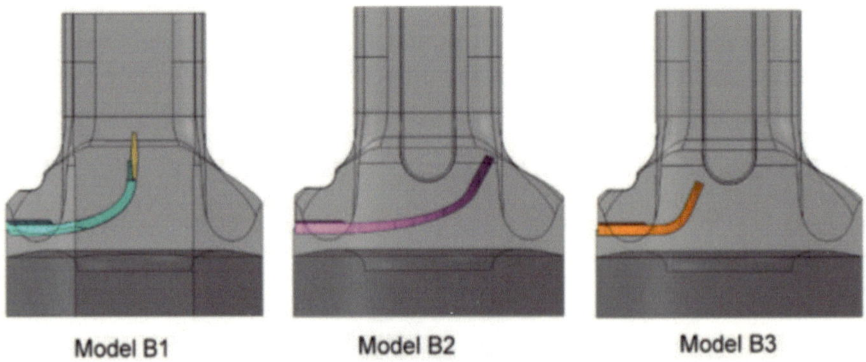

Fig. 10. The design of Model B1~B3 modified the length of the flow channel.

The analytical data is given in Table 2 and pressures obtained from the three designs are compared in Fig. 11. The offset is the difference between maximum pressure inside the housing and mouthpiece, which obtained from simulated data of FLOW-3D®. Our simulations show that when an excessively long flow channel covers the upper part of the suction port, the gas inside the housing will be blocked, unable to enter the suction end, resulting in a pressure rise inside the housing. By the same token, an excessively short flow

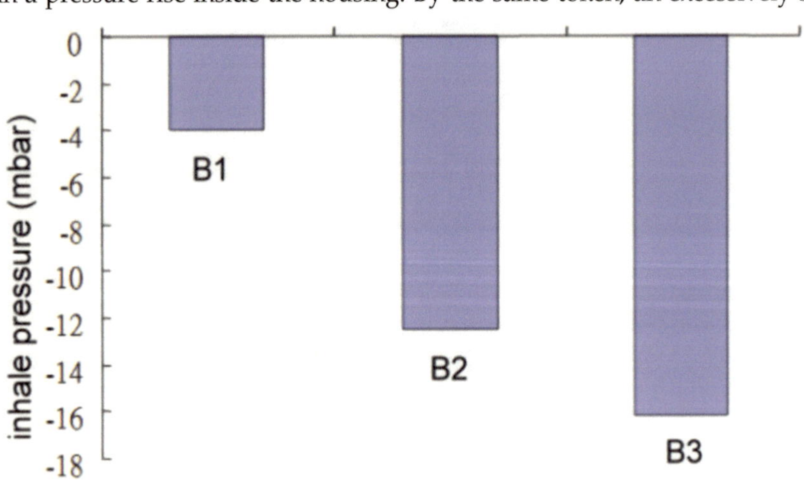

Fig. 11. Inhale pressure of different length of flow channel.

Model	pressure inside the housing (bar)	pressure in the mouthpiece (bar)	Offset (mbar)
B1	6.899	6.931	-4
B2	6.883	6.9065	-12.5
B3	7.0272	7.047	-16.2

Table 2. Comparison of different length of flow channel.

channel will cause the gas inside the flow channels to flow into the housing, creating a pressure drop at the suction end. The offset have to be close zero pressure drop. Our numerical results showed that the original design "B1" obtains the best inhale pressure.

3.2 Angle of the spray holes

After entering the second stage from the spray holes, high-speed gas strikes the diaphragm, which results in a pressure rise inside the diaphragm. A change in the angle of the spray holes will affect the spray whole's pressure shock against the diaphragm. The authors modified the characteristics of actual samples and tested these modified samples on a tester bench. A comparison was made between experimental data and simulation values obtained by FLOW-3D® to confirm that the mode for setting boundary conditions, in which air is vented out at a constant flow rate, is consistent with actual conditions. These modifications mainly aim at the characteristics of the dimensions of period, line and depth. After confirming that the data obtained from numerical model is consistent with experimental data.

Table 3 shows the characteristics of the five groups of samples were varied across experiments (A1 to A5) in order to verify that numerical simulation data and experimental data reliable. These modifications are mainly aimed at the characteristics of deflector holes and spray holes.

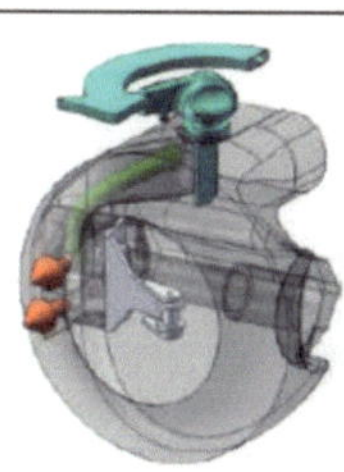

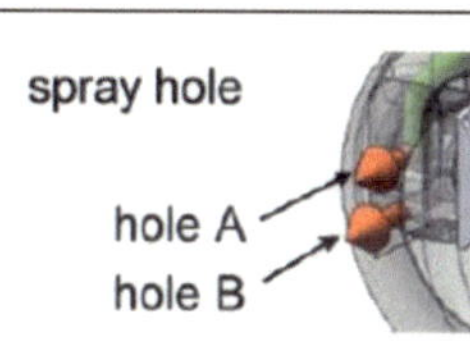

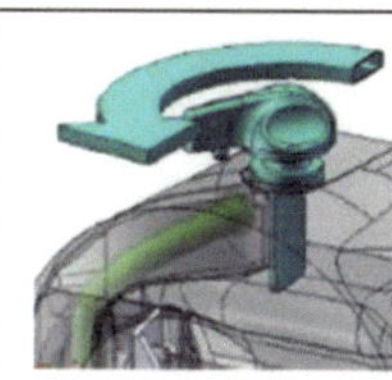

	spray open=O, close=X		The angle of the flow channel
	A	B	
A1	O	O	0°
A2	X	O	0°
A3	X	X	0°
A4	X	X	22.5°
A5	X	X	45°

Table 3. The test parameters of the regulator.

As we have seen, a comparison of the experimental data and numerical simulation results is given in Table 4. And the comparison of the experimental data and the simulated data of inhale pressure are shown in Fig. 12. The results show that the simulated data are acceptably close to the experimental data. After confirming that the data obtained from numerical model is consistent with the experimental data, the authors changed the characteristics of the second stage mechanisms and then determined how the shape characteristics affect regulator performance, using numerical simulation.

Model	pressure inside the housing (bar)	pressure in the mouth piece (bar)	Offset (mbar) (Simulated Data)	Inhale pressure measured from ANSTI test machine (mbar) (Experimental Data)
A1	7.0905	7.111	-16.4	-15.5
A2	6.899	6.931	-5	-4
A3	8.468	8.515	8.43	11
A4	6.655	6.696	5.56	5
A5	6.654	6.683	-6.89	-7

Table 4. The experimental results VS. The results of simulation.

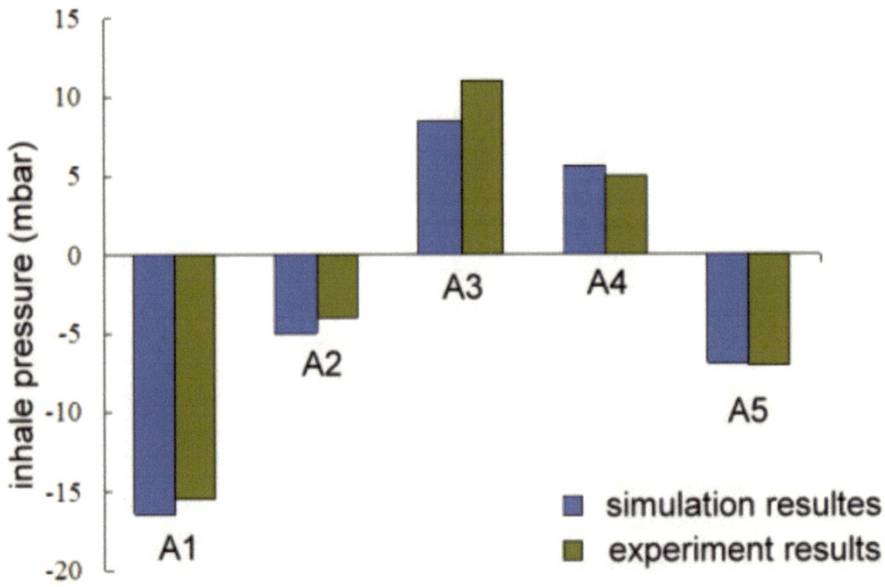

Fig. 12. Simulation VS experimental results in inhale pressure.

In our model design, Model A2 is used as a control, and the angle of its lower spray hole B is varied. Three designs are adopted, designated C1, C2 (Fig.13) and C3 (Fig.14). The key point in the C3 design is to prevent the gas from directly striking the housing and instead allow it to pass along the housing edge when gas enters the housing. The analytical data is summarized in Table 5 and a comparison is given in Fig.15. Our results are also confirmed in the stream line diagram as shown in Fig.16. At this condition, the internal pressure value can be reduced to about one-third of the pressure of the original design of Model A2. However, if the gas is not applied to the diaphragm at all after entering the housing, as in C2, the inhale pressure will become a positive value, which increases respiratory impedance. This must be avoided by the designer. Fig. 17 depicts a stream line diagram of C2.

Model	pressure inside the housing (bar)	pressure in the mouthpiece (bar)	Offset (mbar)
A2	6.4685	6.5025	-2
C1	6.6745	6.7095	-1
C2	6.623	6.665	6
C3	6.7395	6.775	-0.5

Table 5. Comparison of different angle of spray hole.

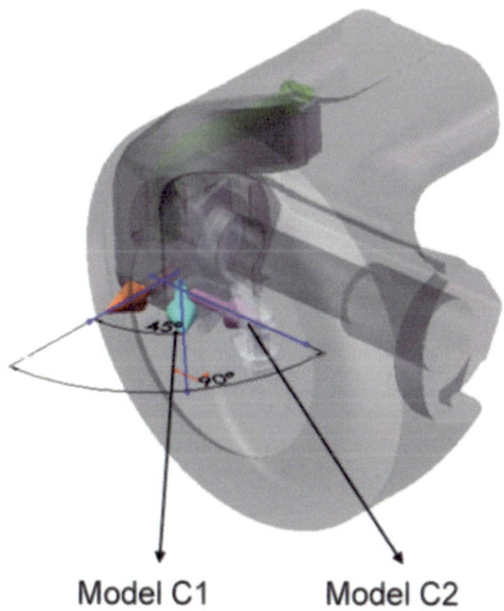

Fig. 13. Model C1 and C2 modified the direction of the spray hole.

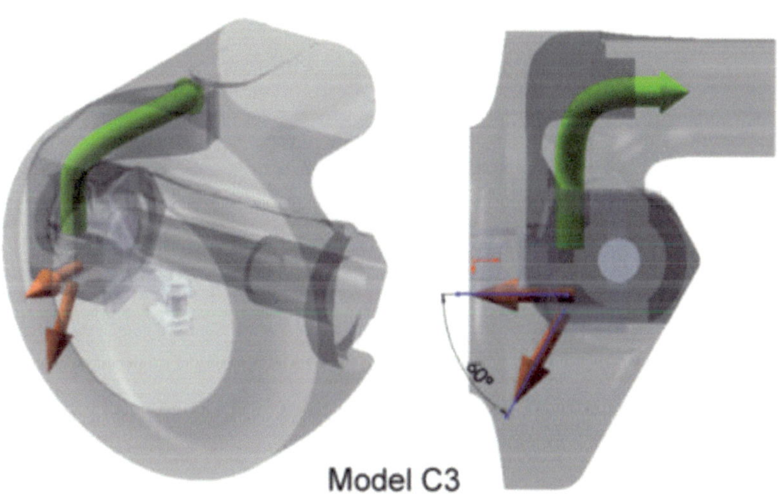

Fig. 14. Model C3 modified the direction of the spray hole.

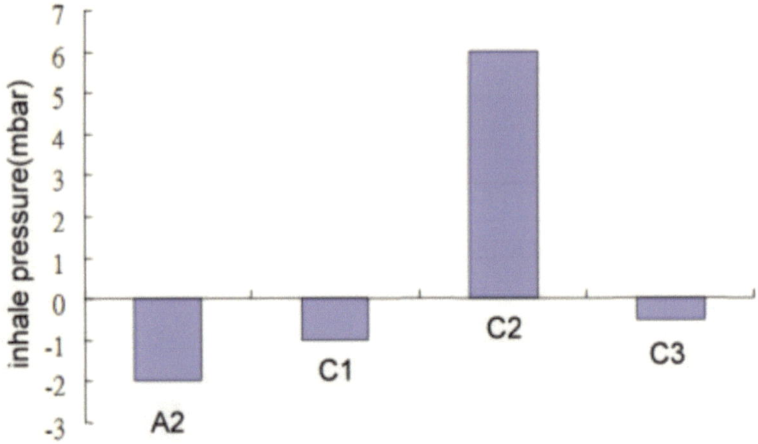

Fig. 15. The inhale pressure of different angle of spray hole.

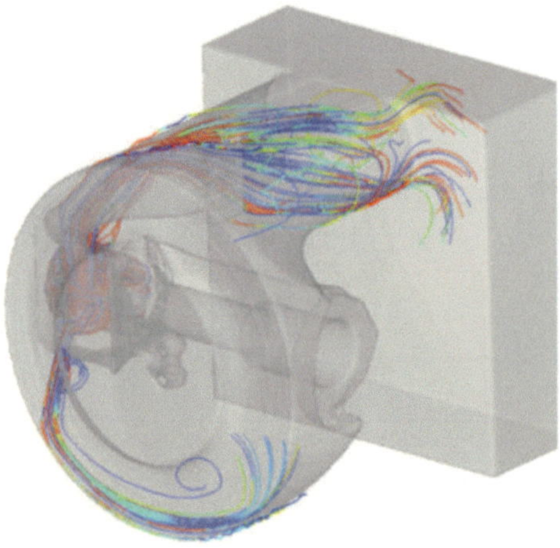

Fig. 16. The stream lines of Model C3.

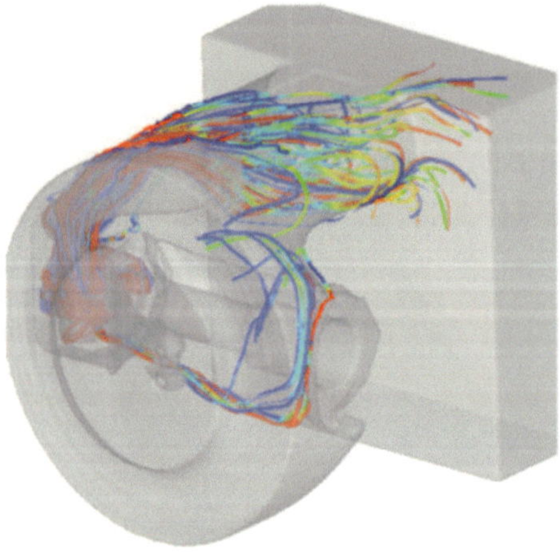

Fig. 17. The stream lines of Model C2.

3.3 Size of spray holes

The size of the spray holes will affect the flow velocity of the gas into the second stage, resulting in a change in internal pressure. The spray hole affects the pressure inside the diaphragm and housing to a great extent. When high-speed gas from the spray holes directly strikes the diaphragm, the pressure inside the diaphragm will rise. The size of the spray holes affects the gas velocity inside the housing. In this research, three groups of models, namely D1, D2 and D3, are used for testing. The position and diameters of the spray holes are shown in Fig. 18.

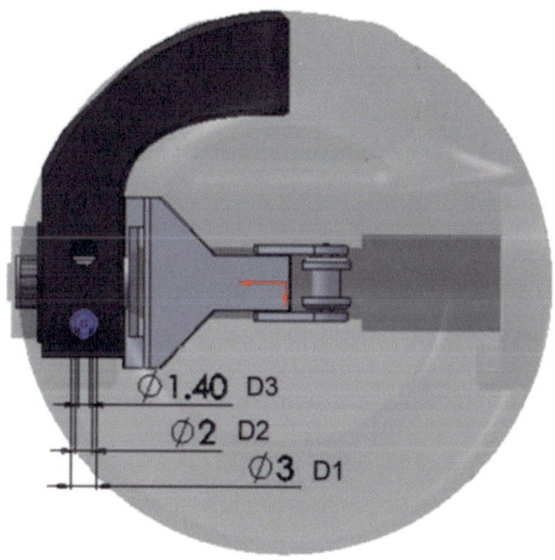

Fig. 18. Modify the size of spray hole (Model D1, D2, and D3).

The analytical data in Table 6 and Fig.19 shows that the pressure inside the housing is proportional to the hole diameter. The larger the hole diameter is, the higher the pressure inside the housing, which in turn reduces the differential pressure between the internal pressure and pressure at the suction end. However, when the hole diameter is too small (D3), the increasing differential pressure will cause the inhale pressure to change from a negative value to a positive value, thus reducing respiratory impedance. After several tests by simulation, when the spray hole diameter decreases from 2mm to 1.8mm, the inhale pressure will be reduced to one-third of the inhale pressure of the original design. In last, these experimental results can be simulated by numerical analysis software FLOW-3D®.

Model	pressure inside the housing (bar)	pressure in the mouthpiece (bar)	Offset (mbar)
D1	6.899	6.931	-4
D2	6.4685	6.5025	-2
D3	6.3835	6.3435	+4

Table 6. Comparison of different size of spray hole.

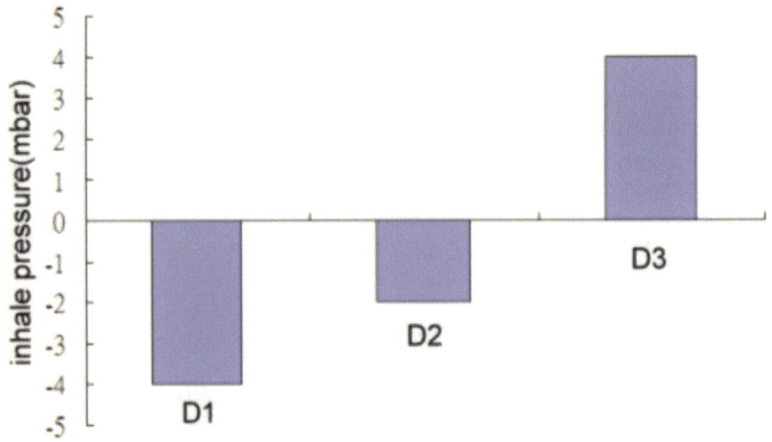

Fig. 19. The inhale pressure of different size of spray hole.

4. Conclusions

This study proved that, among existing insert molding process, the temperature of inserts has impact on the final assembly strength of product. In this chapter, the local heating mechanism of vapour chamber can control the molding temperature of inserts; and the assembly strength can be improved significantly if the temperature of inserts prior to filling can be increased over the mold temperature, thus allowing the local heating mechanism to improve the weld line in the insert molding process. A VC_RHCS for injection molding can effectively reduce the welding lines of the transparent plastic products. The heating and cooling injection molding system associated with vapour chamber can raise the tensile strength and reduce the defect of the welding lines of a plastic product because of VC_RHCS rapid-uniform heating and cooling cycle. The results show that the plastic products with two opposite gates was found increasing by 6.8% and 10% of tensile strength compared with the conventional one, and the other plastic product with eight holes plate is decreased from 12μm to 0.5μm of the depth of the welding line. And the performance of the breathing regulator can be accurately confirmed using numerical simulation software. The angle and diameter of the spray holes are key parameters affecting the performance of breathing regulators. Further, the diameter of the internal spray holes is inversely proportional to inhale pressure. These results indicate that the product formed by the VC_RHCS can effectively achieve high material strength and reduce weld line.

5. Acknowledgments

This chapter originally appeared in these References and is a minor revised version. Some of the materials presented in this chapter were first published in these References. The authors gratefully acknowledge Prof. R.-Q. Hsu and his MPDB Lab. and Dr. Y.-P. Tsai for guidance their writings to publish and permission to reprint the materials here. The work and finance were supported by National Science Council (NSC), National Taiwan Ocean University (NTOU), National Taiwan Normal University (NTNU) and National Defense University (NDU). Finally, the authors would like to thank all colleagues and students who contributed to this study in the Chapter.

6. References

Belloni A. (2001). Regulator with bypass tube, US Patent 6,279,575.

Brown, R.I. ; Brown, D.S. (2000). System and method to prevent the transmission of pathogenic entities between the multiple users of second stage regulators, US Patent 6,089,225.

Brown, R.I. ; Brown, D.S. (2002). Diving regulator with valved mouthpiece, US Patent 6,354,291.

Christianson T. (1989). Regulator second stage for scuba, US Patent 4,862,884.

Ferguson A.R. (1997). Adjustment mechanism for a scuba second stage airflow regulator, US Patent 5,660,502.

Garraffa D.R. (1997). Breathing regulator apparatus having automatic flow control, US Patent 5,678,541.

Garraffe, D.R. (1996). Second stage scuba diving regulator, US Patent 5,549,107.

Hansen H.R. ; Lingenfelter T.A. (1987). Breathing regulator mouthpiece, US Patent 4,683,881.

Houston, C.E. (1981). Underwater breathing apparatus, US Patent 4,245,632.

Toth D.J. (1985). Diaphragm assembly for scuba diving regulator, US Patent 4,508,118.

Tsai, Y.-P. ; Wang, J.-C. & Hsu, R.-Q. (2011). The Effect of Vapour Chamber in an Injection Molding Process on Part Tensile Strength. *EXPERIMENTAL TECHNIQUES,* Vol. 35, No 1, January/February, 2011, pp.60-64.

Wang, J.-C. & Chen, T.-C. (2009). Vapour chamber in high performance server. *Microsystems IEEE 2010 Print ISBN: 978-1-4244-4341-3, Packaging Assembly and Circuits Technology Conference (IMPACT), 2009 4th International,* Taipei, October, 2009, pp.364-367.

Wang, J.-C. & Huang, C.-L. (2010). Vapour chamber in high power LEDs. *IEEE 2011 Print ISBN: 978-1-4244-9783-6, Microsystems Packaging Assembly and Circuits Technology Conference (IMPACT), 2010 5th International,* Taipei, October, 2010, pp.1-4.

Wang, J.-C. & Tsai,Y.-P. (2011). Analysis for Diving Regulator of Manufacturing Process. *Advanced Materials Research,* Vol. 213, February, 2011, pp.68-72.

Wang, J.-C. & Wang R.-T. (2011). A Novel Formula for Effective Thermal Conductivity of Vapour Chamber, *EXPERIMENTAL TECHNIQUES,* Vol. 35, No 5, September/ October, 2011, pp.35-40.

Wang, J.-C. (2010). Development of Vapour Chamber-based VGA Thermal Module. *International Journal of Numerical Methods for Heat & Fluid Flow,* Vol. 20, No. 4, June, 2010, pp.416-428.

Wang, J.-C. (2011a). Applied Vapour Chambers on Non-uniform Thermo Physical Conditions. *Applied Physics,* Vol. 1, April, 2011, pp.20-26.

Wang, J.-C. (2011b). Investigations on Non-Condensation Gas of a Heat Pipe. *Engineering,* Vol. 3, April, 2011, pp.376-383.

Wang, J.-C. (2011c). L-type Heat Pipes Application in Electronic Cooling System. *International Journal of Thermal Sciences,* Vol. 50, No. 1, January, 2011, pp.97-105.

Wang, J.-C. (2011d). Thermal Investigations on LEDs Vapour Chamber-Based Plates. *International Communication in Heat and Mass Transfer,* Vol.38, No 9, November, 2011, pp. 1206-1212.

Wang, J.-C. ; Wang, R.-T. ; Chang, C.-C. & Huang, C.-L. (2010b). Program for Rapid Computation of the Thermal Performance of a Heat Sink with Embedded Heat Pipes. *Journal of the Chinese Society of Mechanical Engineers,* Vol. 31, No. 1, January/February, 2010, pp.21-28.

Wang, J.-C. ; Wang, R.-T.; Chang, T.-L. & Hwang, D.-S. (2010a). Development of 30 Watt High-Power LEDs Vapour Chamber-Based Plate. *International Journal of Heat and Mass Transfer,* Vol. 53, No. 19/20, September, 2010, pp.3900-4001.

Wang, J.-C.; Li, A.-T.; Tsai,Y.-P. & Hsu, R.-Q. (2011a). Analysis for Diving Regulator Applying Local Heating Mechanism of Vapour Chamber in Insert Molding Process. *International Communication in Heat and Mass Transfer,* Vol.38, No. 2, February, 2011, pp.179-183.

Wang, R.-T. ; Wang, J.-C. and Chang T.-L. (2011b). Experimental Analysis for Thermal Performance of a Vapour Chamber Applied to High-Performance Servers, Journal of Marine Science and Technology-Taiwan, Vol.19, No.4, July/August, 2011, pp.353- 360.

10

Thermoplastic Matrix Reinforced with Natural Fibers: A Study on Interfacial Behavior

Mohammad Farsi
Department of Wood and Paper Science, Sari Branch, Islamic Azad University, Iran

1. Introduction

The composites and their constituent components and structures have to meet increasingly development during recent decades. Some important concerns such as increasing price of petroleum and the impending depletion of fossil fuels and the interest in reducing the environmental impact of polymers is leading to the development of newer materials that can reduce stress on environment. Current developments and likely future trends are covered across key areas of the natural fibers reinforced polymer industry, together with existing and potential opportunities for the innovative use of plastic and bio-based fibers products. The challenges facing the world, such as environmental requirements and the need for recycling of plastic materials, are also included. Hence, the attention is increasingly being given to the use of natural fibers as reinforcement filler in low melting thermoplastic matrix manufactured by conventional plastic process such as extrusion and injection molding process. Injection molding is one of the most widely used processes for manufacturing molded parts from reinforced thermoplastic materials. Short natural fiber reinforced composites can be processed into complex shaped components using standard thermoplastic injection molding equipment.

Although the use of bio-based fillers is not as popular as the use of mineral or inorganic fillers, natural fiber-derived fillers have several advantages over traditional fillers and reinforcing materials such as low density, flexibility during the processing with no harm to the equipment, acceptable specific strength properties and low cost per volume basis.

The worldwide markets show the increased demands for natural and bio-based fibers. In 1967, the USA demand for fillers by the plastic industry was 525,000 tons; filler use had grown to 1,925,000 tons by 1998 (Eckert, 1999) and the projected use of fillers by the USA plastic industry in 2010 is to 8.5 billion pounds, of which 0.7 billion pounds (8%) was estimated to be bio-based fibers. It has been also summarized major markets for natural fibers in plastic composites as fig. 1a, on a weight basis (Eckert, 2000). Based on Fig. 1a, the main application areas of bio-based fibers filled composites are the building products in which they are used in structural applications as fencing, decking, roofing, railing, cladding and siding, park benches and etc. as shown in Fig. 2.

Most bio-fiber plastic additives are derived from wood that incorporated in Wood-plastic composites (WPCs). WPCs contain wood (fiber or flour) and polymer as matrix. The

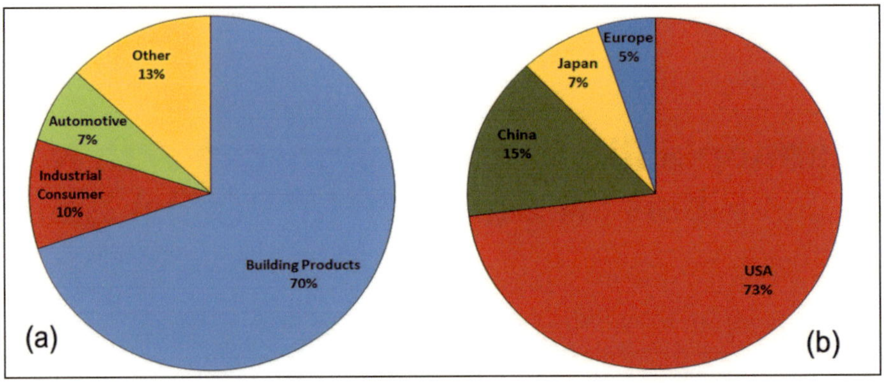

Fig. 1. Application areas and countries share for wood- and natural fiber-thermoplastic composites.

Fig. 2. Applications of natural fiber reinforced plastics in building products.

elasticity of wood fiber is approximately forty times greater than that of neat polyethylene and the overall strength is about twenty times higher (Oksman and Selin, 2004). However, other natural fibers, such as flax or wheat straw are finding their way into the fiber/plastic industry.

WPCs represent one of the rapidly growing markets within the plastics industry. More than 1.5 million tons of WPCs are produced worldwide, with the bulk of composites manufactured in the U.S., as shown in Fig. 1b.

Based on Freedonia Group report on 2010, Demand for WPC and plastic lumber is projected to advance 9.2 per cent per year to US$5.3 billion in 2013, creating a market for 3.3 billion pounds of plastic. WPC lumber will post more rapid gains than plastic lumber, advancing 10.5 per cent annually to US$2.4 billion in 2013. Growth in demand for plastic lumber is forecast to rise 8.2 per cent per year to US$2.9 billion in 2013. Demand for WPC and plastic lumber to reach $5.3 billion in 2013 (Centre Magazine, 2010).

In the USA, the wood- and natural fiber-plastic composites market has been dominated by rail and decking products, while in Europe more emphasis has been placed on automotive applications. Applications for these composites include interior panels, headliners, dashboard; car roofs, seat panels, parcel shelves, and acoustic panels (Fig. 3,4).

Fig. 3. Some applications of natural fibers in the automotive field.

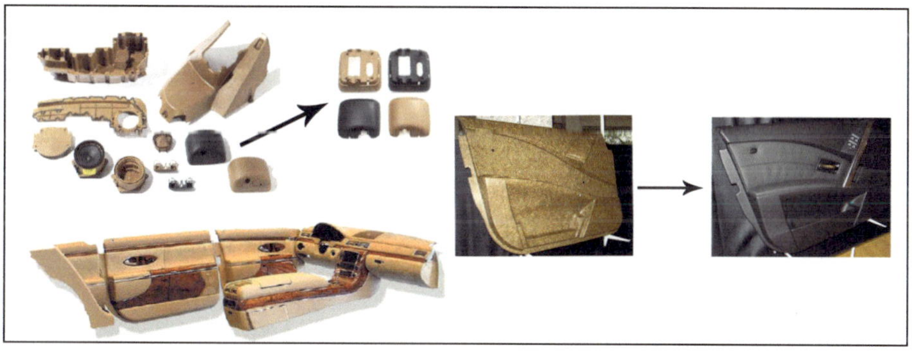
Fig. 4. Interior automotive parts made of wood and natural fiber reinforced plastics.

The WPCs global market for automative applications is estimated at 0.3 million metric tons in 2011 and is expected to increase at 17.1% compound annual growth rate to reach nearly 0.8 million metric tons in 2016 (BCC Research Report, 2011).

In China and other parts of Asia, wood and natural fiber reinforced plastics are beginning to experience considerable growth along with a wider variety of product offerings including pallets, doors, and architectural moldings. While other countries are manufacturing these products, international trade is still limited, in part due to the weight of the finished products.

Despite all the advantages there are also serious concerns. For explanation of these, must be said that:

1. the density of WPCs is almost twice that of solid lumber (Li and Matuana, 2003). The concept of creating cellular foamed structures has been shown to greatly reduce the weight of WPCs.
2. Exposure to ultraviolet (UV) radiation during outdoor use is of particular concern for WPCs.
3. the main disadvantages of using natural fibers as fillers is their high water absorption or desorption when subjected to changes in the relative humidity of the environment.

4. Thermoplastics typically perform poorly in long-term loading because linear polymer molecules exhibit a strong time and temperature dependent response. Addition of wood filler to the polymer matrix decreases creep response during loading (Bengtsson et al., 2005), but it is still a problem.
5. The processing temperature of the cellulosic fibres in thermoplastics is limited due to the potential fibre degradation at higher temperatures. This limits their application with plastics of low melting temperatures.
6. Low impact strength is another disadvantages of WPCs, due to the presence of natural fibers in the polymeric matrix provides points of stress concentrations, thus providing sites for crack initiation and potential composite failure.

But there is more important drawback in natural fiber-thermoplastic composites that limit its applications. The highly hydrophilic nature of natural fibers causes compatibility problems with the hydrophobic thermoplastics in composites. The mechanical properties of composites depend strongly on the interfacial adhesion between components. This can be maximized by improving the interaction and adhesion between the two phases in final composites. There are two approaches to improve the interfacial adhesion: polymeric matrix and lignocelluloses fiber modification. Different coupling agents have been used to modify the polymeric matrix and improve the interfacial strength and subsequently the mechanical properties of the products. Maleic anhydride grafted styrene-ethylene-butylene-styrene (SEBS-*g*-MA)(Oksman et al., 1998) and maleic anhydride grafted polyolefin such as HDPE-*g*- MA (Polec et al., 2010), PP-*g*-MA (Farsi, 2010), and LDPE-*g*-MA (Tasdemir et al., 2009) are the most common examples of reported works in the literature.

Another approach for enhancement of interfacial adhesion in natural fiber reinforced thermoplastic matrix is fiber treatment before mixing with polymer. Some of these treatments have physical nature and some of them are of chemical nature. Plasma and corona treatments of the fibers as physical methods have been reported in some papers (Gassan & Gutowski, 2000; Yuan et al., 2004).Their results show enhanced polymer-matrix adhesion.

Natural fibers have a good potential for chemical treatment due to presence of hydroxyl groups in lignin and cellulose. Reaction of hydroxyl groups can change the surface energy and the polarity of the natural fibers. Many studies have been undertaken to modify the performance of natural fibers. Different surface treatment methods such as alkali treatment (Chang et al., 2009), isocyanate treatment (Maiti et al., 2004,) acrylation (Huda et al., 2008), benzoylation (Mohanty et al., 2001), latex coating (Sreekala, 2000), permanganate treatment (Joseph, 2000), acetylation (Larsson-Brelid et al., 2008), silane (Bouza et al., 2008) and peroxide treatment (Sapieha et al., 1990) have been applied on the fiber to improve its strength, size and its shape and the fiber-matrix adhesion. The aforementioned methods have their own merits and demerits, e.g., alkali treatment improves the fiber-polymer adhesion due to the removal of natural and artificial impurities and changes in chemical composition of the fiber by removing the cementing substances like lignin and hemicelluloses. Generally, mechanism of the performance of these methods is different and is depended on the chemical structure of the reagent.

There are also several other challenges presented by natural fibers filled polymer composites such as large variability of mechanical properties (Sydenstricker et al., 2003) lower ultimate strength, lower elongation, problems with nozzle flow in injection molding machines, bubbles in the product (Toriz et al., 2002), and poor resistance to weathering (Pavithran, 1981).

In this chapter, we focus on structure and composition of natural fiber and review some research on optimization of interfacial adhesion between natural fibers and thermoplastic matrix witch published during the past decades and then report the effect of various chemical modifications of wood fiber and contemporaneous effect of PP-g-MA coupling agent on the interfacial strength of wood polypropylene injection molded composites. The effects of chemical treatments and PP-g-MA coupling agent on the interfacial behavior were evaluated by mechanical properties and dynamic mechanical analysis and so determined tensile strength, adhesion factor, storage modulus and loss factor.

2. Natural fibers

2.1 Characteristics of natural fibers

Natural fibers are subdivided based on their origins, coming from plants, animals or minerals. All plant fibers are composed of cellulose while animal fibers consist of proteins (hair, silk and wool). Natural fibres can be classified according to which part of the plant they are obtained from, as shown in Fig. 5.

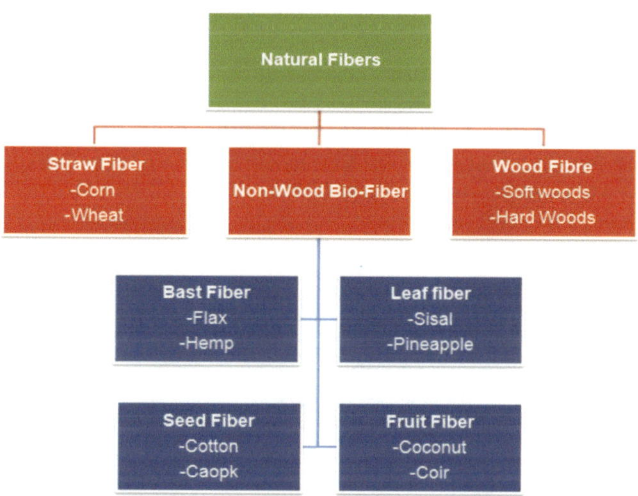

Fig. 5. Classification of natural fibres.

The strength characteristics of fiber depend on the properties of the individual constituents, the fibrillar structure and the lamellae matrix (Joseph et al., 2000).

Natural fibers exhibit considerable variation in diameter along with the length of individual filaments. Quality and other properties of fibers depend on factors such as size, maturity and processing methods adopted for the extraction of fiber (Mohanty et al., 2001). Properties such as density, electrical resistivity, ultimate tensile strength and initial modulus are related to the internal structure and chemical composition of fibers (Mohanty et al., 2001).

The structure, microfibrillar angle, cell dimensions, defects, and the chemical composition of fibres are the most important variables that determine the overall properties of the fibres.

Long term supply of resources is important point about natural fibers that can influence on natural fiber–thermoplastic composites production. The world consumption of natural fibers as illustrated in Table 1. The data presented in Table 1 suggests that wood with 68.5 % of the total world consumption, will continue to be a major source of bio-based fibers.

Fiber	**Quantity(*10³ tons)**	**% of the total**
Wood fiber	1,750,000	68.5
Rice straw	700,000	27.4
Rice husks	70,000	2.8
Cotton	18,645	0.75
Bamboo	10,000	0.39
Jute	3,630	0.14
Kenaf	970	0.04
Flax	830	0.03
Sisal	380	0.01
Hemp	220	0.009
Ramie	110	0.004
Coir	100	0.0039

Table 1. Annual world production of natural fibers (Kandachar, 2000; Bolton, 1995).

2.2 Structure of natural fibers

The cell wall in a fiber is not a homogenous membrane. Each fibril has a complex, layered structure consisting of a thin primary wall that is the first layer deposited during cell growth encircling a secondary wall (Fig. 6).

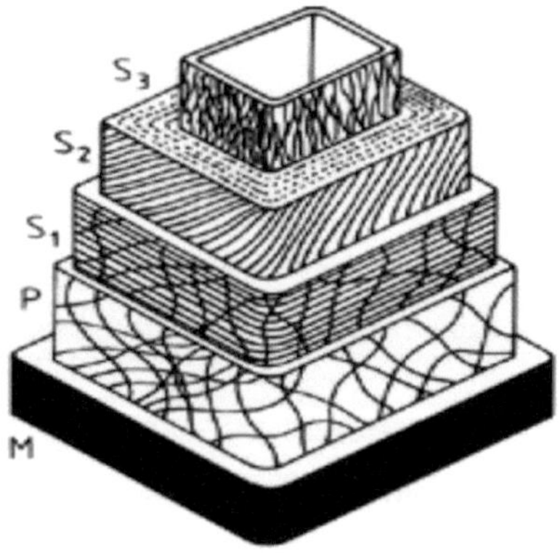

Fig. 6. Positioning of the cellulose fibrils in wood fibres. M) Middle lamella; P) Primary wall; S) Secondary wall; S1) Secondary wall I ; S2) Secondary wall II; S3) Secondary wall III.

The secondary wall is made up of three layers and the thick middle layer determines the mechanical properties of the fiber. The middle layer consists of a series of helically wound cellular microfibrils formed from long chain cellulose molecules: the angle between the fiber axis and the microfibrils is called the microfibrillar angle. The characteristic value for this parameter varies from one fiber to another. The outer secondary cell wall (Sl) is comparable in thickness to the primary wall and consists of four to six lamellae which spiral in opposite directions around the longitudinal axis of the tracheid. The main bulk of the secondary wall is contained in the middle secondary cell wall (S2), and may be as little as 1 pm thick in early woods and up to 5 pm in summer wood. The microfibrils of this part of the wall spiral steeply about the axial direction at an angle of around 10 to 20°. The inner secondary wall (S3), sometimes also known as the tertiary wall, is not always well developed, and is of no great technological importance. The orientation of the microfibrils within the S2 layer has an important bearing on mechanical properties of the fibre such as its modulus of elasticity.

Mechanical properties are determined by the cellulose content and microfibril angle. A high cellulose content and low microfibril angle are desirable properties of a fiber to be used as reinforcement in polymer composites (Williams & Wool, 2000).

Selective removal of non-cellulosic compounds constitutes the main objective of fiber chemical treatment. Both the hemicellulosic and pectic materials play important roles in fiber bundle integration, fiber bundle strength and individual fiber strength as well as water absorbency, swelling, elasticity and wet strength. The production of individual fibers without the generation of kink bands will generate fibers with much higher intrinsic fiber strength which is very useful for composite application (Mooney et al., 2001).

2.3 Composition of natural fibers

2.3.1 Cellulose

The reinforcing efficiency of natural fiber is related to the nature of cellulose and its crystallinity. Cellulose is a natural linear homopolymer (polysaccharide), in which D-glucopyranose rings are connected to each other with β- (1-4)-glycosidic linkages. The structure of cellulose units is shown in Fig. 7.

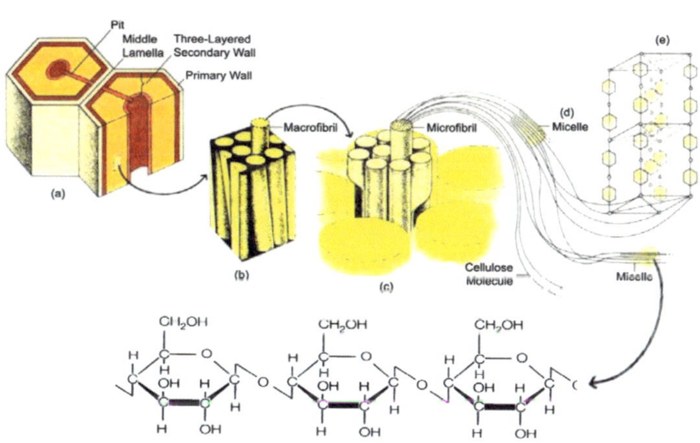

Fig. 7. Molecular structure of cellulose (Raven et al., 1999)

It is thus a 1, 4-β-D-glucan. The cellobiose polymer chains are ordered in three-dimensional levels, which give the supramolecular structure of cellulose. The linear polymeric chains (one dimension) form sheets that are held together with hydrogen bonds (second dimension). Then, these sheets are connected by Van der Waals bonds generating microfibril crystalline structures (third dimension). The crystal nature (monoclinic sphenodic) of naturally occurring cellulose is known as cellulose I. Cellulose is resistant to strong alkali (17.5 wt%) but is easily hydrolyzed by acid to water-soluble sugars. Cellulose is relatively resistant to oxidizing agents. Native cellulose (cellulose I) has two crystalline allomorphs, Iα and Iβ. The main difference between the two crystalline phases is the relative position of the chains to each other. These hydroxyl groups and their ability to hydrogen bond play a major role in directing the crystalline packing and also govern the physical properties of cellulose. Solid cellulose forms a microcrystalline structure with regions of high order, i.e. crystalline regions, and regions of low order, i.e. amorphous regions. Although the chemical structure of cellulose from different natural fibers is the same, the degree of polymerization (DP) varies. Its degree of polymerisation is typically between 10000 and 15000 glucose residues depending upon source and it is never found in a completely crystalline form, but occurs as a partly crystalline.

2.3.2 Hemicellulose

Hemicelluloses are another component of plant fibers. Hemicelluloses are polysaccharides and differ from cellulose in that they consist of several sugar moieties, are mostly branched, and have lower molecular mass with a degree of polymerization (DP) of 50 - 200. They are not, as the name seems to imply, biosynthetic precursors of cellulose. The two main types of hemicelluloses are xylans and glucomannans. Fig. 8 shows partial structure of hemicelluloses with a combination of 5-ring carbon ring sugars.

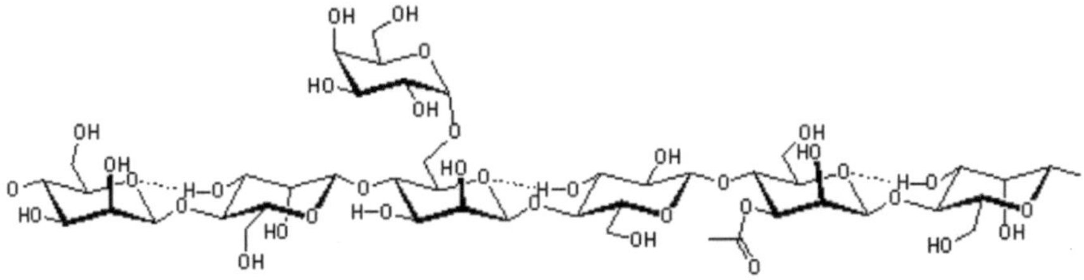

Fig. 8. Structure of softwood galacto-glucomannan (Bledzki & Gassan, 1999).

Hemicllulose differs from cellulose in three aspects. Firstly, they contain several different sugar units whereas cellulose contains only 1,4-b-D-glucopyranose units. Secondly, they exhibit a considerable degree of chain branching containing pendant side groups giving rise to its noncrystalline nature, whereas cellulose is a linear polymer. Thirdly, DP of hemicellulose is around 50–300, whereas that of native cellulose is 10–100 times higher than that of hemicellulose.

2.3.3 Lignin

After cellulose, lignin is the most abundant natural organic polymer. Its content is higher in softwoods (27–33 %) than in hardwoods (18–25 %) and grasses (17–24 %). Lignin is totally amorphous and hydrophobic in nature. It is the compound that gives rigidity to the plants. It is thought to be a complex, three-dimensional copolymer of aliphatic and aromatic constituents with very high molecular weight. Lignin is a randomly branched polyphenol, made up of phenyl propane (C9) units and it is the most complex polymer among naturally occurring high-molecular-weight materials. Due to its lipophilic character, lignindecreases the permeation of water across the cell walls, which consist of cellulose fibres and amorphous hemicelluloses, thus enabling the transport of aqueous solutions of nutrients and metabolites in the conducting xylem tissue. Secondly, lignin imparts rigidity to the cell walls and, in woody parts, together with hemicelluloses, functions as a binder between the cells generating a composite structure with outstanding strength and elasticity. Finally, lignified materials effectively resist attacks by micro organisms by impeding penetration of destructive enzymes into the cell walls. When incorporated in a plastic, lignin, due to its phenolic base structure, could improve the mechanical properties (Thielemans et al., 2002).

Fig. 9 shows building blocks of lignin. There is a wide variation of structure within different plant species (Alder, 1977). Lignin is considered to be a thermoplastic polymer exhibiting a glasstransition temperature of around 90°C and melting temperature of around 170°C (Olesen & Plackett, 1999). It is not hydrolyzed by acids, but soluble in hot alkali, readily oxidized, and easily condensable with phenol.

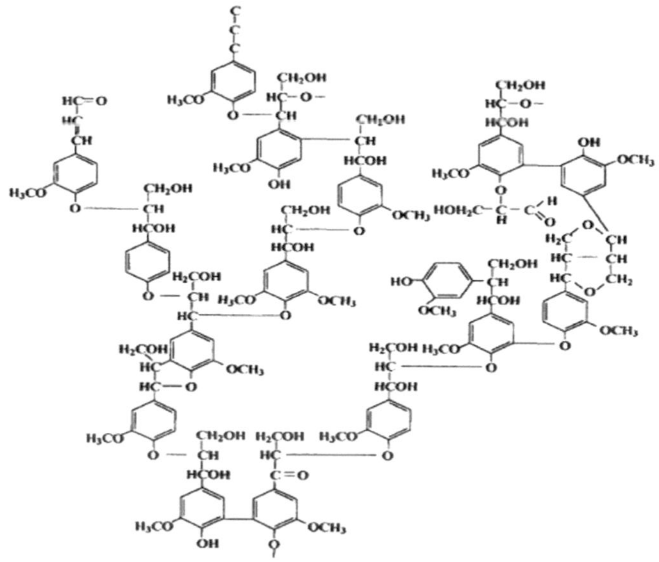

Fig. 9. Partial structure of a spruce lignin fragment (Alder, 1977).

3. Natural fiber for use in injection molded composite

Injection molding is a widely used technique for mass producing articles with a high degree of geometrical complexity. Injection molding has many advantages, such as short product cycle, excellent surface of the product and easily molded complicated shapes. The characteristics of the product are easily affected by the flow type of the melt, the heat transfer effect, the material properties and the specific geometry of the mold. Thus, different injection molding conditions will induce different fiber orientation. The degree of fiber orientation depends on the fiber characteristics, the variation of the cross-sectional area for flow, and the injection molding conditions. Composite materials for use in injection molding applications must be capable of fluid-like flow during processing and thus usually consist of short fibers with a relatively low fiber fraction. Natural fiber compounds offer numerous advantages over other injection molding compounds, for instance, low wear of manufacturing tools, often reduced cycle time, and ease of recycling. The final properties of natural fiber thermoplastic composites manufactured by injection molding depend not only on the properties of raw materials used and their compositions, but also on processing methods. Injection molding is a very suitable procedure to process natural fiber reinforced polymers into sophisticated 3-dimensional parts. In this case, the use of granular material, which already includes natural fibers and coupling agent, proves a success. But the incorporation of natural fibers to thermoplastics leads to flow limitation, which are increased by the incompatibility of natural fibers and thermoplastics. A lot of methods have been used to solve this incompatibility problem, and natural fiber chemical modification and coupling agent have a better result that is explained below.

4. Natural fiber surface modification: Case study outline

Natural fibers are amenable to modification as they bear hydroxyl groups from cellulose and lignin. The hydroxyl groups may be involved in the hydrogen bonding within the cellulose molecules thereby reducing the activity towards the matrix. Chemical modifications may activate these groups or can introduce new moieties that can effectively interlock with the matrix. Interfaces play an important role in the physical and mechanical properties of composites (Joseph et al., 2000). Simple chemical treatments can be applied to the fibers with the aim of changing surface tension and polarity of fiber surface.

4.1 Most important surface modification

The different surface chemical modifications of natural fibers have achieved various levels of success in improving fiber strength, fiber fitness and fiber-matrix adhesion in natural fiber composites. Brief descriptions of some important fiber chemical modifications witch applied in this research are summarized in the following sub-sections.

4.1.1 Alkaline treatment

The important modification done by alkaline treatment is the disruption of hydrogen bonding in the network structure, thereby increasing surface roughness. This treatment removes a certain amount of lignin, wax and oils covering the external surface of the fiber cell wall, depolymerizes cellulose and exposes the short length crystallites (Li et al., 2000) Addition of aqueous sodium hydroxide (NaOH) to natural fiber promotes the ionization of the hydroxyl group to the alkoxide (Agrawal et al., 2000). Thus, alkaline processing directly influences the cellulosic fibril, the degree of polymerization and the extraction of lignin and hemicellulosic compounds (Jahn et al., 2002). Alkali treatment increases surface roughness resulting in better mechanical interlocking and the amount of cellulose exposed on the fiber surface. This increases the number of possible reaction sites and allows better fiber wetting. The following reaction, takes place as a result of alkali treatment.

$$\text{Fibre} - \text{OH} + \text{NaOH} \longrightarrow \text{Fibre} - \text{O}^-\text{Na}^+ + \text{H}_2\text{O}$$

Fig. 10. Alkaline treatment reaction (Pothan et al., 2006).

Many attempts have been done by researchers to increase mechanical properties of biocomposites by alkaline treatments. For example, the effect of alkali treatment on properties of hybrid fiber biocomposite was reported by John et al., (2008). It has been reported that alkali treatment leads to fiber fibrillation i.e. breaking down of fiber bundles into smaller fibers which increases the effective surface area available for contact with the matrix.

Partial removal of lignin and hemicellulose on the alkali modification of cellulose fibers was reported by Sreekala et al., (1997). Mukherjee et al., (1993) reported that the removal of hemicellulose produces less dense and less rigid interfibrillar region. Kokot et al., (1995) also noted that as lignin is removed, the middle lamella joining the ultimate cells is expected to be more plastic as well as homogeneous, due to the gradual elimination of microvoids. Alkaline treatment increases the amount of crystalline cellulose and removes natural and artificial impurities, producing a rough surface topography.

4.1.2 Silane treatment

Silane is a chemical compound with chemical formula SiH4. Silanes are used as coupling agents to let glass fibers adhere to a polymer matrix, stabilizing the composite material. Silane coupling agents may reduce the number of cellulose hydroxyl groups in the natural fiber-matrix interface. Silanes undergo hydrolysis, condensation and the bond formation stage. The reaction scheme is given as follows:

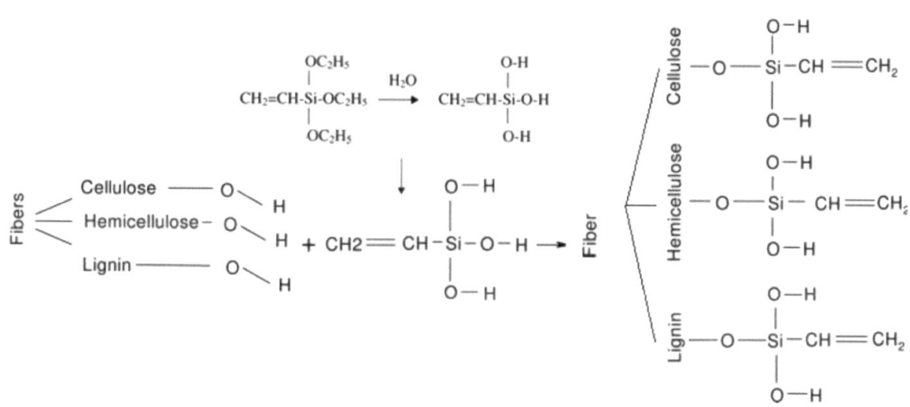

Fig. 11. Hydrolysis of silane and hypothetical reaction of fibers and silane (Sreekala et al., 2000).

Bledzki et al., (1996) concluded that in the process of interaction between natural fibers and silanes, alkoxy silanes are able to form bonds with hydroxyl groups. In presence of moisture hydrolyzable alkoxy group of silans leads to the formation of silanols. Silanols can form polysiloxane structures by reaction with hydroxyl group of the fibers.

John et al., (2008) and Mathew et al., (2004) investigated the effect of silane treatment on the mechanical properties of biocomposites. They observed a marked improvement in the properties after chemical modification. Sreekala et al., (1997) also suggested that the silane treated cellulose fiber composite showed an increase in nucleation density compared the untreated fiber composite. The increased nucleation yielded smaller crystals that result in a transcrystalline interphase region, with improved bonding between the fiber and the matrix (Agrawal et al., 2000).

4.1.3 Benzoylation treatment

Benzoylation is an important transformation in organic synthesis (Paul et al., 2003). Benzoyl chloride is most often used in fiber treatment. The Benzoyl clride forme an ester linkage to the natural fibers, reducing its hydrophylicity and making it more compatible with matrix. The reaction between the cellulosic-OH group of natural fiber and benzoyl chloride is shown as follows:

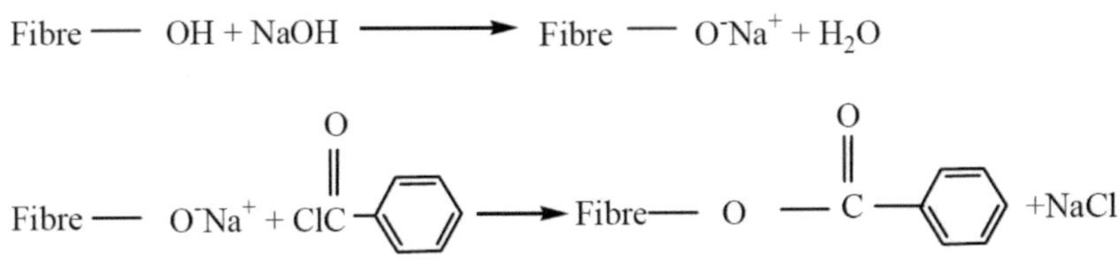

Fig. 12. Reaction between cellulosic-OH groups and benzoyl chloride (Joseph et al., 2000).

Abu bakar & Baharulrazi (2008) also indicated that the benzoylated oil palm empty fruit bunch was able to improve the tensile properties, impact strength and the increase of water resistance, and the reduction of glass transition temperature of composites when compared to the untreated fiber.

4.1.4 Acrylation treatment

Acrylation reaction is initiated by free radicals of the cellulose molecule. Cellulose can be treated with high energy radiation to generate radicals together with chain scission (Bledzki & Gassan, 1999). The reaction was accomplished between OH groups and acrylic acid as follows:

$$\text{Fiber} - \text{OH} + \text{CH2} = \text{CH} - \text{COOH} \rightarrow \text{Fiber} - \text{O} - \text{CH2} - \text{CH2} - \text{COOH}$$

Fig. 13. Reaction between cellulosic-OH groups and acrylic acide (Mohanty et al., 2001).

Sreekala et al., (2000, 2002) used acrylic acid in natural fiber surface modification. acrylation led to strong covalent bond formation and thereby, the tensile strength and Young's modulus of treated fibers were improved marginally.

4.1.5 Polyolefin-g-MA coupling agents

Coupling agents such as PP grafted with maleic anhydride (PP-g-MA) and PP grafted with acrylic acid (PP-g-AA) are usually employed to improve interfacial properties. Esterification reaction and H-bond interactions may take place at the interface of the cellulosic filler and the PP-g-MA as suggested in Fig. 14. The PP chain permits maleic anhydride to be cohesive and produce maleic anhydride grafted polypropylene (PP-g-MA). Then the treatment of cellulose fibers with hot PP-g-MA copolymers provides covalent bonds across the interface. After this treatment, the surface energy of cellulose fibers is increased to a level much closer to the surface energy of the matrix. This results in better wettability and higher interfacial adhesion of the fiber.

Many of maleic anhydride grafted polyolefin such as HDPE-*g*-MA, PP-*g*-MA, and LDPE-*g*-MA studied by Polec et al., (2010), Farsi, (2010) and Tasdemir et al., (2009) respectively. they observed strong interfacial strength by incorporation of the coupling agents.

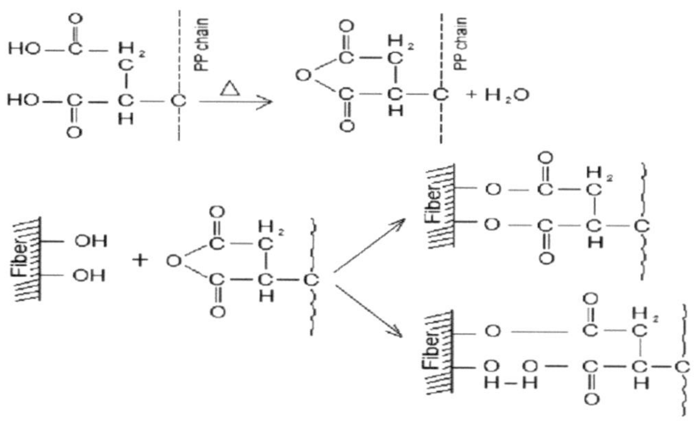

Fig. 14. PP-g-MA reaction with natrural fiber (Bledzki et al., 1996).

4.2 Role of chemical treatment on thermo-mechanical behavior

One of the most important methods for studying the effect of natural fiber chemical surface treatment on the interfacial strength of the composites is dynamic mechanical thermal analysis (DMTA). The effect of wood chemical surface modification on the interfacial strength was tracked using adhesion factor. This parameter is obtained from DMTA data. Effect of PP-g-MA as coupling agent on the interface adhesion of WPC investigated by Correa et al., (2007). They mentioned a method based on a simplified single rule of mixtures aiming to compare differences in interface adhesion in the presence of PP-g-MA is proposed in terms of relaxation spectra of polypropylene-wood composites obtained by DMTA. DMTA has been widely used to investigate the structures and viscoelastic behaviors of composite materials as determined by their storage modulus (E'), loss modulus (E'') and loss factor (tanδ). This analysis technique can provide information on the stiffness of the composites (Kim et al., 2005). Relaxation peaks (α, β and γ) are observed for the tanδ curves, which are caused by the onset of the various motions of the chain molecules. The dominant β peak represents the glass-to-rubber transition of the amorphous portion in composites and is assigned to the glass transition temperature.

For determination of adhesion factor as interfacial intraction criterion, Correa et al., (2007) has been used equation originated from Kubat et al., (1990) work, about high density polyethylene filled with 20 Vol. % glass fibers. They assumed that the mechanical loss factor ($\tan\delta_c$) of the composite can be written:

$$Tan\delta_c = \Phi_f Tan\delta_f + \Phi_i Tan\delta_i + \Phi_p Tan\delta_p \tag{1}$$

Where the subscript f,i and p denotes filler ,interphase, and matrix respectively and Φ is the corresponding volume fraction. By considering $\delta_f \approx 0$ and since the volume fraction of the interphase is rather small, above equation can be rearranged as follows:

$$\frac{Tan\delta_c}{Tan\delta_p} = (1+\Phi_f)(1+A) \tag{2}$$

$$A = \left[[\frac{\Phi_i}{(1+\Phi_f)}] \right] (\frac{Tan\delta_i}{Tan\delta_p}) \tag{3}$$

Where Equation (3) can be rewritten as:

$$A = \left[[(\frac{1}{(1+\Phi_f)})\ (\frac{Tan\delta_c}{Tan\delta_p})] \right] -1 \tag{4}$$

With calculating A factor from DMAT data, one can interpret the interaction in the interphase, where there is strong interaction between wood fiber and polymer matrix due to reduction of macromolecular mobility in the vicinity of the filler surface, A factor decreases. In other words, a low value of A factor is an indication of good adhesion or high degree of interaction between two phases. This factor presents a macroscopic quantitative measure of interfacial adhesion during dynamic loading.

5. Experimental

5.1 Materials

Polypropylene of Arak Petrochemical Company in Iran (Trade Name of V30S) with a density of 0.9 g/cm^3, and the melt flow index (MFI) of 16 g/10 min was used in this study as matrix. 60-mesh virgin wood flour was used as filler. The characteristics of the natural fibres are shown in Fig. 15.

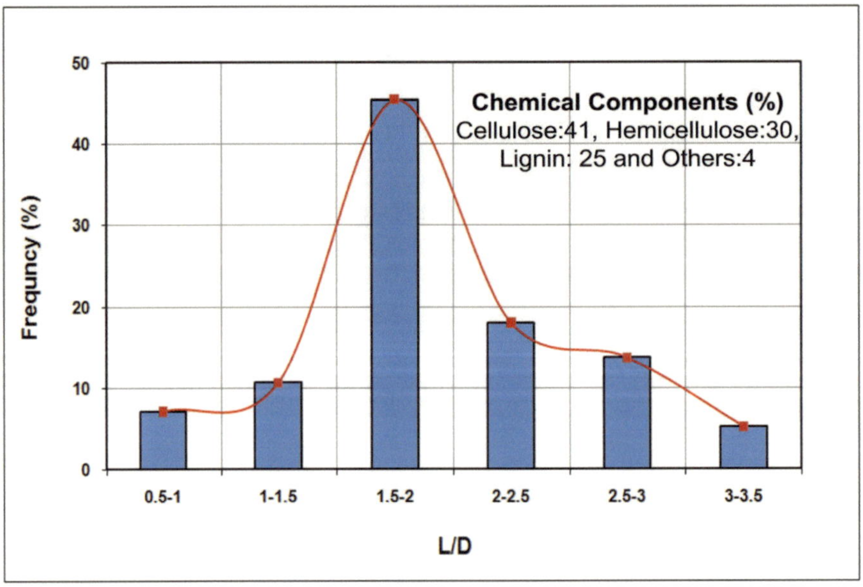

Fig. 15. Characteristics of the wood flour.

The chemical structure of natural fibres was measured according to TAPPI standard and aspect ratio was obtained using optical microscopy. Triethoxy vinyl silane, acrylic acid; benzoyl chloride and sodium hydroxide were from Merck Co, Germany.

5.2 Preparation of natural fibers

5.2.1 Modification with sodium hydroxide

The Wood flour was introduced in a stainless steel vessel and 2 wt.% solution of NaOH was added into the vessel and stirred well. Wood flour was immersed for half an hour. Once this time was over, the fillers were separated from the solution and washed with distilled water containing a few percent of acetic acid to remove residual of alkali. The washed fiber was then dried in the oven at 80° C for 24 hours.

5.2.2 Modification with benzoyl chloride

Wood flour was dipped in solution of 5% of NaOH and benzoyl chloride for 15 minutes. Ethanol solution was used to remove of extra benzoyl chloride for one hour. Finally the fiber was washed with distilled water and dried in oven at 80° C for 24 hours.

5.2.3 Modification with acrylic acid

Wood flour (which was treated with NaOH) was immersed to acrylic acid solution at 50° C for half an hour and was then washed and dried similar to previous steps.

5.2.4 Modification with silane

The silane used was Triethoxy vinyl silane. 1% of the respective silane was prepared by mixing with an ethanol/water mixture in the ratio 60/40 and was allowed to stand for 1 h. The pH of the solution was maintained between 3.5- 4 with the addition of acetic acid. Wood flour dipped in this solution and was allowed to stand for 1.5 h. The ethanol/water mixture was drained out and the washing and drying steps were repeated as mentioned in previous treatments.

5.3 Processing of the composites

The modified and unmodified wood flour was dried at 103±2 °C to constant weight before mixing process. PP and the modified and unmodified wood flour were blended in a batch mixer (Haake Buchler) at 190°C and 60 rpm for 8 min. In all cases, the weight ratio of fiber and polymer was 40:60 (Table 2). From the compounds which had been granulated, specimens were injection molded into ASTM standard by an injection molder at a molding temperature of 190°C and injection pressure was 3 Mpa.

5.4 Measurements

5.4.1 Thermo-mechanical test

Dynamic mechanical thermal analysis (DMTA) was carried out by using Triton instrument, Model Tritic 2000 made by UK in triple-point bending mode. The dimension of each sample was 5×1×2 cm. The range of testing temperature was from -50 to 150°C and the experiments were performed at 1 Hz frequency and heating rate of 2°C/min. During testing DMTA parameters of storage modulus and loss factor were recorded as function of temperature. Then based on equation (4) data analyzed for determining A factor.

Code*	Polypropylene content (Wt. %)	Wood flour content (Wt. %)	PP-g-MA (Wt. %)
PP	100	0	0
UW-P	60	40	0
UW-P-M	60	40	3
AW-P	60	40	0
AW-P-M	60	40	3
BW-P	60	40	0
BW-P-M	60	40	3
CW-P	60	40	0
CW-P-M	60	40	3
SW-P	60	40	0
SW-P-M	60	40	3

*PP: Polypropylene, W:Wood Flour, U:Unmodified, A:Alkline, B:Benzoylation, C:Acrylication, S:Silane, M:PP-g-Ma

Table 2. Composition of the Studied WPCs.

5.4.2 Mechanical test

Tensile strength tests of the specimens were carried out according to ASTM D-638 by Instron 6025 model from UK at crosshead speed of 5 mm/min. For each test and type of the composite, five specimens were tested and the average values are reported.

5.4.3 Fourier Transform Infrared Spectroscopy (FTIR)

The infrared spectra of raw and treated wood flour were recorded on a Bomem, 150-MB series model Spectrophotometer to characterize the chemical change upon treatment of the wood flour with chemical components.

5.4.4 Scanning Electron Microscopy (SEM)

The morphology of the wood modified-PP composites and interfacial bonding between the filler and the PP matrix was examined using a scanning electron microscope (JXA-840) supplied by JEOL Company Limited, Japan. The samples were viewed perpendicular to the fractured surfaces.

6. Results and discussion

6.1 FTIR

FTIR Spectrum in Fig. 16 shows a spectrum of modified and unmodified wood flour samples. As can be seen, the intensity of the peak around 3400 cm^{-1}, which is evidence of OH band, is decreased after treatment of fibers. The intensity of the band around 1730 cm^{-1}

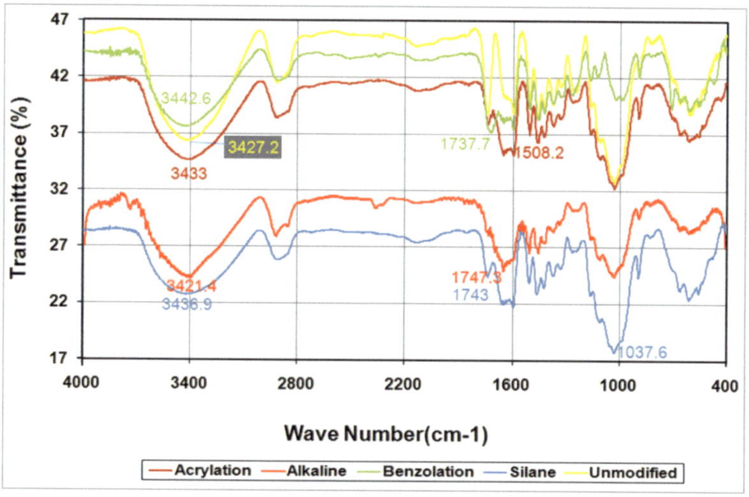

Fig. 16. FTIR spectra of the chemical treated wood flour.

increased due to formation of ester band from the reaction between OH group and bezoyl chloride. The intensity peak for aromatic ring at 1508.2 cm^{-1} is decreased after surface modification due to removal of lignin. A strong peak at 1730 cm^{-1} in the FTIR spectrum indicates the presence of acetyl group in the fiber. The intensity peak at 1037.6 cm^{-1} is increased after silane absorbance, which is an overlap of si-o-si band and c-o stretching of fiber (Lu & Drazel, 2010). In the presence of moisture, hydrolysable alkoxy group leads to the formation of silanols. The silanol then reacts with the hydroxyl group of the fiber, forming stable covalent bonds to the cell wall that are chemisorbed onto the fiber surface (Agrawal et al., 2000).

6.2 Adhesion factor

Results for the adhesion factor as an evaluation parameter for fillers-polymer interactions versus temperature is presented in Fig. 17 for different chemical modification.

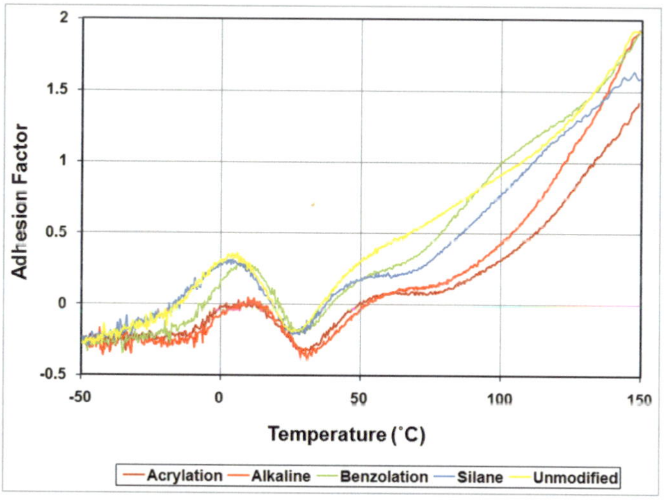

Fig. 17. Adhesion factor versus temperature for the treated WPCs.

Adhesion factor presents a macroscopic quantitative measure of interfacial adhesion during dynamic loading and at high levels of interface adhesion, the molecular mobility surrounding the filler is reduced, and consequently low values of the adhesion factor suggest improved interactions at the matrix-filler interface (Kubat et al., 1990). As can be seen, below of glass transition temperature, the maximum amount of adhesion factors are related to untreated samples which means the weakest interface has been formed for samples containing untreated wood flour. It seems the chemical modification facilitates the interaction between fillers polymer and decreases the adhesion factor. It is important to note that adhesion factor seems to be very sensitive to glass transition temperature of samples. Around this temperature (around 22°C), slop of curves changes and the adhesion factor passes through a maximum due to more polymer chains mobility. According to Kubat et al., (1990) by increasing the temperature there was a release of the thermal stresses at the filler surface and reduced filler-matrix friction and should be related to a more cohesive matrix-filler interface (lower A). In other words, a strong interfacial adhesion i.e. samples which are treated with silane, restricts the chain mobility at the filler matrix interface, therefore the adhesion factor decreases and its maximum shifts to higher temperature. At the higher end of the temperature range, the most curves converge.

6.3 Storage modulus

The variation of the storage modulus (E') value of the composites as a function of temperature is shown in Fig. 18 for different chemical treatment. E' Value determines relevant stiffness of WPCs (Kim et al., 2005).

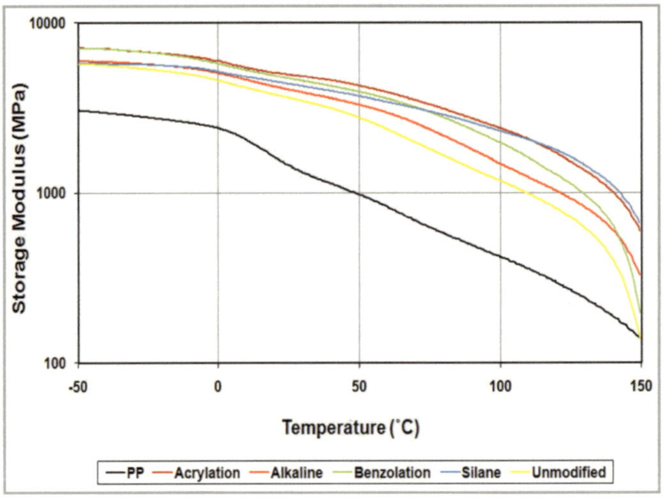

Fig. 18. Storage modulus versus temperature for the treated WPCs.

The stiffness of the composites is greater than that of the neat PP in the whole temperature range, and this trend is more significant in the higher temperature range. Fillers play an important role in increasing the storage modulus of polymeric materials. As can be seen, a general decrease trend was also observed over the entire range of temperature and with incorporation of wood flour to PP, significant increase in the E' values of composites is clearly seen. This is probably due to increase in the stiffness of the matrix with the reinforcing effect imparted by the fiber, which allowed a greater degree of stress transfer at the interface (Jain et al., 1992). As the temperature is increased, relaxation process of the molecular matrix is initiated. Also, thermal expansion occurs which decreases the intermolecular forces (George et al., 1999). An appreciable improvement in the storage modulus was observed for the treated composite, due to the increase in the interfacial stiffness brought about by the more intense filler-matrix interaction. The composite modified by silane improves the interfacial adhesion more than other composites and this more lessens the molecular mobility in the interfacial region. In Storage modulus plots, around temperature 22°C, slop of most curves change that can be considerable as β transition. At the higher end of the temperature range, the curves of PP and unmodified composites converge.

6.4 Tan δ

Fig. 19 shows the loss factor (tan δ) versus temperature for wood-PP composites and their corresponding different chemical treatment. The tan δ peak was shifted to higher temperature for filled samples in comparison to neat polypropylene. Relaxation peaks for treated and untreated samples are present for the tan δ curves in the vicinity of -40°C (γ), 22°C (β) and 100°C (α) which are caused by the onset of the various motions of the chain molecules. The dominant β-peak represents the glass-to-rubber transition of the amorphous portion in PP and is assigned to the glass transition temperature.

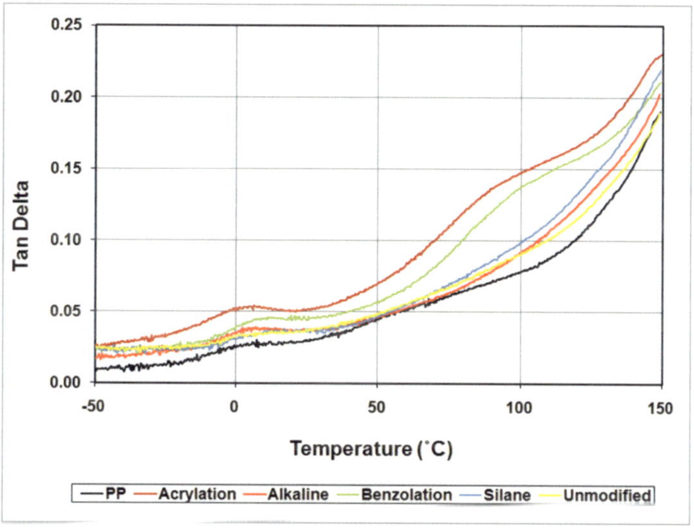

Fig. 19. Tan δ versus temperature for the treated WPC.

Table 3 depicts the shift in glass transition temperature of the sample which is taken from tan δ curves. Depending on the nature of lignocelluloses filler and filler/matrix interaction, glass transition of the composites shifts to higher temperature. In other words, during stress transfer at the interface the strong bonding causes the fiber constraint and the poor bonding leads to dissipation energy. Among composites, those containing unmodified samples have the lowest values, whereas those containing acrylic acid treated samples have the highest tan δ values compared with the other samples. The difference between various chemical modifications becomes more pronounced at higher temperatures.

Sample	Shift in Tg (°C)
PP	-
Unmodified	2
Acrylation	7
Benzoylation	8
Alkaline	6
Silane	13

Table 3. Shift in glass transition temperature of the treated WPC.

The effects of PP-g-Ma coupling agent on the storage modulus and tan δ of the silane treated wood flour /PP composites are shown in Fig. 20. As can be seen, the addition of PP-g-MA improved E' of the composites, that was because PP-g-MA could lead to the creation of a thin and irregular polymer layer, which could assist formation of plastic deformation zone around the fiber (Hristov et al., 2004). Further, incorporation of compatibilizer to composites containing silane treated wood flour decreased tan δ due to the more intense filler-matrix interaction. This results indicated that, simultaneous use of silane and coupling agent on storage modulus and tan delta had synergic effect.

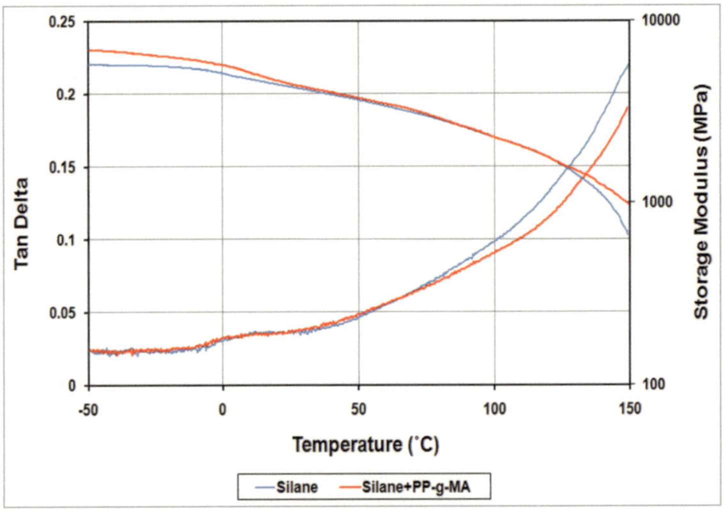

Fig. 20. Simultaneous effect of compatibilizer and silane on storage modulus and loss factor of the wood flour/PP composite.

6.5 Tensile properties

Results for the tensile strength and modulus of composites as function of chemical modification are presented in Fig. 21. Incorporation of wood flour in PP matrix significantly increased strength and modulus of composites. The increase in tensile strength or modulus is primarily attributed to the presence of fiber, which allowed a uniform stress distribution from continuous PP matrix to dispersed fiber phase (Coutinho et al., 1997).

Benzoylation treatment of fiber enhancement the tensile modulus of composites, but all other modulus has no significant variation together. Wood flour after modification are somewhat leached, leading to dissolution of hemicellulose, lignin and pectin. The removal of surface impurities can make the fiber cleaner and rougher than before (Liu et al., 2009). Wood flour reinforced plastic composites often showed enhancement in tensile strength upon different modification owing to the increased fiber-matrix adhesion. Optimum strength is observed for alkali treated composite.

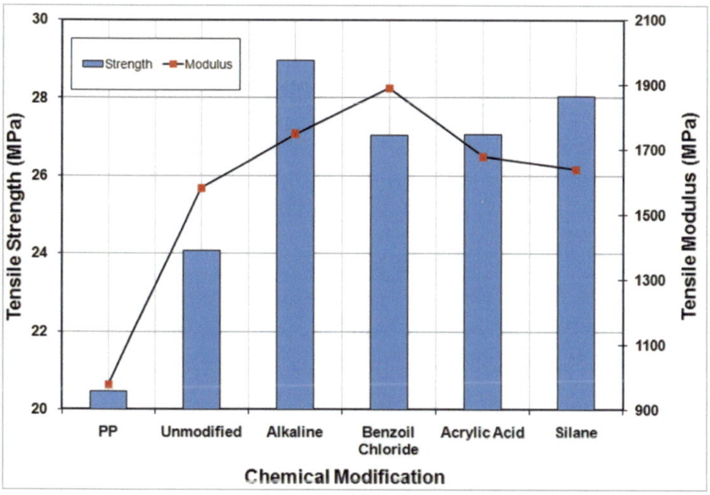

Fig. 21. Tensile strength and modulus of neat PP and the ones reinforced with modified and unmodified fiber composites.

Fig. 22 illustrated the result of tensile strength conducted on the modified specimens with and without PP-g-MA. The tensile strength of the composites increased with using coupling agent; because of improve the bonding strength between wood flour and PP matrix. So, the tensile strength of composites increased with conjunction of use both modified fiber and coupling agent.

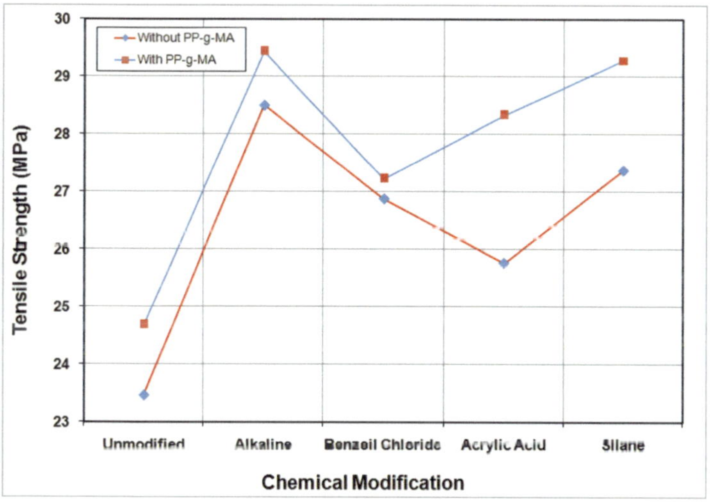

Fig. 22. Tensile strength of modified and unmodified fiber composites with and without coupling agent.

6.6 Surface morphology

It is also clear from the SEM images in Fig. 23a that the wood fibers in unmodified sample are pulled out easily and some holes are noticed around the fibers which imply that there are weak interactions between the filler and polymer. As it can be seen in Fig. 23e, there is a better polymer-filler adhesion with the silane treatment than in the composite prepared with untreated wood flour, which implies an increase in the thickness of the interface between the particles and polymers. In samples undergone alkali treatment (Fig. 23b), fibers removed from pp matrix and broken, but not the isolated fibrils were observed, which means that the interactions between the phases are not strong enough. Similar trend is also observed for samples containing acrylic acid (Fig. 23c) and benzoyl (Fig. 23d) treated fibers. As in the case of adhesion factor, the best encapsulation of wood fibers with polymer matrix can be seen in samples with silane treatment. This explanation is similar to that of adhesion factor results.

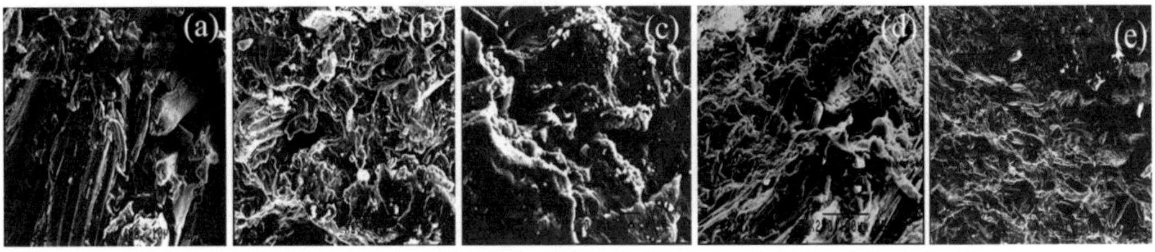

Fig. 23. SEM Micrographs of modified wood polymer composites with: (b) alkali, (c) acrylic acid, (d) benzoyl chloride, (e) silane, and (a) unmodified samples.

7. Conclusion

Due to the growing competition in producing cost-effective products, in addition to the excreted environmental limitations, natural fibers have recently become attractive to researchers because of their advantages over conventional mineral fillers. However, several limitations must be overcome in order to exploit the full potential of natural fibres. The incompatibility between natural fiber and polymer matrix is a major problem for interfacial adhesion between these two component materials, which is of critical importance for the mechanical properties of the composite. In the present chapter, we focused on the natural fiber-thermoplastic composites and its worldwide application markets, reviewed some influence factors on the injection molding process to produce the natural fibers thermoplastic compound and introduced some research on interfacial adhesion strength between natural fibers and thermoplastic matrix and then reported the effect of various chemical modifications of wood fiber on the interfacial strength of wood polypropylene injection molded composites. The results showed that, at first, chemical modification of natural fiber is necessary and respectively silane and alkaline treatment of wood flour improved interfacial adhesion and so increased the mechanical performance of these composites. Secondly, simultaneous use of chemical modification and coupling agent on the properties had synergic effect. Thirdly, however, adhesion factor is a good way for understanding interfacial behavior of WPCs, further evidences of improved matrix–filler interactions is observed by SEM. Finally, future areas of interest should be focused on developing new coupling agents and new class of natural fiber modification such as enzymatic treatment.

8. References

(2010). Demand for wood-plastic composite and plastic lumber to reach $5.3 billion in 2013, In:, *Centre Magazine*, March 2010, Available from: <http://www.centremagazine.com/news/demand-for-wood-plastic-composite-and-plastic-lumber-to-reach-5-3-billion-in-2013/1000369238/>

(1999). Wood-Plastic Composites: Technologies and Global Markets, In: *BCC Research Report*, June 2011, Available from: <http://www.bccresearch.com/report/wood-plastic-composites-tech-markets-pls034b.html>

Abu Bakar, A. & Baharulrazi, N. (2008). Mechanical properties of benzoylated oil palm empty fruit bunch short fiber reinforced poly (vinyl chloride) composites. *Polymer-Plastics Technology and Engineering*, Vol.47, No.10, (September 2008), pp. 1072-1079, ISSN 0360-2559

Agrawal, N.S.; Saxena, K.B.; Sharma, Thomas S. & Sreekala, M.S. (2000). Activation energy and crystallization kinetics of untreated and treated oil palm fiber reinforced phenol formaldehyde composites. *Materials Science and Engineering: A*, Vol.277, No.1-2, (January 2000), pp. 77–82. ISSN 0921-5093

Alder, E. (1977). Lignin chemistry—past, present and future. *Wood Science and Technology*, Vol.11, No.3, (September 1977), pp. 169-218, ISSN 0043-7719

Bengtsson, M.; Gatenholm, P. & Oksman, K. (2005). The effect of crosslinking on the properties of polyethylene/wood flour composites. *Composites Science and Technology*, Vol.65, No.10, (August 2005), pp. 1468–1479, ISSN 0266-3538

Bledzki, A.K.; Reihmane, S. & Gassan, J. (1996). Properties and modification methods for vegetable fibres for natural fibre composites. *Journal of Applied Polymer Science*, Vol.59, No.8, (February 1996), pp. 1329–1336, ISSN 0021-8995

Bledzki, A.K. & Gassan, J. (1999). composites reinforced with cellulose based fibres. *Progress in Polymer Science*, Vol.24, No.2, (May 1999), pp.221-274, ISSN 0079-6700

Bolton, A.J. (1995). The Potential of Plant Fibres as Crops for Industrial Use. *Outlook on Agriculture*, Vol. 24, No.2, (June 1995), pp.85-89, ISSN 0030-7270

Bouza, R.; Lasagabaster, A.; Abad, M.J. & Barral, L. (2008). Effects of vinyltrimethoxy silane on thermal properties and dynamic mechanical properties of polypropylene-wood flour composites. *Journal of Applied Polymer Science*, Vol.109, No.2, (July 2008), pp. 1197-1204, ISSN 0021-8995

Chang, W.P.; Kim, K.J. & Gupta, R.K. (2009). Moisture absorption behavior of wood/plastic composites made with ultrasound-assisted alkali-treated wood particulates. *Composite Interfaces*, Vol.16, No.7-9, (October 2009), pp. 937- 951, ISSN 0927-6440

Correa, C.A.; Razzino, C.A. & Hage JR, E. (2007). Role of Maleated Coupling Agents on the Interface Adhesion of Polypropylene—Wood Composites. *Journal of Thermoplastic Composite Materials*, Vol.20, No.3, (May 2007), pp.323-339, ISSN 0892-7057

Coutinho, F.M.B.; Costa, T.H.S. & Carvalho, D.L. (1997). Polypropylene-wood fiber composites: Effect of treatment and mixing conditions on mechanical properties. *Journal of Applied Polymer Science*, Vol.65, No.6, (August 1997), pp.1227-1235, ISSN 0021-8995

Eckert, C.H. (1999). Functional fillers for plastics: outlook to the year 2005, *Proceedings of the Fifth International Conference on Woodfiber-plastic Composites*, pp. 10-22, Proceedings No.7263, Madison, USA, May 26-28, 1999

Eckert, C. (2000). Opportunities for natural fibers in plastic composites, *Proceedings of the Conference on Progress in Woodfibre-plastic Composites,* University of Toronto, Canada, May 25-26, 2000

Farsi, M. (2010). Wood–plastic composites: influence of wood flour chemical modification on the mechanical performance. *Journal of Reinforced Plastics and Composites,* Vol.29, No.24, (December 2010), pp. 3587–3592, ISSN 0731-6844

Gassan J. & Gutowski V.S. (2000). Effects of corona discharge and UV treatment on the properties of jute-fibre epoxy composites, *Composites Science and Technology,* Vol.60, No.15, (November 2000), pp. 2857-2863, ISSN 0266-3538

George, J.; Bhagawan, S.S. & Thomas, S. (1999). Effect of Strain Rate and Temperature on the Tensile Failure of Pineapple Fiber Reinforced Polyethylene Composites. *Journal of Thermoplastic Composite Materials,* Vol.12, No.6, (November 1999), pp. 443-446, ISSN 0892-7057

Hristov, V.N.; Krumova, M.; Vasileva, S. & Michler, G.H.*J.* (2004). Modified polypropylene wood flour composites. II. Fracture, deformation, and mechanical properties. *Journal of Applied Polymer Science,* Vol.92, No.2, (April 2004), pp. 1286–1292, ISSN 0021-8995

Huda, M.S.; Drzal, L.T.; Mohanty, A.K. & Misra, M. (2008). Effect of fiber surface treatments on the properties of laminated biocomposites from poly(lacticacid) (PLA and kenaf fibers. *Composites Science and Technology,* Vol.68, No.2, (February 2008), pp. 424-432, ISSN 0266-3538

Jahn, A.; Schroder, M.W.; Futing, M.; Schenzel, K. & Diepenbrock, W. (2002). Characterization of alkali treated flax fibres by means of FT Raman spectroscopy and environmental scanning electron microscopy. *Spectrochimica Acta Part A: Mol Biomol Spectrosc,* Vol.58, No.10, (August 2002), pp. 2271-2279, ISSN 1386-1425

Jain, S.; Kumar, R. & Jindal, U.C. (1992). Mechanical behaviour of bamboo and bamboo composite. *Journal of Materials Science,* Vol.27, No.17, (September 1992), pp. 4598-4604, ISSN 0022-2461

John, M.J.; Francis, B.; Varughese, K.T. & Thomas, S. (2008). Effect of chemical modification on properties of hybrid fiber biocomposites. *Composites Part A: Applied Science and Manufacturing,*Vol.39, No.2, (February 2008), pp. 352-363, ISSN 1359-835X

Joseph, K.; Mattoso, L.H.C.; Toledo, R.D.; Thomas, S., de Carvalho, L.H.; Pothen, L.; Kala, S. & James, B. (2000). Natural fiber reinforced thermoplastic composites. In: *Natural Polymers and Agrofibers Composites,* Frollini, E.; Leao, A.L. & Mattoso, L.H.C. (Eds), 159-201, Embrapa Instrumentacao Agropecuaria, ISBN 858646306X, San Carlos, Brazil

Kandachar, P.V. (2000). Designing with natural fibre composites, *Proceedings of the International Workshop on the Development of natural Polymers and Composites in East Africa,* pp. 187–202, Arusha Tanzania, September 2000

Kim, H.S.; Yang, H.S.; Kim, H.J.; Lee, B.J. & Hwang, T.S. (2005). Thermal properties of agro-flour-filled biodegradable polymer bio-composites. *Journal of Thermal Analysis and Calorimetry,* Vol.81, No.2, (July 2005), pp. 299-306, ISSN 1388-6150

Kokot, S. & Stewart, S. (1995). An Exploratory Study of Mercerized Cotton Fabrics by DRIFT Spectroscopy and Chemometrics. *Textile Research Journal,* Vol.65, No.11, (November 1995), pp. 643-651, ISSN 0040-5175

Kubat, J.; Rigdahl, J. & Welander, M. (1990). Characterization of interfacial interactions in high density polyethylene filled with glass spheres using dynamicmechanical analysis. *Journal of Applied Polymer Science*, Vol.39, No.7, (April 1990), pp.1527-1539, ISSN 0021-8995

Larsson-Brelid, P.; Walinder, M.E.P.; Westin, M. & Rowell, R.M. (2008). Ecobuild- a center for development of fully biobased material systems and furniture applications. *Molecular Crystals & Liquid Crystals*, Vol.484, No.1, (April 2008), pp. 623- 630, ISSN 1542-1406

Li, Y.; Mai, Y.W. & Ye, I. (2000). Sisal fibre and its composites: a review of recent developments. *Composites Science and Technology*, Vol.60, No.11, (August 2000), pp. 2037–2055, ISSN 0266-3538

Li, Q. & Matuana, L.M. (2003). Foam extrusion of high density polyethylene/wood-flour composites using chemical foaming agents. *Journal of Applied Polymer Science*, Vol.88, No.14, (June 2003), pp. 3139–3150, ISSN 0021-8995

Liu, L.; Yu, J.; Cheng, L. & Qu, W. (2009). Mechanical properties of poly(butylene succinate) (PBS) biocomposites reinforced with surface modified jute fibre. *Composites Part A: Applied Science and Manufacturing*, Vol.40, No.5, (May 2009), pp. 669-674, ISSN 1359-835X

Lu, J. & Drazel, L.T. (2010). Microfibrillated cellulose/cellulose acetate composites: Effect of surface treatment. *Journal of Polymer Science Part B: Polymer Physics*, Vol.48, No.2, (January 2010), pp. 153-161, ISSN 0887-6266

Maiti, S.N.; Subbaro, R. & Ibrahim, M.N. (2004). Effect of wood fibers on the rheological properties of i-PP/wood fiber composites. *Journal of Applied Polymer Science*, Vol.91, No.1, (January 2004), pp. 644-650, ISSN 0021-8995

Mathew, L.; Joseph, K.U. & Joseph, R. (2004). Isora fibres and their composites with natural rubber. *Progress in rubber, plastics and recycling technology*, Vol.20, No.4, (- 2004), pp. 337–34, ISSN 1477-7606

Mohanty, A.K.; Misra, M. & Drzal, L.T. (2001). Surface modifications of natural fibers and performance of the resulting biocomposites: An overview. *Composite Interfaces*, Vol.8, No.5, (October 2001), pp. 313-343, ISSN 0927-6440

Mooney, C.; Stolle-Smits, T.; Schols, H. & de Jong, E. (2001). Analysis of retted and non retted flax fibers by chemical and enzymatic means. *Journal of Biotechnology*, Vol.89, No.2- 3, (August 2001), pp. 205-216, ISSN 0168-1656

Mukherjee, A.; Ganguly, P.K. & Sur, D. (1993). Structural mechanics of jute: the effects of hemicellulose or lignin removal. *Journal of the Textile Institute*, Vol.84, No.3, (-1993), pp. 348–353, ISSN 0040-5000

Oksman, K.; Lindberg, H. & Holmgren, A. (1998). The nature and location of SEBS-MA compatibilizer in polyethylene- wood flour composites. *Journal of Applied Polymer Science*, Vol.69, No.1, (July 1998), pp. 201-209, ISSN 0021-8995

Oksman, K. & Selin, J.F. (2004). Plastics and composites from polylactic acid. In: *Natural fibers, plastics and composites*, Wallenberger, F.T. & Weston, N. (Eds), pp. 149-166, Kluwer Academic Publishers, ISBN 1-4020-7643-6, Boston

Olesen, P.O. and Plackett, D.V. (1999). Perspectives on the Performance of Natural Plant Fibres, *Proceedings of Natural Fibres Performance Forum Plant Fibre Products-Essentials for the Future*, pp. 1-7, Copenhagen Denmark, 27-28 May 1999

Paul, S.; Puja, N. & Rajive, G. (2003). PhCOCl-Py/Basic Alumina as a Versatile Reagent for Benzoylation in Solvent-Free Conditions. *Molecules*, Vol.8, No.3, (March 2003), pp. 374-380, ISSN 1420-3049

Pavithran, C.; Gopakumar, K.; Prasad, S.V. & Rohatgi, P.K. (1981). Copper coating on coir fibres. *Journal of Materials Science*, Vol.16, No.6, (June 1981), pp. 1548–56. ISSN0022-2461

Polec, I.; Hine, P.J.; Bonner, M.J.; Ward, I.M. & Barton, D.C. (2010). Die drawn wood polymer composites. I. mechanical properties. *Composites Science and Technology*, Vol.70, No.1, (January 2010), pp. 45-52, ISSN 0266-3538

Pothan, L.A.; Thomas, S. & Groeninckx, G. (2006). The role of fiber matrix interaction on the dynamic mechanical properties of chemically modified banana fiber/polyester composites. *Composites Part A: Applied Science and Manufacturing*, Vol.37, No.9, (September 2006), pp. 1260-1269, ISSN 1359-835X

Raven, P.H.; Evert, R.F. & Eichhorn, S.E. (1999). *Biology of Plants* (6th edn), W.H. Freeman and Company, ISBN 0716710072, New York

Sapieha, S.; Allard, P. & Zang, Y.H. (1990). Dicumyl peroxidemodified cellulose/LLDPE composites. *Journal of Applied Polymer Science*, Vol.41, No.9-10, (November 1990), pp. 2039-2048, ISSN 0021-8995

Sreekala, M.S.; Kumaran, M.G. & Thomas, S. (1997). Oil palm fibers: morphology, chemical composition, surface modification, and mechanical properties. *Journal of Applied Polymer Science*, Vol.66, No.5, (October 1997), pp. 821–835, ISSN 0021-8995

Sreekala, M.S.; Kumaran, M.G.; Joseph, S.; Jacob, M. & Thomas, S. (2000). Oil palm fiber reinforced phenol formaldehyde composites: influence of fiber surface modifications on the mechanical performance. *Applied Composite Materials*, Vol.7, No.5-6, (November 2000), pp. 295-329, ISSN 0929-189X

Sreekala, M.S.; Kumaran, M.G. & Thomas, S. (2002). Water sorption in oil palm fiber reinforced phenol formaldehyde composites. *Composites Part A: Applied Science and Manufacturing*, Vol.33, No.6, (June 2002), pp. 763-777, ISSN 1359-835X

Sydenstricker, T.H.; Mochnaz, S. & Amico, S.C. (2003). Pull-out and other evaluations in sisal-reinforced polyester biocomposites. *Polymer Testing*, Vol.22, No.4, (June 2003), pp. 375–80, ISSN 0142-9418

Tasdemir, M., Biltekin, H., Caneba, G. & Gerald, T. (2009). Preparation and characterization of LDPE and ppwood fiber composites, *Journal of Applied Polymer Science*, Vol.112, No.5, (June 2009), pp. 3095-3102, ISSN 0021-8995

Thielemans, W.; Can, E.; Morye, S.S. & Wool, R.P. (2002). Novel applications of lignin in composite materials, *Journal of Applied Polymer Science*, Vol.83, No.2, (January 2002), pp. 323-331, ISSN 0021-8995

Toriz, G.; Denes, F. & Young, R.A. (2002). Lignin-polypropylene composites. Part 1: composites from unmodified lignin and polypropylene. *Polymer Composites*, Vol.23, No.5, (October 2002), pp. 806–11, ISSN 0272-8397

Williams, G.I. & R.P. Wool. (2000). Composites from natural fibers and soy oil resins. *Applied Composite Materials*, Vol.7, No.5-6, (November 2000), pp. 421-432, ISSN 0929-189X

Yuan, X.; Jayaraman, K. & Bhattacharyya, D. (2004). Effects of plasma treatment in enhancing the performance of wood fibre-polypropylene composites. *Composites Part A: Applied Science and Manufacturing*, Vol.35, No.12, (December 2004), pp. 1363-1374, ISSN 1359-835X

11

Properties of Injection Molded High Density Polyethylene Nanocomposites Filled with Exfoliated Graphene Nanoplatelets

Xian Jiang and Lawrence T. Drzal
Michigan State University, Composite Materials and Structures Center, Department of Chemical Engineering and Materials Science, East Lansing, Michigan, USA

1. Introduction

This book chapter investigated the potential of using exfoliated graphene nanoplatelets, GNP, as the multifunctional reinforcement in high density polyethylene (HDPE) matrix. HDPE/GNP nanocomposites were fabricated by the conventional compounding method of melt-extrusion followed with injection molding. The mechanical properties, crystallization behaviors, thermal stability, thermal conductivity, and electrical conductivity of the resulting HDPE/GNP nanocomposites were evaluated as a function of GNP concentration. Results showed that HDPE/GNP nanocomposites exhibit equivalent flexural modulus and strength to HDPE composites filled with other commercial reinforcements such as carbon fibers (CF), carbon black (CB) and glass fibers (GF). But they have superior impact strength. By investigating the crystallization behavior of HDPE/GNP nanocomposites, it was found that GNP is a good nucleating agent at low loading levels and as a result can significantly increase crystallization temperature and crystallinity of HDPE. At high GNP loadings, however, the close proximity of GNP particles retards the crystallization process. The thermal stability and thermal conductivity of HDPE/GNP nanocomposites were significantly enhanced as a function of GNP concentration due to the excellent thermal properties of GNP. Meanwhile, results indicated that the percolation threshold of these nanocomposites prepared by the conventional melt - extrusion and injection molding is relatively high at around 10-15 vol% GNP loading. The high percolation threshold is mainly due to the sever GNP aggregation and platelets alignment during the processing conditions as verified by the morphology. To enhance their electrical conductivity and lower the percolation threshold, a wax coating method was introduced in this study which is proved to be efficient in improving the dispersion of GNP in HDPE which is responsible for the better electrical and mechanical properties in the resulting nanocomposites.

Polymeric nanocomposites have attracted research interest both in industry and in academia in recent years, they can be found useful in many applications such as electromagnetic interference (EMI) shielding devices, low power rechargeable batteries, electronic devices, light emitting diodes (LEDs), gas sensors, super capacitors and photovoltaic cells (Hussain et al., 2006; Vaia, 2003). Polymeric nanocomposites have represented a radical alternative to conventional filled polymers or polymer blends. The difference between the conventional fillers and nano-fillers can be explained that nano-reinforcements must have at least one dimension in the nanometer range.

The advantages of using nano-fillers have been summarized in the work by Griffith and Weibull (Griffith, 1920; Weibull, 1951). They claimed that the smaller the reinforcement is, the stronger it becomes. It is believed that the failure of macroscopic specimens is mainly due to the presence of critical size defects. As the material size decreases, the probability of critical size flaws also reduces which allows the material to approach its intrinsic strength. Piggot and Hussain (Hussain et al., 2006; Piggott, 1980) concluded that nano-fillers are more effective reinforcements than their conventional counterparts because a smaller amount of nanoparticles could result in a larger enhancement in the mechanical, electrical, and thermal properties of the polymer matrix.

Recently, carbon nanotubes (CNTs) have been extensively explored as the nano-reinforcement in polymers due to their exceptional mechanical, electrical and thermal properties (Hermant et al., 2009). Many papers have appeared in literature discussing the CNTs-filled polymeric nanocomposites with excellent mechanical, electrical and thermal properties for numerous applications (Balasubramanian & Burghard, 2006; Kim et al., 2005; Singh et al., 2006; Yao et al., 2006). However due to the poor yield and costly fabrication and purifying process, the price of CNTs in the market is still high, which limits the commercial applications of CNTs to date (Kim & Drzal, 2009).

To search for an alternative nano-filler which exhibits the superior properties of CNTs but have a low cost and easy processing, graphite based materials are gaining more and more research attention. Polycrystalline graphite is a material that consists of extended networks of sp^2-hybridized carbons in a planar layered structure (graphene), leading to excellent thermal and electrical conductivity within this graphitic basal plate. It is found that exfoliation of these graphite layers and dispersion into polymers offers the potential to bring multifunctionality to the host polymers. Furthermore, research has shown that fully exfoliated graphite nanosheets are as effective in conductivity enhancement as CNTs due to their two-dimensional lattice of sp^2-bond carbon and extremely high aspect ratio (Xie et al., 2008). Based on this principal, a new form of graphite based nano-filler, exfoliated graphene nanoplatelets, has been under investigation in the Drzal group for several years (Fukushima, 2003; Kalaitzidou, 2006). Research has shown that this nano-particle is a potential alternative to other nano-reinforcements such as nano-clays and CNTs since it combines the low cost and layered structure of nano-clays and the superior thermal and electrical properties of CNTs (Jiang & Drzal, 2010,2009,2011; Kalaitzidou et al., 2007). The objective of this book chapter is to: (1) determine the mechanical properties, i.e., flexural strength, flexural modulus, and impact strength of HDPE/GNP nanocomposites made by melt-extrusion and injection molding and their comparison to the HDPE composites reinforced by commercially available fillers such as glass fiber, carbon fiber and carbon black; (2) investigate the crystallization behavior of HDPE with the presence of GNP; (3) explore the thermal stability, thermal conductivity, and electrical conductivity of injection molded HDPE/GNP nanocomposites; (4) observe the morphology of HDPE/GNP nanocomposites to determine the dispersion and orientation of the nano-reinforcement under the melt-extrusion and injection molding conditions; (5) enhance the dispersion of

GNP in HDPE by a wax coating method which is suitable for melt-extrusion and injection molding.

2. Experimental

2.1 Materials

In this book chapter, HDPE pellets with the trade name Marlex® HXM 50100 (Density 0.948 g/cm³, MW~ 230,000) were obtained from Chevron Phillips Chemical Company. Paraffin wax (max C30, Density 0.92 g/cm³, MW~ 500) with the melting point of 55°C was purchased from Sigma-Aldrich. GNP nanoplatelets were obtained from XG Science, Inc (www.xgsciences.com). There are two kinds of GNP particles used in this study. GNP-1 has the thickness of around 5-10nm and a platelet diameter of 1 μm; while GNP-15 has the same thickness but the diameter is around 15 μm.

Several commercial reinforcements and fillers were also combined with HDPE to make composites for comparison to the HDPE/GNP nanocomposites. They are: (1) CF-PAN based carbon fiber (PANEX 33 MC Milled Carbon Fibers, Zoltek Co), (2) CB-nanosize 'High Structure' carbon black (KETJENBLACK EC-600 JD, Akzo Novel Polymer Chemicals LLC), and (3) GF-chopped glass fiber (StarStran® LCF, Johns Manville Co.) . The physical properties of these materials are detailed in Table 1.

Filler	Length (μm)	Diameter (μm)	Aspect Ratio	Surface Area (m^2/g)	Density (g/cm^3)
GNP-1	<0.01	1	<100	100	2.1
GNP-15	<0.01	15	~1500	40	2.1
PAN CF	175	7.2	~24	16	1.8
GF	51 (mm)	13	~4000	NA	2.6
CB	0.4 - 0.5	0.4 - 0.5	1	1400	1.8

Table 1. Geometrical and surface characteristics of various fillers.

2.2 Processing methods

Melt - extrusion of HDPE/GNP nanocomposites was carried out in a DSM Micro 15cc Compounder, (Vertical, co-rotating, twin-screws micro-extruder) operating at 220°C for 5 minutes at a screw speed of 100rpm. The composite melt was then transferred to a Daca Micro injector with the T_{barrel}=220°C and T_{mold}=90°C. The injection pressure applied for the injection molding of flexural coupons was at 0.6MPa. Round disks (thickness ~1.5mm, diameter ~25mm) were also injection molded for thermal conductivity test under the same injection conditions. The melt - extrusion and injection molding systems are shown in the Fig. 1.

To enhance the dispersion of GNP in the HDPE matrix, a wax coating technique was applied in this study, which uniformly coated the surface of GNP with wax. For this method, wax was first dissolved in xylene at around 60°C and GNP was added afterwards. Sonication was then applied for 30 minutes at 100W to initially break down the GNP aggregates and to ensure a uniform wax coating. The resultant mixture was poured into an aluminum pan and left in a hood at room temperature to evaporate the solvent. After xylene was completely evaporated, the wax coated GNP was further dried in a vacuum oven

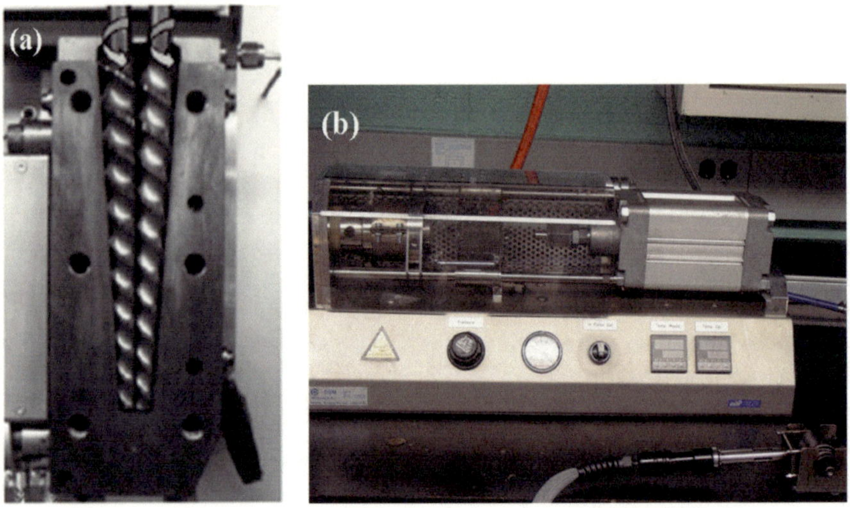

Fig. 1. (a) A DSM Micro 15cc Compounder, (Vertical, co-rotating, twin-screws microextruder); (b) A Daca Micro injector.

overnight at 30°C. In this study, four different wax coated GNP-15 samples were prepared having wax to GNP-15 ratios of 5:95, 10:90, 20:80, and 30:70wt%. This procedure of producing wax coated GNP is schematically shown in the Fig. 2.

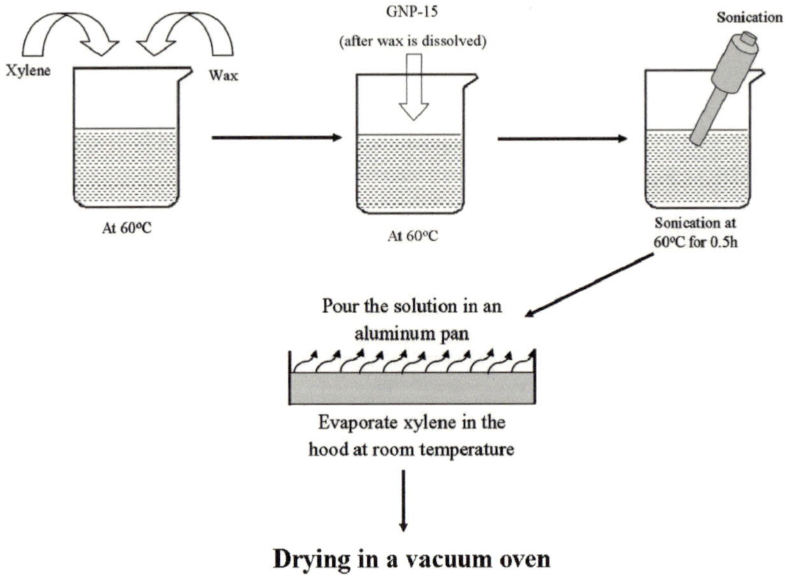

Fig. 2. The procedure of producing wax coated GNP-15.

Then wax coated GNP-15 was re-dispersed in HDPE by melt - extrusion in the DSM Micro 15cc Compounder with the same processing parameters described above. The actual loading of GNP-15 in the final nanocomposites was kept at 5vol%. After extrusion, the resulting

HDPE/wax coated GNP-15 (HDPE/WaxGNP-15) composite melts were injection molded into flexural coupons for mechanical and electrical properties test. The sample nomenclature for HDPE/WaxGNP-15 nanocomposites with different wax to GNP-15 ratios is: 5vol% HDPE/WaxGNP-15 (20:80wt%) means the actual GNP-15 loading in this nanocomposite is 5vol% and the weight ratio between wax and GNP-15 is 20:80.

To exclude any artifact due to the sonication, GNP-15 was added into xylene with the same sonication time of 30 minutes but without the addition of wax. After complete evaporation of xylene, GNP-15 was melt mixed with HDPE by the same melt - extrusion and injection molding process to obtain the HDPE/GNP-15 (sonic.) nanocomposites with GNP-15 loading at 5vol% as a reference.

2.3 Characterization techniques

Flexural tests were performed with a UTS SFM-20 machine (United Calibration Corp.) at room temperature by following the ASTM D790 standard test method (3-point bending mode). The test was performed at a flexural rate of 0.05 in/min. Impact strength tests (Izod impact type) were performed following the ASTM D256 standard test method.

The crystallinity and crystallization temperature were measured by using Dynamic Scanning Calorimetry (DSC Q2000, TA instrument). The samples used were 5–10 mg and non-isothermal crystallization was studied using the following experimental conditions: the sample was heated to 160°C at a rate of 20°C/min. The prior thermal history of the sample was erased by maintaining isothermal conditions at 160°C for 5 minutes. Then the sample was cooled down to 40°C at a rate of 20°C/min, held isothermally for 5 minutes and reheated at 20°C/min to 160°C and cooled back to 40°C again. The data of melting enthalpy (ΔH_m) and crystallization peak temperature (T_c) were collected during the second cycle. The degree of crystallization was calculated by the following equation, where χ% is the percent crystallinity of the matrix, wt% is the weight percentage of GNP, and ΔH_m^0 is the theoretical melting enthalpy of the polymer matrix if it is 100% crystalline.

$$\chi\% = \frac{1}{1-wt\%}\frac{\Delta H_m}{\Delta H_m^0} \tag{1}$$

The thermal stability of HDPE/GNP nanocomposites was determined from the thermogravimetric analysis (TGA), which was carried out on a TA instrument (TGA 2950) at a heating rate of 20°C/min under nitrogen from 30°C to 800°C. Thermal diffusivity (a, m^2/s) of GNP nanocomposites (round disks) was measured by a LFA Nanoflash 447 Light flash system. To calculate the thermal conductivity, the bulk density of the samples (ρ, kg/m^3) was obtained by dividing the mass over the volume, and the specific heat capacity (C_p: J/(kg·K)) was measured through the Dynamic Scanning Calorimetry (DSC Q2000, TA instrument). The thermal conductivity (κ, W/(m·K)) of GNP samples was then calculated by the following equation:

$$\kappa = a \times \rho \times C_p \tag{2}$$

The electrical resistivity of HDPE/GNP nanocomposites was measured both along the material flow direction (in-plane resistivity) and through the sample thickness direction (through-plane resistivity, normal to the flow direction), using the impedance spectroscopy by applying a two-probe method at room temperature. Samples with dimensions of 10.0 x 3.2 x 12.2mm (Length x Thickness x Width) were cut from the middle portion of flexural coupons. The two surfaces connected to the electrodes were first treated with O_2 plasma (14mins, 375W) in order to remove the top surface layers which are rich in polymer and then conductive silver paste was applied to the surface to ensure a good contact with the electrodes. The resistance of samples was measured and converted to resistivity by taking the sample dimensions into account.

The preparation of SEM samples in this study included epoxy mounting, grinding, polishing and etching steps. First, specimens were mounted with epoxy in cylindrical sample holders to maintain a flat surface over the entire grinding and polishing area. After epoxy was fully cured, samples were carefully grounded and polished. O_2 plasma etching (25mins, 375W) was then applied at the last step to remove the polymer in top surface allowing the GNP platelets to stand out under the SEM observation. A JEOL (model JSM-6400) SEM was then used to characterize the dispersion of GNP in HDPE. Samples were also gold coated to avoid charging.

3. Results and discussion

3.1 Flexural and impact properties of HDPE/GNP nanocomposites made by melt-extrusion and injection molding and the comparison of GNP to other reinforcements

The flexural strength and flexural modulus of HDPE/GNP nanocomposites and their comparison to HDPE composites filled with other reinforcements are presented in the Fig. 3 and Fig. 4 (Jiang & Drzal, 2010) respectively. All filler concentrations are from 0vol% to 15vol% except CB. HDPE/CB composites with CB concentration higher than 5vol% are not included because at the given processing conditions, the viscosity of these composites increases to the

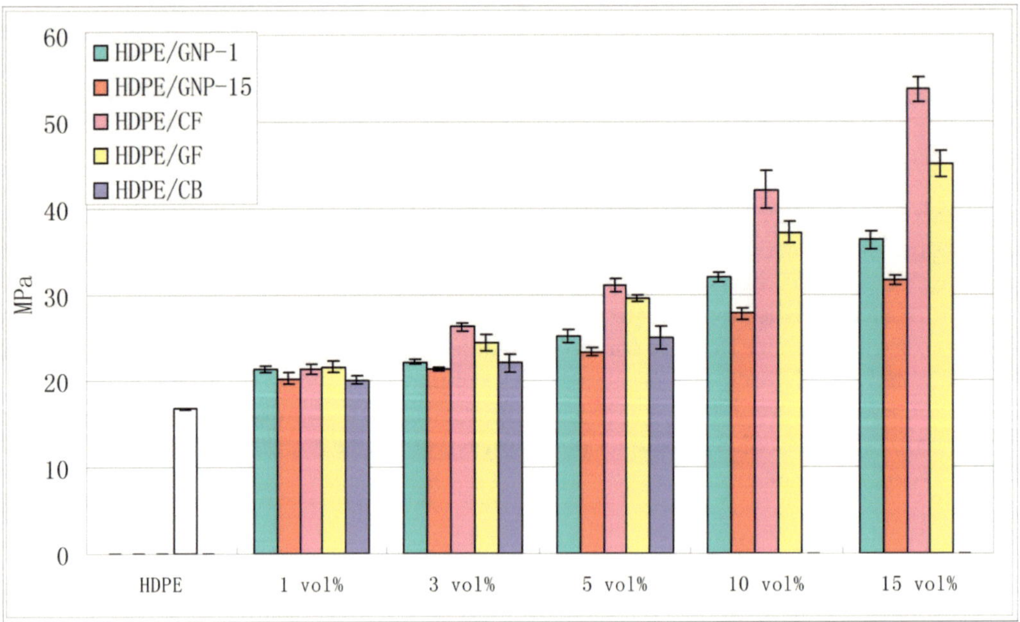

Fig. 3. Flexural strength of various HDPE composites.

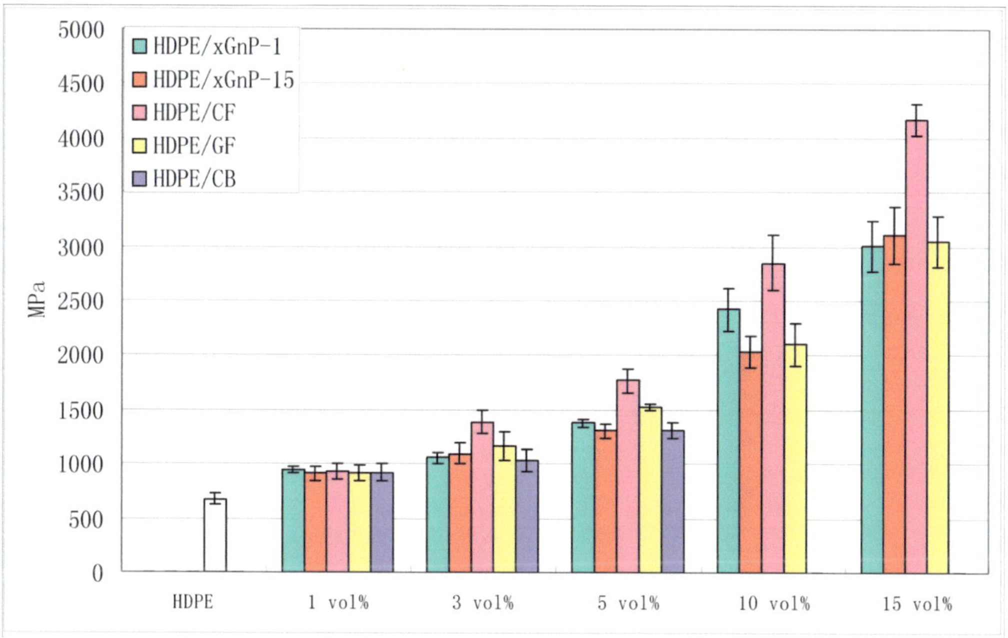

Fig. 4. Flexural modulus of various HDPE composites.

level where the DSM extruder could not generate sufficient pressure to extrude the mix properly. For the flexural strength shown in the Fig. 3, HDPE/CF composites exhibit the highest improvement at all filler loading levels followed by HDPE/GF composites. At the highest loading of 15vol%, HDPE/CF and HDPE/GF result in ~220% and ~170% improvement in flexural strength compared to the neat HDPE respectively. The great enhancement in the flexural strength for HDPE/CF and HDPE/GF composites is largely due to the high aspect ratio and excellent flexural properties of these carbon and glass fibers (Mai et al., 1994). HDPE/GNP nanocomposites also exhibit a significant increase in flexural strength with the increasing of GNP content. At 15vol% GNP loading, HDPE/GNP-1 and HDPE/GNP-15 nanocomposites result in ~116% and ~90% improvement in flexural strength respectively. Meanwhile, it is detected that GNP-1 nanocomposites are superior to GNP-15 counterparts in flexural strength at every GNP loading. At low CB concentrations up to 5vol%, HDPE/CB composites have the flexural strength value close to those of HDPE/GNP-1 samples. And as seen for the flexural modulus of these composites presented in the Fig. 4, HDPE/CF composites display the greatest enhancement in flexural modulus. HDPE/GNP nanocomposites are competitive to their HDPE/GF and HDPE/CB (up to 5vol%) counterparts.

The Izod impact strength of various HDPE composites up to a filler loading of 15vol% is displayed in the Fig. 5. (Jiang & Drzal, 2010). A reduction in impact strength is observed in all HDPE composites compared to the neat HDPE which is the case normally accompanies incorporation of a rigid filler into a relatively tough polymer (Wakabayashi et al., 2008). However, it is noted that HDPE/GNP nanocomposites exhibit the smallest reduction, which implies the advantage of using GNP as the reinforcement. At the filler loadings from 1vol% to 15vol%, overall HDPE/GNP-1 nanocomposites have the highest impact strength

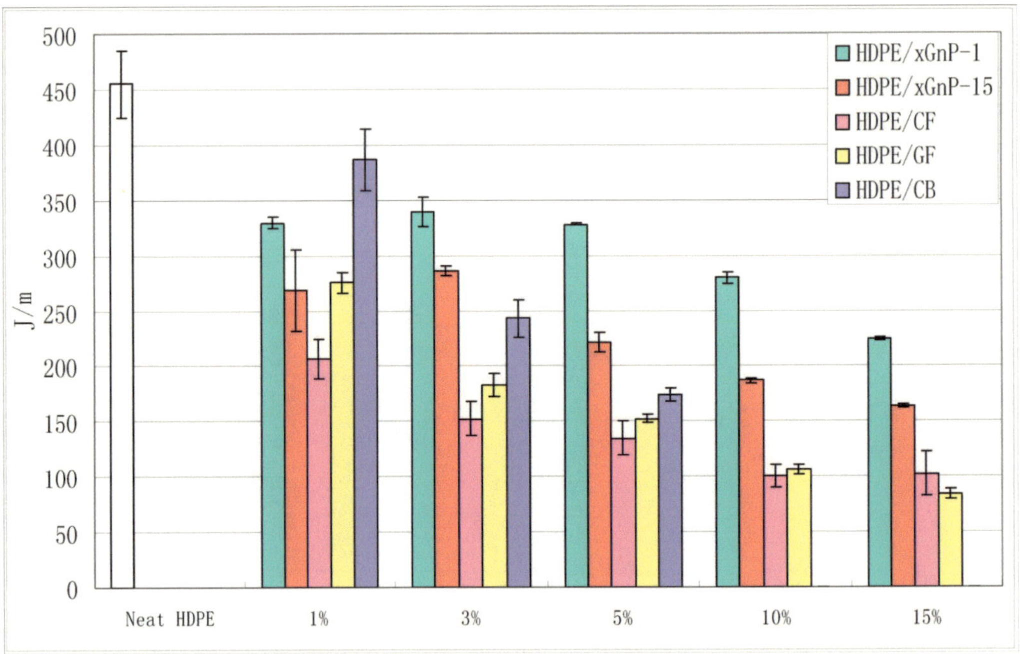

Fig. 5. Impact strength of various HDPE/GNP nanocomposites.

followed by HDPE/GNP-15 nanocomposites and then HDPE/CB composites (up to 5vol%). HDPE/CF and HDPE/GF show inferior performance compared to the other composites. The difference in impact strength may result from the difference in reinforcement size and aspect ratio, difference in dispersion of reinforcements in the polymer matrix and the difference in the adhesion between the reinforcements and the polymer matrix (Kim & Mai, 1991). Those differences eventually result in the different energy absorbing mechanisms at the impact fracture surface.

3.2 Crystallization behavior of HDPE/GNP nanocomposites made by melt-extrusion and injection molding

The crystallization temperature of HDPE/GNP nanocomposites with the GNP loading up to 15vol% is shown in the Fig. 6. It is concluded that incorporation of GNP has a significant effect on the crystallization temperature of HDPE, that is, the crystallization temperature increases as the GNP content increases. A relatively large temperature increase can be detected (around 3°C) even at a low GNP loading of 1vol%. The increased crystallization temperature suggests that the presence of GNP particles acts as nucleating agents which facilitate the crystallization process of HDPE (Peneva & Minkova, 2006).

The total percent of crystallinity of HDPE/GNP-1 and HDPE/GNP-15 nanocomposites obtained from equation [1] after the non-isothermal crystallization process is presented in the Fig. 7. It is interesting to note that the crystallinity value of GNP-1 and GNP-15 nanocomposites first increases with the GNP content and then it drops to an almost constant value. The highest crystallinity of HDPE/GNP-1 and HDPE/GNP-15 nanocomposites both occurs at 3vol% GNP loading. This interesting phenomenon is believed to be the result of a

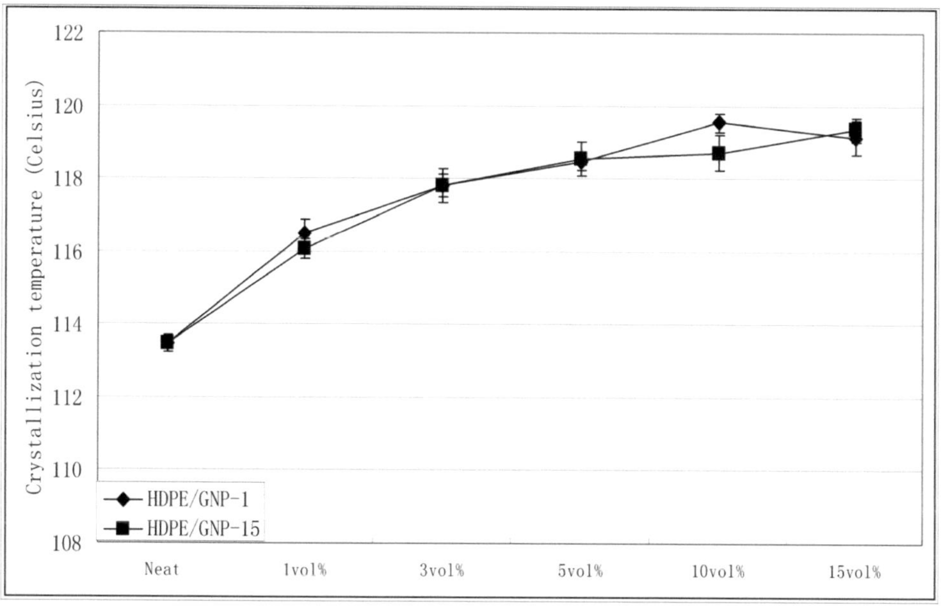

Fig. 6. Crystallization temperature of HDPE/GNP nanocomposites.

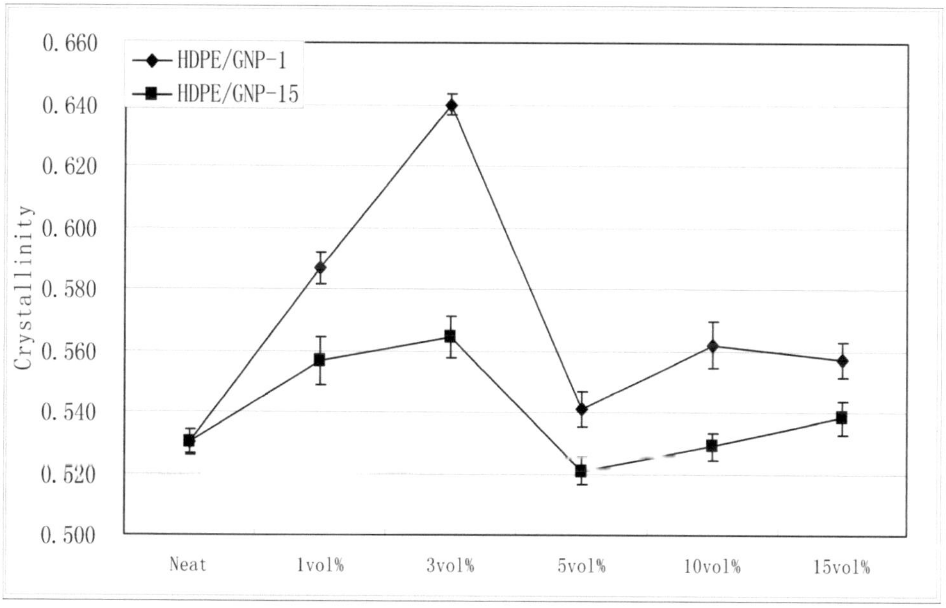

Fig. 7. Total percent of crystallinity of HDPE/GNP nanocomposites.

compromise between the nucleating and retarding effects of GNP on the host polymer during the non-isothermal crystallization (Di Maio et al., 2004). When the GNP content is relatively low, GNP particles exhibit a strong nucleating effect and act as heterogeneous nucleating sites at the GNP-HDPE interfaces, thus increasing the total percent of crystallinity of HDPE. And if the GNP concentration reaches a high level, the presence of the abundant GNP particles significantly hinders the diffusion of polymer chains to the growing crystallites. The crystallinity is therefore reduced. From the Fig. 7, it is also seen that HDPE/GNP-1 nanocomposites always show higher crystallinity than HDPE/GNP-15 counterparts at the same GNP loading, which indicates the nucleating efficiency between GNP-1 and GNP-15 is different. The superior nucleating efficiency of GNP-1 is due to the fact that the absolute number of GNP-1 particles is much larger than GNP-15 at the same filler concentration which provides more heterogeneous nucleating sites to initiate the polymer crystallization (Jiang & Drzal, 2011).

3.3 Thermal stability and thermal conductivity of HDPE/GNP nanocomposites made by melt-extrusion and injection molding

Fig. 8 and Fig. 9 present the TGA curves for HDPE/GNP-1 and HDPE/GNP-15 nanocomposites with different GNP loadings in nitrogen respectively. From the thermogravimetric curves, it is seen the pure HDPE is relatively thermal stable until around 300°C and after the onset degradation temperature of around 420°C, the degradation rate accelerates. However, for HDPE/GNP-1 nanocomposites, the onset degradation temperature increases from 420°C for neat HDPE to around 500°C for the sample at 15vol% GNP loading. Significant increase in onset degradation temperature suggests that the thermal stability of HDPE/GNP-1 nanocomposites is substantially enhanced, which is due to the high thermal stability of GNP-1 and the shielding effect of GNP-1 on the diffusion of combustion gases into and out of the polymer matrix during its thermal decomposition (Yang et al., 2007). A largely enhanced thermal stability can also be observed in HDPE/GNP-15 nanocomposites, where the onset thermal decomposition temperature is increased by around 70°C from neat HDPE to the HDPE/GNP-15 sample at 15vol% filler loading.

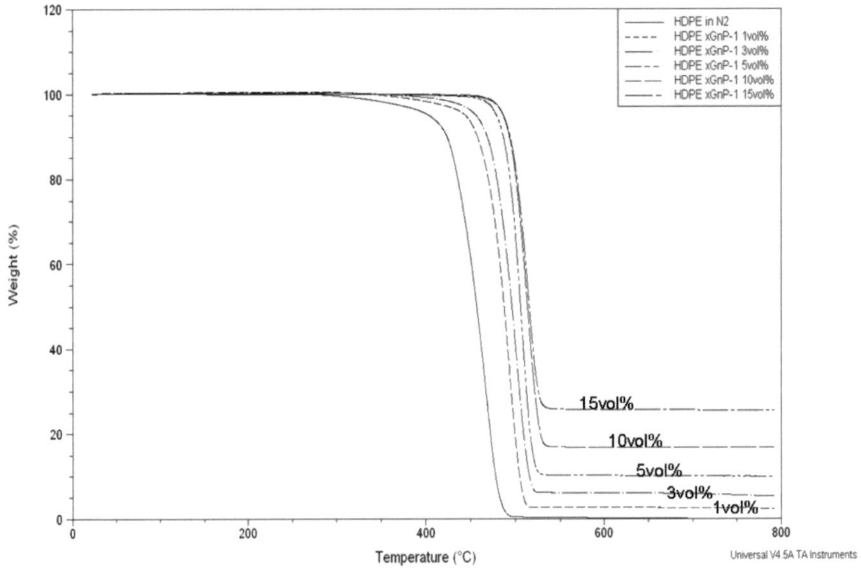

Fig. 8. TGA curves of HDPE and HDPE/GNP-1 nanocomposites.

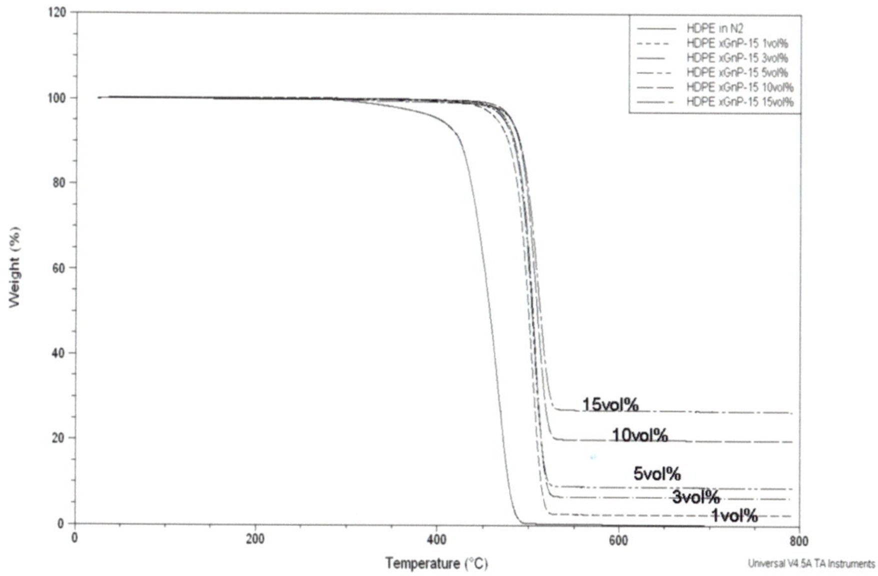

Fig. 9. TGA curves of HDPE and HDPE/GNP-15 nanocomposites.

The in-plane as well as the through-plane thermal conductivity of HDPE/GNP nanocomposites calculated by the equation [2] is characterized in the Fig. 10. As the GNP loading increases, it is detected that both the in-plane and through-plane thermal conductivity of HDPE/GNP nanocomposites undergo a substantial increase. At relatively

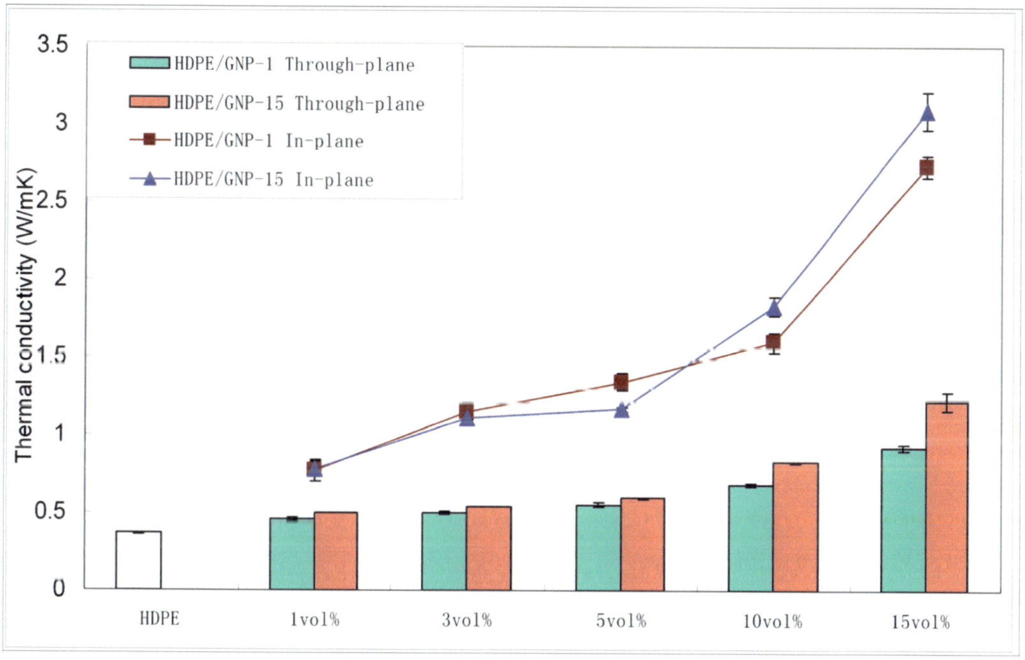

Fig. 10. Thermal conductivity of HDPE/GNP nanocomposites.

low GNP contents (1-5 vol%), HDPE/GNP-15 nanocomposites exhibit a thermal conductivity that is almost the same as that of HDPE/GNP-1 counterparts. When the GNP loading is higher than 5vol%, however, HDPE/GNP-15 nanocomposites show both superior in-plane and through-plane thermal conductivity. The higher thermal conductivity in GNP-15 samples is due to the larger aspect ratio of GNP-15 particles, which makes them more easily to connect with each other in forming conductive pathways thus reducing the thermal contact resistance and increasing the thermal conductivity of the resulting HDPE/GNP-15 nanocomposites.

If we compare the in-plane conductivity with the through-plane conductivity, it is found HDPE/GNP nanocomposites exhibit a much higher in-plane thermal conductivity at every GNP loading. For injection molded HDPE/GNP samples, higher in-plane conductivity is resulted from the preferential GNP alignment during the injection molding process, which will be fully addressed later in the morphology section and the anisotropic thermal property of GNP itself (The in-plane thermal conductivity of GNP platelet is 3000W/mK, and the through-plane conductivity is only 10W/mK due to its anisotropic structure (Kalaitzidou, 2006)).

3.4 Electrical conductivity of HDPE/GNP nanocomposites made by melt-extrusion and injection molding

The in-plane and through-plane electrical resistivity of HDPE/GNP nanocomposites made by melt-extrusion and injection molding are shown in the Fig. 11. For both in-plane and through-plane resistivity, it is seen that between 10vol% and 15vol% GNP content, there is a large decrease in the resistivity. This concentration range (10-15vol%) is thus noted as the percolation threshold for HDPE/GNP nanocomposites. The percolation threshold is defined as the concentration where a connected assembly of conductive particles

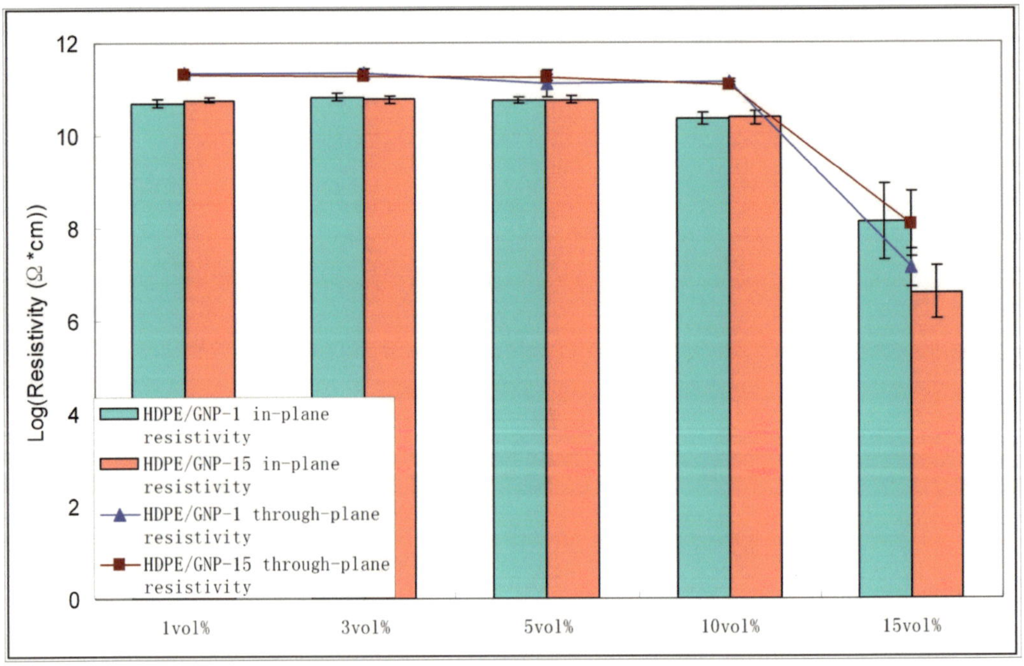

Fig. 11. In-plane and through-plane electrical resistivity of HDPE/GNP nanocomposites.

is formed within a polymer matrix. This assembly penetrates throughout the sample in forming conductive paths for electron transportation. At this concentration of the conductive filler, the electrical conductivity of the nanocomposites significantly increases and the nanocomposites become electrical conductive (I. Krupa et al., 2004).

Interestingly, at the GNP concentration of 15vol% (higher than the percolation threshold), HDPE/GNP-15 nanocomposites exhibit a lower in-plane electrical resistivity than HDPE/GNP-1 samples, which means HDPE/GNP-15 nanocomposites are more electrical conductive. However, at this GNP loading, HDPE/GNP-1 samples show a slightly higher through-plane electrical conductivity. By comparing the in-plane resistivity with the through-plane resistivity, it is concluded that HDPE/GNP-1 and HDPE/GNP-15 nanocomposites both exhibit a higher in-plane electrical conductivity, which is due to the platelet structure of GNP particles and the anisotropic property of injection molded composites as described above for the thermal conductivity. During injection molding, GNP platelets are aligned parallel to each other along the material flow direction, and at certain loading levels will they start intersecting with each other to form a conductive path. Because of the platelet morphology, aligned GNP particles are more difficult to get contact with each other through the thickness direction, thus leading to higher through-plane resistivity or lower electrical conductivity.

From the literature, it is noted that GNP nanocomposites fabricated by other processing methods such as solution compounding, pre-mixing, solid state ball milling, and solid state shear pulverization tend to have lower percolation threshold or higher electrical conductivity than the samples made by melt-extrusion and injection molding in this study (Jiang & Drzal, 2011; Kalaitzidou et al., 2007). The high percolation threshold is a result of the GNP aggregation during melt - extrusion and preferential platelets alignment in the injection molding process due to the large aspect ratio and planar shape of GNP, which will be fully explored in the next section.

3.5 Morphology of HDPE/GNP nanocomposites made by melt-extrusion and injection molding

The morphology of HDPE/GNP-1 and HDPE/GNP-15 nanocomposites fabricated by melt-extrusion and injection molding is presented in the Fig. 12. The samples at 5vol% GNP loading were taken as an example. Images (a) and (c) show the injection molded morphology near the mold wall or at the edge, where the shear forces under the injection molding conditions is high. And images (b) and (d) represent the injection molded morphology in the center of the composites, where the shear force is small. From the images of (a) and (c), it is concluded that all the GNP-1 and GNP-15 particles align parallel along the material flow direction and they are totally separated by the polymer matrix. Meanwhile, large GNP aggregates can be observed in these two images, especially in the GNP-15 sample, which is the indication of insufficiency of DSM extrusion for good GNP separation and dispersion. The presence of these large GNP aggregates drastically reduces the number of GNP platelets available as 'effective' reinforcing particles and significantly decreases the probability of interconnections between GNP platelets in forming electrically conductive pathways. Therefore, the high percolation threshold or low electrical conductivity in melt-extrusion and injection molded HDPE/GNP nanocomposites could be attributed to the sever GNP aggregation and preferential GNP alignment during the melt-extrusion and

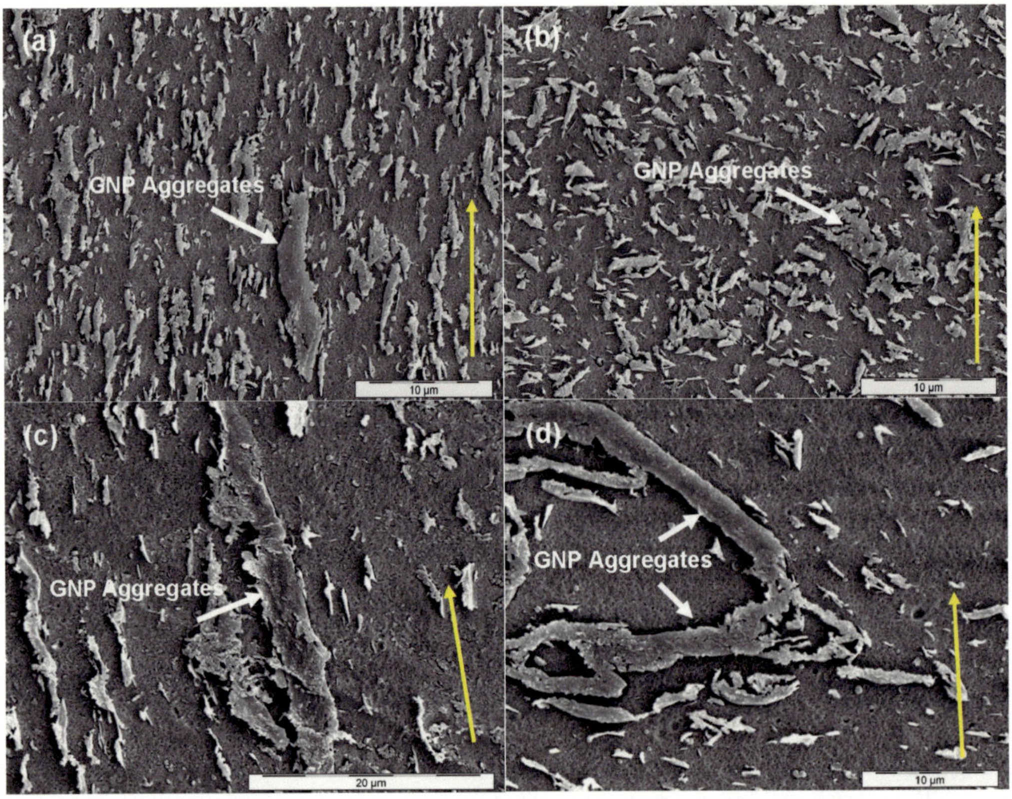

Fig. 12. Morphology of HDPE/GNP nanocomposites made by melt-extrusion and injection molding. The GNP loading is 5vol%. (a) GNP-1 sample at the edge; (b) GNP-1 sample in the center; (c) GNP-15 sample at the edge; (d) GNP-15 sample in the center. The yellow arrow on the right bottom indicates the material flow direction during injection molding.

injection molding process. The aligned structure of GNP platelets in the resulting nanocomposites also leads to much higher in-plane thermal and electrical conductivity as described above. Meanwhile, it is detected that the number density of GNP-1 particles are much larger than that of GNP-15 at the same GNP loading due to their smaller size, which is considered as a major reason of higher flexural strength, impact strength, and crystallinity in the resulting GNP-1 nanocomposites.

Compared with the morphology at the edge, it is found that GNP platelets in the center of the composites (images (b) and (d)) do not exhibit preferential alignment along the material flow direction. They are randomly oriented because of the minimum shear forces encountered during injection molding. However, large GNP aggregates can also be observed which further confirm the insufficient shear force attainable in the DSM extrusion to break down the GNP aggregates and to achieve a uniform GNP dispersion.

In summary, GNP nanoplatelets tend to align along the material flow direction at the edge of injection molded nanocomposites while they are randomly oriented at the center. The presence of large GNP aggregates constrains the fully translation of superb mechanical and electrical properties of GNP particles into good mechanical and electrical properties of the resulting nanocomposites.

3.6 Wax coating method to improve the dispersion of GNP in HDPE

To improve the dispersion of GNP particles in melt-extrusion and injection molded HDPE/GNP nanocomposites, the wax coating method was applied which is based on the steric repulsion force between the wax coated GNP nanoplatelets in preventing the aggregation of GNP during the processing conditions. Take the GNP-15 nanocomposites for example, the in-plane and through-plane electrical resistivity of HDPE/WaxGNP-15 nanocomposites with different wax to GNP weight ratios are displayed in the Fig. 13. (Jiang & Drzal, 2011). The GNP loading in all these nanocomposites is kept at 5vol%. Interestingly, with the presence of wax coated on the surface of GNP, the previous non-conductive 5vol% HDPE/GNP-15 nanocomposite (control sample) becomes electrical conductive along the material flow direction. And the in-plane resistivity is continuously decreased as the wax content increases. A more than 5 orders of magnitude decrease in resistivity is seen from the control sample to the 5vol% HDPE/WaxGNP-15 (30:70wt%) nanocomposite, suggesting that the electrical conductivity of HDPE/WaxGNP-15 nanocomposites is significantly enhanced. Therefore, the percolation threshold of injection molded HDPE/GNP-15 nanocomposites is reduced from the previous 10-15vol% GNP loading to less than 5vol%. However, the through-plane electrical resistivity of these nanocomposites remains high and it is unaffected by the addition of the wax.

Meanwhile, if we compare the resistivity value of the sonicated sample (5vol% HDPE/GNP-15 (sonic.)) with the control sample, no significant difference in resistivity is observed. This implies that the application of sonication alone without the use of wax is insufficient to improve the electrical conductivity of the resulting nanocomposites.

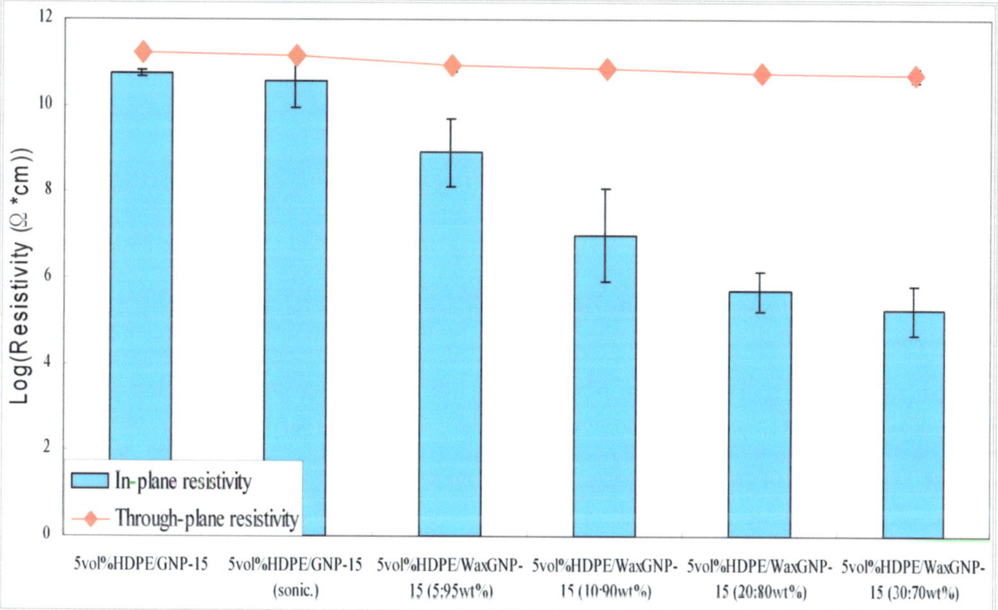

Fig. 13. In-plane and through-plane electrical resistivity of various HDPE/WaxGNP-15 nanocomposites at 5vol% GNP loading made by melt-extrusion and injection molding.

Fig. 14 (Jiang & Drzal, 2011) gives the flexural properties of various HDPE/WaxGNP-15 nanocomposites at 5vol% GNP loading. It is found that as the amount of wax coating on the

GNP surface increases, both the flexural strength and flexural modulus of the resulting nanocomposites firstly increase and then the trend reverses. The largest enhancement in the flexural properties is detected at the 5vol% HDPE/WaxGNP-15 (20:80wt%) sample, of which the flexural strength is increased by 12% and the flexural modulus is improved by 20%. Higher flexural strength and modulus is a strong indication of an improved GNP dispersion in HDPE (Kashiwagi et al., 2007). And the reduction in mechanical properties in the 5vol%HDPE/WaxGNP-15 (30:70wt%) nanocomposite could be explained as the addition of too much low molecular weight polyethylene (wax) into the nanocomposite, which compromises the positive effect of enhanced GNP dispersion on the mechanical strength (Jiang & Drzal, 2011). Again, we do not see any enhancement in flexural properties for the sonicated sample.

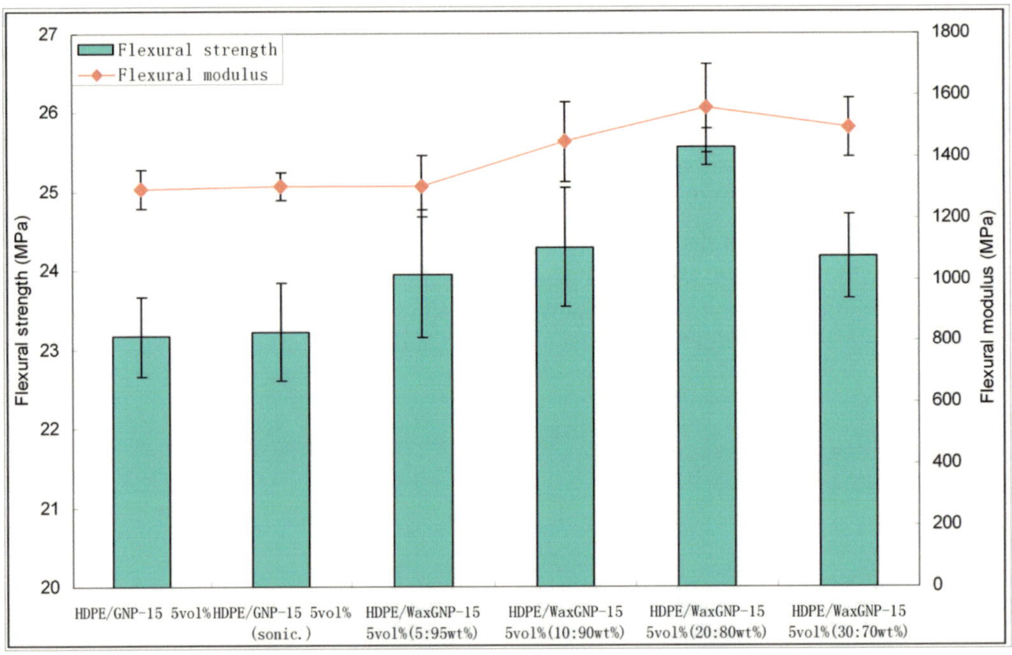

Fig. 14. Flexural strength and flexural modulus of various HDPE/WaxGNP-15 nanocomposites at 5vol% GNP loading made by melt-extrusion and injection molding.

The morphology of these nanocomposites at the edge is presented in the Fig. 15 (Jiang & Drzal, 2011), which helps to explain the substantial improvement in electrical and mechanical properties for HDPE/GNP-15 nanocomposites made by the wax coating method. As discussed in the previous section, GNP-15 platelets in the control sample are aligned along the material flow direction and large GNP-15 aggregates are present (Fig. 12(c))). From the Fig. 15(a), which is the morphology of the sonicated sample, a similar morphology can be seen, that is, there exhibit preferential GNP-15 alignment and large GNP-15 aggregates. This similarity in morphology between the sonicated sample and the control sample suggests that sonication alone without wax is not sufficient to improve the dispersion of GNP-15 in HDPE. Although the technique of sonication is proved as an efficient method to break down the agglomerates or aggregates of nano-particles (Bang & Suslick, 2010) in solution, once incorporated in polymers, these nano-particles which were initially separated would like to re-aggregate again during the melt processing conditions.

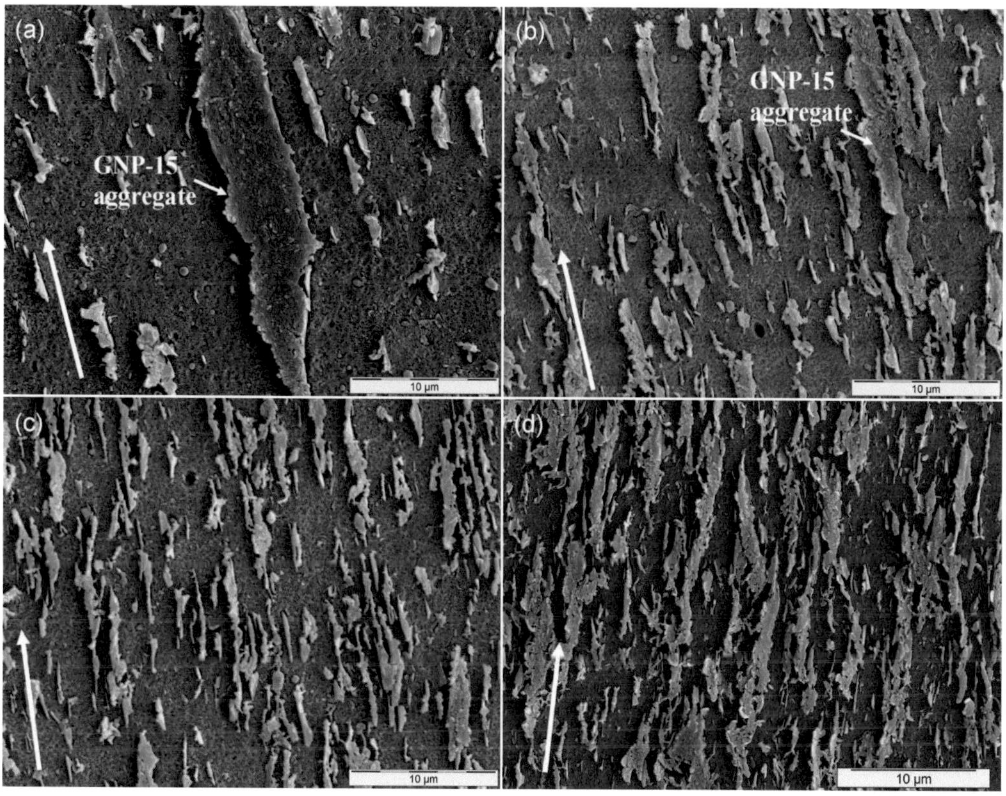

Fig. 15. Morphology of various HDPE/WaxGNP-15 nanocomposites at 5vol% GNP loading: (a) sonicated HDPE/GNP-15; (b) HDPE/WaxGNP-15 (10:90wt%); (c) HDPE/WaxGNP-15 (20:80wt%); (d) HDPE/WaxGNP-15 (30:70wt%). The arrow on the left bottom indicates the material flow direction during injection molding.

Fig. 15(b) to Fig. 15(d) compare the morphology of 5vol% HDPE/WaxGNP-15 nanocomposites with increasing wax content. It is clear to see that the number density of GNP-15 platelets is increasing as the wax content increases and big GNP-15 aggregates can no longer be detected. The disappearance of GNP aggregates as well as the increased GNP number density indicates that the re-aggregation of GNP-15 platelets is indeed prevented by the addition of the proper amount of wax coating on the surface. In this case, individual GNP-15 platelets can be dispersed uniformly in the polymer matrix. Meanwhile, the presence of well dispersed GNP-15 platelets significantly improves their interconnections in forming conductive pathways along the material flow direction, which is mainly responsible for the enhanced in-plane electrical conductivity of HDPE/WaxGNP-15 nanocomposites.

The morphology of the 5vol% HDPE/WaxGNP-15 (20:80wt%) nanocomposite in the center is presented in the Fig. 16. Compared with the morphology of the control sample (Fig. 12(d)), enhanced GNP-15 dispersion in HDPE is further confirmed, which results in disappearance of GNP-15 aggregates and higher GNP number density in HDPE. In this case, it is concluded that the wax coating method is capable of enhancing GNP dispersion in HDPE which leads to better mechanical and electrical properties in the resulting HDPE/GNP nanocomposites.

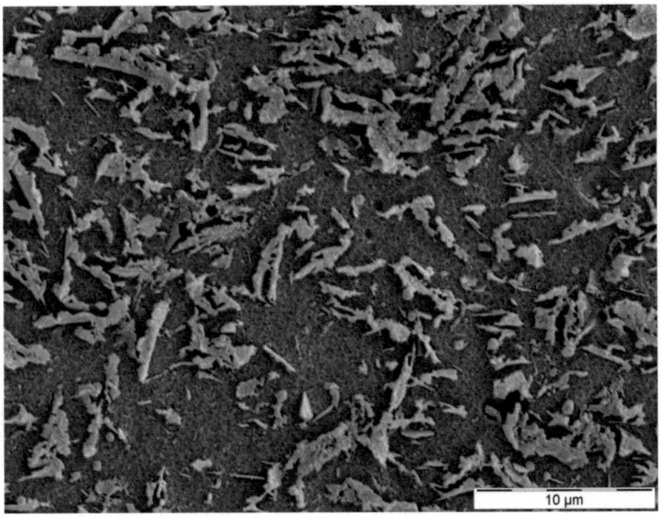

Fig. 16. Morphology of 5vol%HDPE/WaxGNP-15 (20:80wt%) sample in the center.

4. Conclusions

Various properties of HDPE/GNP nanocomposites made by melt-extrusion and injection molding were explored and analyzed in this book chapter, including mechanical properties, crystallization behaviors, thermal stability, thermal conductivity, and electrical conductivity. Results show some unique features of these injection molded nanocomposites. First of all, HDPE/GNP nanocomposites exhibit anisotropic thermal and electrical conductivity, that is, in-plane thermal and electrical conductivity were found to be much higher than the through-plane conductivity. Higher in-plane thermal and electrical properties of these nanocomposites are on the account of the alignment of GNP platelets along the material flow direction during injection molding, which were verified by their morphology. Secondly, it was found that the morphology of injection molded nanocomposites varies from the edge to the center. GNP platelets exhibit strong preferential alignment at the edge where the shear forces under injection molding are maximum while they are randomly orientated in the center where the shear forces are minimum. Additionally, sever GNP aggregation was detected in the melt-extrusion and injection molded samples, which is due to the insufficient shear force attainable during the processing conditions to shear GNP platelets apart and to achieve a uniform GNP dispersion. Regardless of this disadvantage, the technique of melt-extrusion and injection molding still remains as the major processing method used for manufacturing thermoplastics in industry because of its design flexibility, low cost and labor, short cycle time and minimum scrap loss. In order to improve the dispersion of GNP in HDPE, a wax coating technique was reported in this book chapter. It is concluded that the dispersion of GNP in HDPE was dramatically enhanced due to the steric repulsive forces between wax coated GNP platelets, which leads to the better electrical and mechanical properties in the resulting nanocomposites.

5. Acknowledgment

The authors gratefully acknowledgment the support of the Michigan Economic Development Commission, 21st Century Jobs Fund for partial support of this research. And the generous help from Dr. Hiroyuki Fukushima, Dr. Wanjun Liu and Brian Rook in Composite Materials and Structures Center at Michigan State University is also gratefully appreciated.

6. References

GNP is an exfoliated graphene nanoplatelet material obtained from XG Sciences, Inc., East Lansing, MI www.xgsciences.com, last access in 10/31/2011

Balasubramanian, K., & Burghard, M. (2006). Biosensors based on carbon nanotubes *ANALYTICAL AND BIOANALYTICAL CHEMISTRY* 385, 3, pp. (452-468)

Bang, J. H., & Suslick, K. S. (2010). Applications of Ultrasound to the Synthesis of Nanostructured Materials. *Advanced Materials,* 22, 10, pp. (1039-1059), 1521-4095

Di Maio, E., Iannace, S., Sorrentino, L., & Nicolais, L. (2004). Isothermal crystallization in PCL/clay nanocomposites investigated with thermal and rheometric methods. *Polymer* 45, pp. (8893–8900), 0032-3861

Fukushima, H. (2003). *Graphite nanoreinforcements in polymer nanocomposites.* PHD Thesis, Michigan State University, East Lansing, MI, USA

Griffith, A. A. (1920). The phenomena of rapture and flaw in solids. *Phil. Trans. R. Soc. Lond. A,* 221, pp. (163-198)

Hermant, M. C., Klumperman, B., Kyrylyuk, A. V., van der Schoot, P., & Koning, C. E. (2009). Lowering the percolation threshold of single-walled carbon nanotubes using polystyrene/poly(3,4-ethylenedioxythiophene): poly(styrene sulfonate) blends. *Soft Matter,* 5, pp. (878 - 885), 1744-683X

Hussain, F., Hojjati, M., Okamoto, M., & Gorga, R. (2006). Review article: Polymer-matrix Nanocomposites, Processing, Manufacturing, and Application: An Overview. *Journal of Composite Materials,* 40, 17, pp. (1511-1575)

I. Krupa, I. Novák, & Chodák, I. (2004). Electrically and thermally conductive polyethylene/graphite composites and their mechanical properties. *Synthetic Metals,* 145, pp. (245–252)

Jiang, X., & Drzal, L. T. (2010). Multifunctional high density polyethylene nanocomposites produced by incorporation of exfoliated graphite nanoplatelets 1: Morphology and mechanical properties. *Polymer Composites,* 31, 6, pp. (1091-1098), 1548-0569

Jiang, X., & Drzal, L. T. (2011). Improving Electrical Conductivity and Mechanical Properties of High Density Polyethylene through Incorporation of Paraffin Wax Coated Exfoliated Graphene Nanoplatelets and Multi-wall Carbon Nano-tubes. *Composites Part A: Applied Science and Manufacturing,* 42, 11, pp. (1840-1849), 1359-835X

Jiang, X., & Drzal, L. T. (2011). Multifunctional high density polyethylene nanocomposites produced by incorporation of exfoliated graphite nanoplatelets 2: crystallization, thermal, and electrical properties. *submitted to the Journal of Polymer Composites*

Jiang, X., & Drzal, L. T. (2011). Reduction in percolation threshold of injection molded High Density Polyethylene/Exfoliated Graphene Nanoplatelets composites by Solid State Ball Milling and Solid State Shear Pulverization. *Journal of Applied Polymer Science* In Press, Accepted Manuscript

Kalaitzidou, K. (2006). *Exfoliated Graphite Nanoplatelets as Nanoreinforcement for Multifunctional Polypropylene Nanocomposites.* PHD Thesis, Michigan State University, East Lansing, MI, USA

Kalaitzidou, K., Fukushima, H., & Drzal, L. T. (2007). Multifunctional polypropylene composites produced by incorporation of exfoliated graphite nanoplatelets. *Carbon,* 45, pp. (1446–1452)

Kalaitzidou, K., Fukushima, H., & Drzal, L. T. (2007). A new compounding method for exfoliated graphite-polypropylene nanocomposites with enhanced flexural properties and lower percolation threshold. *Composites Science and Technology,* 67, pp. (2045–2051)

Kashiwagi, T., Fagan, J., Douglas, J. F., Yamamoto, K., Heckert, A. N., Leigh, S. D., Obrzut, J., Du, F. M., Lin-Gibson, S., Mu, M. F., Winey, K. I., & Haggenmueller, R. (2007). Relationship between dispersion metric and properties of PMMA/SWNT nanocomposites. *Polymer,* 48, 16, (Jul), pp. (4855-4866), 0032-3861

Kim IH, Kim JH, & KB, K. (2005). Electrochemical characterization of electrochemically prepared ruthenium oxide/carbon nanotube electrode for supercapacitor application *Electrochemical and Solid State Letters,* 8, 7, pp. (A369-A372)

Kim, J.-K., & Mai, Y.-W. (1991). High strength, high fracture toughness fibre composites with interface control—A review *Composites Science and Technology,* 41, 4, pp. (333-378)

Kim, S., & Drzal, L. T. (2009). Comparison of Exfoliated Graphite Nanoplatelets (xGnP) and CNTs for Reinforcement of EVA Nanocomposites Fabricated by Solution Compounding Method and Three Screw rotating Systems *Journal of Adhesion Science and Technology,* 23, pp. (1623–1638)

Mai, K., Li, J., & Zeng, H. (1994). Mechanical property and fracture morphology of fiber-reinforced polysulfone plasticized with acetylene-terminated sulfone. *Journal of Applied Polymer Science,* 52, 9, pp. (1279 - 1291)

Peneva, Y., & Minkova, L. (2006). Non-isothermal and isothermal crystallization of nanocomposites based on functionalized polyethylenes. *Polymer Testing,* 25, 3, pp. (366-376), 0142-9418

Piggott, M. R. (1980). *Load Bearing Fibre Composites,* Pergamon Press

Singh, K. V., Pandey, R. R., Wang, X., Lake, R., Ozkan, C. S., Wang, K., & Ozkan, M. (2006). Covalent functionalization of single walled carbon nanotubes with peptide nucleic acid: Nanocomponents for molecular level electronics. *Carbon,* 44, 9, pp. (1730-1739), 0008-6223

Vaia, R. A. (2003). Polymer nanocomposites open a new dimension for plastics and composites. *AMPTIAC Newsletter,* 6, 1, pp. (17–24)

Wakabayashi, K., Pierre, C., & Dikin, D. A. (2008). Polymer-Graphite Nanocomposites: Effective Dispersion and Major Property Enhancement via Solid-State Shear Pulverization. *Macromolecules,* 41, pp. (1905-1908)

Weibull, W. J. (1951). A statistical distribution function of wide applicability. *Journal of Applied Mechanics,* 18, (September), pp. (293-297)

Xie, S. H., Liu, Y. Y., & Li, J. Y. (2008). Comparison of the effective conductivity between composites reinforced by graphene nanosheets and carbon nanotubes. *Applied Physics Letters,* 92, pp. (243121-3), 1077-3118

Yang, S., Taha-Tijerina, J., Serrato-Diaz, V., Hernandez, K., & Lozano, K. (2007). Dynamic mechanical and thermal analysis of aligned vapor grown carbon nanofiber reinforced polyethylene. *Composites Part B: Engineering,* 38, pp. (228-235), 1359-8368

Yao, X., Wu, H., Wang, J., Qu, S., & Chen, G. (2006). Carbon nanotube/poly(methyl methacrylate) (CNT/PMMA) composite electrode fabricated by in situ polymerization for microchip capillary electrophoresis *Chemistry - A European Journal,* 13, 3, pp. (846 - 853)